Toxic Politics

Toxic Politics

RESPONDING TO
CHEMICAL DISASTERS

Michael R. Reich

Cornell University Press

ITHACA AND LONDON

First published 1991 by Cornell University Press.

International Standard Book Number 0-8014-2434-8
Library of Congress Catalog Card Number 91-55054
Printed in the United States of America
*Librarians: Library of Congress cataloging information
appears on the last page of the book.*

⊚ The paper in this book meets the minimum requirements
of the American National Standard for Information Sciences—
Permanence of Paper for Printed Library Materials, ANSI Z39.48-1984.

For my family

Contents

Acknowledgments

This book combines topics that occupy my professional life: chemicals, politics, victims, and symbols; structures and cultures; the United States, Japan, and Italy. It brings together my past studies of biochemistry, Japanese pollution, death in Japan, comparative politics and policy, social movements, organizations, private business, and public health—to examine the critical problems of chemical disasters and their expression in politics and society.

Many people and institutions have aided me. To research the case studies, I relied on the hospitality of friends on three continents, who provided both material and spiritual support. For the Michigan case: Fred and Beth Fry, Doc Clark, Edie Clark, Stephen M. Soble, Pat Miller. For the Italian case: Letizia Ciotti Miller, Pula Creton, the Cattoretti family, the late Maia Luzzatto, the Scesi family, Alberto Colombi, Peggy Craig. And for the Yusho case: the Nomura family, Watanuki Reiko and Jōji, Isono Naohide, and Kuroda Kōtaro and his parents. (Following East Asian practice, this book places the family name before the given name for Japanese names.) Without the help of these friends, the project would have been impossible.

My research depended on the good will of many individuals who consented to be interviewed and who gave time, documents, and often hospitality. Although I could not include all the materials I gathered, each interview nonetheless contributed to my understanding of the three chemical disasters.

Early on, four members of Yale University's Department of Political Science—Joseph LaPalombara, James C. Scott, Robert Dahl, and William Foltz—provided criticisms and comments that encouraged me to pursue the topic in a more scholarly manner than I might otherwise have done.

LaPalombara was especially helpful with his introductions to the ways of Italian politics and his detailed comments on drafts. Hugh Patrick, then on the Yale Council for East Asian Studies, supported my impulse to extend the study to include a European country.

My deep appreciation goes to friends who reviewed portions of the manuscript. Tom Lifson read early drafts of the analytic chapters and taught me much about organizational theory. Marc J. Roberts deserves special thanks for reading the full manuscript and applying his keen mind to both details and themes, pushing me to improve the quality of the final product. Michael S. Brown read the manuscript on his Medford-to-Boston commuter bus and provided valuable guidance for revision. Isono Naohide reviewed the Yusho chapter and checked my translations from the Japanese. Letizia Ciotti Miller reviewed the Seveso chapter and scrutinized my translations from the Italian.

Other friends supported me through bouts of rewriting: Daniel Einhorn, Richard Diamond, Mike Connor, Elio Riboli, Rashid Shaikh, Peggy Coulson, and Marilyn Arsem. Students in a course I taught one semester at Yale College helped me work through the cases and analysis. Over the years, friends in several quarters provided moral support and encouragement: C. Peter Timmer, Barbara Rosenkrantz, David E. Bell, Arthur Kleinman, Frank Upham, and Sheila Jasanoff. Naomi Pierce provided editorial assistance, improving the flow and removing obstacles; Richard C. Rose edited the final manuscript with great care and expertise. Anonymous reviewers examined major themes and minor details. I am grateful to them all.

Several institutions supported the research. My initial research and travel expenses were provided by a Sumitomo Fellowship from the Yale University Council on East Asian Studies and by an individual grant from the Ford Foundation. The Department of Political Science at Michigan State University graciously allowed me to use an office during the summer of 1978, making my research there more productive and pleasant. In Japan, the National Institute of Public Health provided me with office space and other assistance, for which I thank Dr. Suzuki Takeo, then vice-director, and Dr. Someya Shiro, then director.

I also received support for the writing. The Harvard University Center for European Studies opened its special congenial environment to me and provided office space, through the good efforts of Peter Lange and Abby Collins. My appreciation goes also to the Interdisciplinary Programs in Health at the Harvard School of Public Health, to its director, Donald Horning, and the supporting staff; to the Harvard Business School and its Postdoctoral Research Fellowship Program; and to the Harvard University Center for Population Studies, which provided me with office space through part of the revisions. The Milton Fund awarded me a small grant to assist the rewriting.

Portions of the book have been presented at various institutions, including Princeton University's Center for Environmental and Energy Studies, the Interdisciplinary Programs in Health at the Harvard School of Public Health, and the Triangle East Asian Colloquium "Political and Social Responses to Catastrophe," in North Carolina. I appreciate the permission from two journals to use revisions of materials they published: *Law in Japan: An Annual* 15 (1982) 102–129, on the Yusho case, and the *American Journal of Public Health* 73 (1983) 302–313, on the Michigan case.

Research assistance was provided by Jaquelin Spong, Diana Cooper Weil, Margaret John, and Matsumoto Kiyofumi. Early on, my friend Lucy Leu helped me type parts of the first draft into the computer, transforming black on white into electronic bleeps. Alvin Brass graciously prepared the figures. Peter Agree, Kay Scheuer, and others at Cornell University Press supported me with good humor and guidance in preparing the manuscript for publication.

Last is rarely least. My final thanks go to my mother, the school teacher, who taught me the importance of choosing words carefully; my father, the chemical engineer, who taught me the importance of balancing science with people; and my wife, who struggled with me as I struggled with this book, and who believed in it and me when others expressed doubts.

MICHAEL R. REICH

Boston, Massachusetts

Toxic Politics

Chapter 1

Toxic Chemicals and Toxic Politics

Toxic chemicals have become part of life in the late twentieth century—and also part of politics. On TV and in print in the 1970s, Monsanto announced: "Without chemicals, life itself would be impossible." The statement is true but misleading, a public relations attempt to counter the increasing popular awareness of another truth: that sometimes, because of chemicals, life itself becomes impossible.

Occasionally, toxic substances cause dramatic disasters, creating images that haunt society. These images include the mother holding a daughter deformed by congenital mercury poisoning in Minamata, Japan; the lines of corpses left by the toxic cloud in the middle of the night in Bhopal, India; the worker at Life Sciences, Inc., trembling uncontrollably from Kepone poisoning in Hopewell, Virginia; the carcasses of poisoned cows being buried by machines in pits in Michigan; and anxious parents of children with birth defects at Love Canal, New York, accusing Hooker Chemical of poisoning them and causing deformities in their offspring.

These images embody some of the personal and social problems of chemical disasters. Yet we rarely think about the aftermath for the victims. Their problems involve one of the largest segments of the world economy, the chemical industry, and raise fundamental questions about power, politics, and justice in modern society. A central question, and the main theme of this book, is: How do the victims of chemical disasters obtain redress? I argue that victims of toxic contamination seek redress through politics, in three phases, by making the problem public, organizing group actions, and mobilizing political allies. But that political process of seeking redress creates its own conflicts and costs, its own process of victimization. For the victims, paradoxically, the politics of contamination can become as poisonous as the chemicals themselves. Hence, toxic politics.

This book examines three major chemical disasters and argues that the three political phases in the process of redress are remarkably consistent, despite differences in national cultures and contexts. This conclusion stands in sharp contrast to recent studies that stress national differences in policy responses to similar problems in chemical regulation (Brickman, Jasanoff, and Ilgen) and environmental policy (D. Vogel, 1986). Understanding why the victims of toxic contamination confront similar political obstacles—and what can be done about those obstacles—is the major objective of this book. The first of the three chemical disasters took place in Japan, beginning in 1968, when polychlorinated biphenyls (PCBs) leaked into cooking oil and were consumed directly by humans. The second started in 1973 in Michigan when animal feed was contaminated by polybrominated biphenyls (PBBs) and humans were then poisoned through consumption of animal products. The third occurred in Seveso, Italy, in 1976 when tetrachlorodibenzodioxin (dioxin) erupted from a factory into the air and settled over homes and gardens. The analysis of these cases addresses major questions in political science, although I borrow concepts and approaches from other disciplines.

All three involve similar toxic substances (man-made, halogenated hydrocarbons) with similar toxic effects on humans, including acute skin disease, long-term promotion of cancer, birth defects, disorders in the nervous system, and possibly other health problems (Kimbrough, 1979). The disasters directly affected populations of approximately the same size, ranging from about one to two thousand people, and within the same decade (1968–76). In the three countries, all advanced capitalist societies with democratic forms of government, the disasters came to symbolize for some people the crises of industrial society in the late twentieth century.

All three were unintentional poisonings. The chemicals ended up where they were not supposed to be—in the living environment, in food and feed products, in animals and in humans. In each case, the chemicals entered society from a single source, in a single shot, in what is called an environmental *pulse*. Those chemical pulses produced long-lasting reverberations in the natural and physical orders as well as in the political and social orders.

Despite differences in cultural, historical, and political contexts and in the paths of exposure to toxic chemicals, the analysis reveals basic similarities in the development of the chemical disasters as political phenomena. To understand the common patterns of toxic politics, we first need to look at the nature of toxic contamination.

The Problem of Toxic Contamination

The intentional use of toxic substances for poisoning has a long history. The poisoning of individuals with plant and animal toxins, in hunting, warfare, murder, and suicide, began millennia ago. The death of Socrates by hemlock is a case in point. In the Renaissance, poisoning reached the status of an art, intimately connected to political and social conflicts (Casarett). The first major outbreak of deliberate mass poisoning in warfare occurred early in this century, as lethal gases slaughtered soldiers and civilians in the First World War. Later the Nazis transformed mass poisoning into systematic murder in the gas chambers. Other modern forms of intentional poisoning are chemotherapy for cancer (Tannock) and other serious diseases and the use of lethal injections for capital punishment.

Unintentional poisoning likewise has a long history. The Romans, according to one theory, were poisoned by their lead water pipes and lead-lined cooking wares, a poisoning that may have contributed to the weakening of leadership and the downfall of empire (Gilfillan). By the late 1600s and early 1700s, hazards for certain jobs were especially high, as shown by Ramazzini, the great Italian physician, who compiled and published the first methodical presentation of occupational illnesses (Ramazzini). Both kinds of unintentional poisoning, that affecting the populace and that affecting certain workers, have become increasingly common in the last half of the twentieth century. Industrial society brought deadly infectious diseases under greater control, but also created new diseases from thousands of new chemicals under only haphazard controls.

Humankind has knowingly and unknowingly used toxic substances for centuries. What has changed recently is that many more chemicals with known toxic properties are being deliberately used in industry, that others with unknown toxic properties are being used without proper testing and control, and that many of these substances are much more toxic than others used in the past. The U.S. Environmental Protection Agency listed over sixty-three thousand chemical substances in its 1985 inventory of chemical substances subject to the Toxic Substances Control Act (TSCA) of 1976. That number excludes chemical mixtures, foods, drugs, cosmetics, pesticides, and other materials not covered by TSCA (Toxic Substances Strategy Committee, p. 1). In 1984, a review of toxicity testing concluded that the great majority of chemical substances with potential human exposure lack "data considered to be essential for conducting a health-hazard assessment" (National Academy of Sciences, p. 1).

Production of chemicals has surged worldwide in the postwar period. Global production of chemicals grew annually about 10 percent on average from 1963 to 1969, about 6 percent in the 1970s, and about 3 percent in the early 1980s (Chemical Industries Association, chart 9). According to the same source, global sales of chemicals rose from an estimated total of $96 billion in 1963 to $837 billion in 1986 (table 12). In 1983, the chemical industry accounted for about 13 percent of the world industrial product, exceeded only by the manufacturing sector (Franck, p. 52).

The global growth of the chemical industry is reflected in U.S. statistics. In the United States, production of synthetic organic chemicals in 1941 totaled less than one billion pounds; in 1986, the production of the top fifty organic chemicals reached 188 billion pounds (Reisch, p. 21). The overall chemicals and allied products industry in the United States had more than $200 billion in net sales in 1986. After a slump in the early 1980s, the U.S. industry underwent a drastic reorganization, followed by a period of "high production, rising sales, and strong profits" (Kiefer, p. 28).

The use of toxic chemicals is often accompanied by unanticipated consequences for health and society. Rachel Carson, in her classic volume *Silent Spring*, vividly described the irreversible effects of toxic chemicals on the human environment, in images that foreshadowed the disasters examined in this book. Specific chemicals have long been used without adequate understanding of broader consequences. Vinyl chloride, for example, has been widely used since the 1940s to produce the plastic polyvinyl chloride, but only in 1973 was it shown to produce a rare form of liver cancer in humans (Creech and Johnson). These unintended impacts have occurred for multiple categories of substances—in food additives, pesticides, industrial chemicals, and consumer products—which have infiltrated all aspects of modern life.

The health effects of toxic chemicals, however, are often difficult to understand. Animal experiments on the toxicity of chemical substances do not always indicate the adverse effects on humans. In human populations, environmental exposures are usually low-level and long-term and can interact with other factors. Moreover, epidemiology (the study of disease patterns in populations) is subject to many uncertainties in relating cause and effect, especially when the exposed group is small and the health effects are subtle (MacMahon). Before the full range of deleterious effects are known, the modern distribution system can permit a rapid and thorough contamination of society.

The difficulties in understanding the causes and effects of toxic contamination, while based in scientific uncertainty, derive from other factors as well. Often, political conflict erupts over the definition of causation, with disputes over what specific toxic substance is involved and what human and environmental effects are due to exposure. The resolution of this conflict has real consequences in economic, legal, political, and social dimensions, thereby raising the stakes for winning. A society's decision on causation does not result simply from scientific evidence but from a political and social construction of reality as well.

Incomplete knowledge about toxic contamination interacts with the interests of private and public institutions, suggesting similarity with infectious diseases. In the nineteenth century, vested interests in European and American cities blocked the construction of new systems of water pipes and sewer pipes designed to control and prevent cholera, until approaching epidemics raised public fears of death and compelled public agencies to act (McNeill, pp. 272–74). After Britain established separate water and sewage systems in the mid-nineteenth century in some cities, the idea spread to other European countries, but "not infrequently it took the same stimulus of an approaching epidemic of cholera to compel local vested interests to yield to advocates of sanitary reform" (McNeill, p. 273). Public health measures to control disease often require a redistribution of resources, challenging the interests of public and private institutions. Control efforts therefore can easily result in political conflict for infectious diseases as well as toxic contamination.

In the late 1700s, for example, a major epidemic of yellow fever in Philadelphia erupted into political controversy, as theories of causation split along existing lines of political competition. One side, supporting Hamiltonian ideas of strong central government, believed in the "importation" theory, blaming the epidemic on recent French immigrants from Haiti. The other side, supporting Jeffersonian notions of limited government, promoted a theory of "localism," which pointed to local swamps and unsanitary docks as the source of yellow fever (Eisenberg, p. 244). Even with advances in scientific knowledge, infectious diseases still produce political and social conflicts over causes and cures, as we see today in the tragic example of AIDS (Brandt; Eisenberg; Shilts).

These conflicts can arise with other forms of illness as well. As Leon Eisenberg stressed, "The response to human disability is not something given in the nature of things; the interpretation of disease is a social construction in which class and political values may at times be more decisive than empirical data" (p. 244). The vested interests and powers of social groups and institutions can affect all aspects of the human experience of illness, including the processes of discovery, symptom identification, diagnosis, treatment, and control, at the individual and community levels. Yet relatively little attention has been given to political analysis in studies of how people respond to illness (Zola; McHugh and Vallis).

Industrial society often creates and propagates technology far faster than it can control and respond to the consequences of that technology. Efforts to evaluate technology before its application have been instituted in many fields in the United States; examples include studies by the Office of Technology Assessment in the U.S. Congress, the quantitative risk assessment for pharmaceuticals and medical devices by the Food and Drug Administration, and the regulation of toxic chemicals by the Environmental Protection Agency and the Occupational Safety and Health Administration (Hoel, Merrill, and Perera). But the effectiveness of these assessments is a matter of debate. One review of the carcinogenic risk assessment of formaldehyde showed that "differences in assumptions can lead to orders of magnitude differences in risk estimates," demonstrating "how little has been accomplished in resolving the difficulties . . . inherent in the process of risk estimation" (Cohn, p. 276). The controversy over the regulation of biotechnology shows the continuing tension between those who favor strict controls prior to broad use and those who argue for loose controls with broad use to assess consequences (Krimsky and Ozonoff).

Often, society reexamines the application of new technology only after it is too late, after the device is thoroughly integrated into social institutions, after the device has produced a series of undesirable second- or third-order consequences, or worse, after the device has caused a disaster and a body count. All technologies, indeed all actions, have consequences that extend beyond the primary intended ones. A major challenge for society is to design and implement measures to reduce the second-order consequences that are "unintended, unanticipated or undesirable" (Bauer et al., p. 18). A study of the social history of the machine gun, "for many years the ultimate weapon of mass destruction," in the words of its historian, stressed that American inventors of that machine were motivated not only by the prospects of huge profits but also by a faith in human progress resulting from the application of new technologies (Ellis, pp. 25, 32). As the study concluded, "The technology that was designed to comfort us or to make us feel more secure now seems to threaten us from every side" (p. 178). Similarly, the automobile brought revolutionary mobility to society, but also came to clog the cities, pollute the air, and kill tens of thousands each year in the United States alone. Nuclear fission, once hailed as opening new vistas of man-made power, killed hundreds of thousands of people in the atomic bombs at Hiroshima and Nagasaki and now could obliterate many times more.

For toxic substances, the consequences of inadequate control can involve an insidious, invisible poisoning. In the disasters presented here, as in all toxic contaminations, victims confront multiple problems. The fundamental problem is that what the victims of toxic contamination want—redress—can be achieved only through a process of politicization. Some people may be dismayed that politics enters into chemical disasters and their aftermath. Some may even believe that toxic contamination is a technical problem and should be handled as such. To the contrary, I believe that chemical disasters show aspects that are inherently political and that political action is necessary for the victims of toxic contamination to achieve redress. Moreover, chemical disasters demonstrate political patterns that are predictable, in the sense that the patterns occur with a reasonable degree of probability.

The Process of Redress

People struck by a chemical disaster become suddenly involved in another world of new problems, conflicts, and institutions, even new personal identities. As victims of toxic contamination, they want to be made whole again, to return to their previous existence. But the notion of wholeness is ambiguous. Victims need to articulate their vague desires as specific demands. The victims' demands for redress commonly form around three basic problems of toxic contamination: care, compensation, and cleanup. These problems are common foci of political conflict in the three cases in this book and in chemical disasters generally.

The demand for redress in a chemical disaster can raise ethical questions. One can ask whether the victims should receive redress, whether the original distribution of wealth should be restored. One can imagine cases of toxic contamination in which these ethical questions might pose difficulties—for example, if the chemical affected only the pet animals of a rich family. But in most chemical disasters, the demand for redress poses no major ethical quandary. We intuitively believe that unwitting victims who suffer damage to person and property deserve redress. Without pursuing a full philosophical analysis, this book similarly assumes the ethical importance of redress.

But obtaining what the victims consider appropriate and adequate redress is not automatic. Both public and private institutions are often unwilling to provide immediate or full redress. Much of the burden of obtaining redress therefore depends on the victims themselves, on their identification of their problems as an issue, on their collective mobilization in group action, and on their alliances with other groups in the political arena. Victims use these political and social processes to define and resolve the problems of redress. How they use these processes both reflects and can change the balance of power between victims and social institutions.

Throughout this volume, I use the term *private institution* to refer to private corporations as well as for-profit cooperatives and the term *public institution* to refer to government agencies at the local and national levels. I use *social institution* to refer to both private and public institutions, thereby stressing the similarities of these two kinds of organization. For the sake of convenience, I use *corporation* interchangeably with *private institution,* despite technical differences.

The first phase of the political process of redress focuses on the individual's discovery of a private trouble and the institutional processes that maintain social problems as a *nonissue.* I follow the distinction made by C. Wright Mills and consider a private trouble a problem perceived as concerning an individual or family and a public issue a problem perceived as involving a larger group of people and institutions in society (p. 8). The phase of nonissue lasts until the toxic agent is publicly identified, often through the efforts of the victims and through the mass media. In the second phase, the problem appears on the agenda of society as a *public issue.* While private companies and public administrations generally seek to contain the issue, victims seek to expand the issue, struggling to assert their grievances through organization and through protest. In the third phase, the issue expands to include groups of nonvictims in alliances with victims' groups. These alliances use social conflict around the problem as a *political issue* to pressure private and public institutions to provide redress through a redistribution of resources.

This analysis of the phases of an issue connects to a scholarly tradition concerned with the use of power in communities and how agendas are set in society. The literature on community power and agenda setting examines which problems receive authoritative attention in society, how problems get on the public and political agendas, and especially how some problems are prevented from gaining easy access to the policy arena (Bachrach and Baratz, 1962; Crenson; Cobb and Elder; Kingdon; Nelson, pp. 20–21). This book's investigation of the processes of issue formation gives particular attention to the bottom-up perspective of victims in contrast to other studies (such as Nelson's) that emphasize the top-down perspective of officials. In addition, the three phases I propose in the politics of redress for toxic victims bear strong resemblence to the findings of other analysts who have examined the political evolution of issues in society—for example, the three processes proposed by John Gaventa in his study of protest by poor people in Appalachia (pp. 257–58) and the three phases developed in Stuart Scheingold's analysis of legal rights and political mobilization (p. 131).

E. E. Schattschneider, in *The Semisovereign People,* his classic work on the social processes of conflict in democracy, stressed that the scope of a conflict, especially the number of people involved, determines the out-come (p. 2). "Every change in the scope of conflict has a bias; it is partisan in nature. That is, it must be assumed that every change in the number of participants is about something, that the newcomers have sympathies or antipathies that make it possible to involve them. By defi-nition, the intervening bystanders are not neutral. Thus, in political con-flict, every change in scope changes the equation" (pp. 4–5). Moreover, as more people become involved and the equation changes, the conflict changes its nature, and "the original participants are apt to lose control of the conflict altogether" (p. 4).

Schattschneider's perspective shapes my analysis of the political con-flicts around a chemical disaster. As the scope of the issue expands, the victims incur additional costs and burdens at each phase of the transfor-mation from private to public to political.

Non-Issue

In many cases, victims fall ill without knowing that the cause is toxic contamination. Without a correct diagnosis of their illness as chemical poisoning, victims confront multiple obstacles in getting effective medical care and other assistance. Diagnosis is often difficult and slow because of institutional resistance, conflicts of interest, and the nature of toxic con-tamination. Victims thus begin with little understanding of the social causes of their private trouble and no well-defined common symbols around which to organize. Conflict is contained at the individual or family level. Greater visibility of toxic contamination facilitates public recognition and remedial action by social institutions, as has been shown for other kinds of health problems (Pearson). Victims here face the dilem-ma of going public: they want to keep their illness private to avoid social stigma and its costs, but they also want official recognition of the cause of their problem, which requires going public with its inherent risks and conflicts. Public identification of the cause as a specific chemical contami-nant represents a major turning point, transforming the problem from a private to a public concern. Victims become aware of fellow sufferers and redefine the problem as social and not individual, as controllable and not a private fate or an act of God.

Public Issue

In the second phase, when the problem is a public issue, the victims work to organize themselves, achieve public recognition, and influence public and private institutions. Private companies, as the source of the chemical disaster, commonly respond with measures to contain the issue, for economic and organizational reasons. Most public administrations also seek to contain the issue in attempts to establish "manageable" crises. The victims respond by creating formal organizations and by conducting public protests to counter the power of social institutions over the public issue. Both responses by the victims are efforts at empowerment that expand rather than contain the issue's scope. As a public issue, the problems of care, compensation, and cleanup become fields of public struggle. In this phase, victims confront the dilemma of group action: victims need group action to put pressure on social institutions to provide redress, as dictated by social norms; but both organization and protest create new conflicts and costs for the victims and require that they violate norms and engage in institutional disruptions (Piven and Cloward). Victims overcome this dilemma through a change in consciousness as they come to perceive the prevailing social order as both "unjust and mutable" (p. 12). The transition from the victim's struggle as a public issue to society's conflict as a political issue occurs through another expansion of the issue's scope, often through protest by victims and through the assistance of "issue-entrepreneurs" (Kingdon, pp. 129–30).

Political Issue

In the third phase, when the problem emerges as a political issue, victims become involved in organizations that seek to help them but also to use them. Governments, companies, media, and political groups adopt and adapt the issue into their repertoire of interests. Tension develops between victims' groups that seek redress and outside organizations that pursue other long-range and more general goals. This phase highlights the victims' dilemma of political alliance: their need to ally with other forces and the loss of control produced by alliance. Alliances promise support and power but also produce dissension and weakness, with conflicts following the structure of political competition in the society. For care, compensation, and cleanup, the criteria for "official" victims become foci for bitter political issues. Public and private administrations draw lines to decide who should receive benefits and who should not. Each line-drawing process creates peripheral victims, persons who believe they have been harmed but excluded unjustly from receiving benefits. Those peripheral victims look elsewhere for support from alliances, often outside the political mainstream, to redefine the official boundaries, alter the administrative policies, and change the institutional context.

Not all cases of toxic contamination make the full transformation from private to public to political. In some cases, the epidemiological inference between cause and effect may be difficult to draw, so that victims do not realize their common problem. The victims may be geographically or socially obstructed from organizing group actions; the problem may remain at the individual level, and the victims may remain without redress. Even if the problem becomes a public issue, political groups may not adopt it into their agendas as a political issue. Or public and private institutions may respond in ways that resolve the problem in the private or public sphere.

The three chemical disasters of this book, however, all moved from private to public to political. As the above discussion suggests, the path to political conflict depends on the structures and the strategies of the participants in the disaster but also on the uncertainty of the damage caused by the disaster. Although the politics of redress impose substantial costs on chemical victims, benefits may also result from participation in the political process. An economically "rational" participant would expect the benefits to exceed the costs, as stressed in theories of collective action and rational choice (Olson; Hardin). Some benefits and costs, however, are difficult to calculate. And some toxic victims miscalculate or respond "irrationally," acting not from an economic perspective, but from a political stance of moral outrage, a sense of entitlement, or a desire for revenge. Individual victims deal in different ways with the three dilemmas, and some victims are willing to pay more than others in order to compel social institutions to provide adequate redress.

The Study of Chemical Disasters

Paradoxically, crisis in society creates a window on normality, one that offers a view of underlying political patterns not usually visible (Molotch; R. M. Anderson et al., p. 17). An astute political participant and observer of the Italian chemical disaster at Seveso summed up this characteristic as "the extraordinary event and the ordinary administration" (Conti). In a similar way, I use the crisis of a chemical disaster to view and dissect the ordinary responses of individuals and institutions to extraordinary situations. And I approach the study of chemical disasters with several analytic themes in mind.

The three chemical disasters of this book represent what Charles Perrow deftly termed the "normal accidents" of high-risk technologies. These accidents result from multiple failures within a system characterized by components with complex interactions and tight coupling. In such systems, "multiple and unexpected interactions of failures are inevitable. This is an expression of an integral characteristic of the system, not a statement of frequency. It is normal for us to die, but we only do it once. System accidents are uncommon, even rare; yet this is not all that reassuring, if they can produce catastrophes" (Perrow, p. 5). We therefore need better methods for the prevention and management of such low-probability but high-cost events.

A primary objective of this study of chemical disasters is to understand the problems of power and powerlessness that confront toxic victims. My analysis of these problems is shaped by three "dimensions" of power identified by Steven Lukes and used by John Gaventa in examining patterns of quiescence and rebellion in relatively powerless groups (Lukes; Gaventa).

Analysis of the first dimension of power stresses observable behavior and concrete decisions made by political elites and the use of material resources in political bargaining. This approach to power represents the conventional pluralist school, developed mainly within American political science, especially through the works of Robert A. Dahl (1961) and Nelson Polsby (1963). This one-dimensional view of power emphasizes specific events in which one actor causes another to do something he or she otherwise might not do.

The pluralist view has been criticized on a number of points, particularly its emphasis on problems that have achieved attention within official decision-making arenas and its assumption of an open system with relatively little concern given to the obstacles that keep grievances private (Gaventa, pp. 5–8). Although Dahl protested that "pluralist theory" has been unfairly criticized and that he has indeed recognized the unequal distribution of resources and its political consequences (Dahl, 1982, pp. 207–9), the pluralist approach nonetheless has tended to emphasize the perspective of those with power more than those without.

Inquiry into the second dimension of power pays attention to the processes of agenda formation and emphasizes that some issues and actors are excluded from the decision-making process. Schattschneider wrote, "All forms of political organizations have a bias in favor of the exploitation of some kinds of conflict and the suppression of others because *organization is the mobilization of bias.* Some issues are organized into politics while others are organized out" (italics in original, p. 71). I apply this analytic approach to examine how both public and private organizations shape public debate over chemical disasters by setting the agenda and defining the issues in ways that tend to protect narrow organizational interests more than to serve broader societal purposes.

Analysis of the third dimension of power emphasizes the efforts to influence, shape, and determine the wants and demands of powerless groups (Lukes, p. 23). Whereas the second-dimension approach focuses on "barriers that prevent issues from emerging into political arenas—i.e. that constrain conflict," inquiry into the third dimension analyzes the "use of power to pre-empt manifest conflict at all, through the shaping of patterns or conceptions of non-conflict" (Gaventa, p. 13). This analytic approach requires study of how power holders create a social construction of reality that serves their interests (Berger and Luckmann) and how the powerless can or cannot change that construction.

The political use of symbols plays a central role in the third dimension of power. The ability to have one's symbols accepted by others is an important source of power. Dominant groups use symbols to contain opposing constructions of reality in efforts to maintain the existing distribution of power. But symbols can also be used to advance alternative conceptions of reality, in attempts to transform the distribution of power. The critical question becomes who controls the meaning of these key symbols, a battle with significant economic and social consequences. In this battle, language itself becomes a form of political action, with both power and limitations (Edelman, 1988). For the victims of chemical disasters, the strategic task becomes the development of their own symbolic universe, to reject the definition of reality propagated by social institutions, and to accrue sufficient power to achieve redress.

Grasping all three dimensions of power in the study of chemical disasters requires what Clifford Geertz called "thick description," a notion he borrowed from Gilbert Ryle, to "aid us in gaining access to the conceptual world in which our subjects live so that we can, in some extended sense of the term, converse with them" (Geertz, p. 24). For toxic chemicals, the assessment of risk often assumes that hazards can be objectively measured through expert calculations of exposures and probabilities of responses and then translated into economic gains and losses. A number of analysts reject this thin technocratic approach and adopt a thick description of risk (Perrow, p. 328; Douglas and Wildavsky). Dorothy Nelkin and Michael S. Brown, for example, in *Workers at Risk: Voices from the Workplace*, argued that subjective, contextual, and experiential aspects must be included to understand the political and social realities of toxic chemicals and other risks.

In this book, the thick description of chemical disasters is achieved through case studies. This method allows one to appreciate the ethnographic detail of a few instances, bringing alive the experiences and perspectives of toxic victims. But the small number of cases in the sample also reduces the reliability in a statistical sense and raises problems of representativeness. These problems are addressed through the use of discrete, middle-level hypotheses in analytic chapters, to examine similarities and differences among the cases and argue that some patterns occur more generally. Although single-case studies exist for a number of chemical disasters for Love Canal (Levine), Bhopal (Shrivastava), Seveso (Conti; Whiteside), and Michigan's PBBs (Chen; Egginton)—none adopts an explicitly comparative approach and none places the issues of power and powerlessness at the center of analysis.

This study of chemical disasters thus depends on the comparative analysis of both politics and policies. This subfield of political science surged in popularity in the 1980s (Hancock), including various comparative studies of environmental and occupational health policies (Enloe; Lundquist; Kelman; Badaracco; Brickman, Jasanoff, and Ilgen; D. Vogel, 1986). Yet, the three cases in this book probably do not add up to what Adam Przeworski and Henry Teune called the "most similar systems" design for comparative analysis. Even with many similarities, moreover, that design does not allow strong conclusions about causal relations. The conclusions are weaker when the number of causal factors being explored equals or exceeds the number of systems being explored. But the comparative approach used in this book does allow an exploration of important variables and an assessment of relations between variables and outcomes, thereby lending plausibility to hypotheses. Most important, the three case studies show a remarkably similar pattern in the political evolution of chemical disasters as issues in Japan, the United States, and Italy, despite significant differences in cultural context. Comparative analysis can explain why.

This comparative analysis of chemical disasters is intended as well to indicate more effective public policies and more effective political strategies. Both public and private institutions currently lack means of sufficient complexity to predict or to cope with chemical disasters, disasters those organizations helped create. The tragic example of the chemical disaster in Bhopal, India, reminds us of the terrible destructive potential of modern technology and the difficulties of control. What initially appear as simple accidents turn into complex disasters, involving much uncertainty and varied interests, patterns of unexplored problems, unanticipated risks, and ambiguous responsibilities. Comparative analysis can help identify commonalities, providing lessons for policy makers in both public and private institutions on ways to prevent chemical disasters and to manage the consequences when prevention fails.

Comparative analysis can also help correct tendencies toward ethnocentric perspectives found in writings on both Japan and Italy. For Japan, many scholars have used the United States as the basis of comparison, while virtually ignoring European countries. Yet Japan's political institutions, ideological spread, geophysical constraints, and other factors more closely resemble those of such nations as France, Germany, Britain, and Italy than those of the United States. Various scholars have observed similarities in the political economies of Italy and Japan, but few studies have been completed (McNelly, p. 23; Leiserson, p. 80). And Italy, according to a leading American specialist on that country's politics, has generally been isolated from comparative political studies and treated as an anomaly or exception to various models and hypotheses (Lange, 1980b, pp. 1–2). One objective of this book is to widen those perspectives.

Finally, this book's comparative analysis allows us to explore the role of culture in chemical disasters. Over the past two decades, political analysis has generally discounted cultural variables in favor of economic factors as expressed in models of rational choice. Recently, renewed interest has arisen in using theories of political culture to explain political systems (Inglehart) and political change (Eckstein). A tendency nonetheless remains to consider culture either as an independent causal variable or as irrelevant. That dichotomy is simplistic. Cultural variables play a complex role in politics generally and in the politics of chemical disasters specifically (Laitin and Wildavsky).

Culture here is defined as a system of interacting symbols that give shared meanings to a social group. In Geertz's symbolic approach culture "is not a power, something to which social events, behaviors, institutions, or processes can be causally attributed; it is a context, something within which they can be intelligibly—that is, thickly—described" (p. 14). Geertz's call for "thick description" stresses that the analysis of culture is "not an experimental science in search of law but an interpretive one in search of meaning" (p. 5).

Defined in this way, culture plays an essential role in our three chemical disasters, providing people with a universe of meaning for understanding, for interpreting, for communicating, for acting. Although its role is essential, culture does not, I argue, influence human behavior as immediately or as directly as structure. In some cases, culture provides conflicting guides and contradictory propensities. Furthermore, culture is not static, but allows for choice, for redefinition, for change. Cultures and cultural variables, then, never operate alone. They serve as intervening variables.

Despite differences of national culture in the three cases, victims and organizations responded in common patterns, which suggest both the complex role of culture and the overwhelming importance of structural variables in shaping behavior, especially the distribution of power in society, the organization of private corporations, the organization of public administrative agencies, and the organization of the mass media. For these reasons, the analysis emphasizes structural more than cultural aspects. And from these structural variables, I argue, emerge the common political patterns of chemical disasters.

The chapters on the three disasters are based on field research conducted from the summer of 1978 to the fall of 1979, with subsequent follow-up. In the field research, I collected primary and secondary documents and conducted interviews with approximately eighty major participants in each incident, selecting prominent persons who represented different perspectives in seven categories: victims, politicians, journalists, civil servants, businessmen, private and public interest activists, and legal and medical specialists. Inevitably, gaps remain in the stories, due to inherent complexities, limitations in space, and difficulties in access to some information. Books have been written about each of the three disasters. In summarizing the case in a single chapter, I have selected material most relevant to the politics of redress.

This book first reconstructs the three cases chronologically in separate chapters. The second section consists of three chapters of comparative analysis, one chapter for each phase as an issue. The final chapter draws conclusions for the policies and politics of chemical disasters and for the politics of relatively powerless groups in advanced industrial society.

Chapter 2

Kanemi PCB Contamination

In spring 1968, life turned bitter for the family of Kamino Ryūzō. Kamino had recently retired, after thirty years as an office worker at a coal mine in northern Kyūshū, Japan's westernmost main island. That March, his eldest son, home from university on spring vacation, began working to earn spare money. He soon ran into problems. On the first day of work, he lost his appetite and felt fatigued. His eyelids puffed and his legs and arms swelled. Sweat drenched his work clothes. After five days of work, he could go no further. Other members of the Kamino family developed similar problems: headaches, painful limbs, total exhaustion. The family tried all sorts of medicines, Western drugs as well as Chinese herbs, but with no relief. The symptoms only worsened. In April, their eyesight began to weaken (Kamino, 1972, p. 14).

No one knew the cause of the strange symptoms, but some imagined a resemblance to syphilis. Tension soared between Kamino's eldest daughter and her husband, as each suspected the other of some sex-related disease. The wife finally visited a gynecologist, who assured her the illness was not syphilis. But as health problems persisted, so did marital tension (p. 15).

The family's health continued to deteriorate in May. They lost all desire to eat and had to force food down. Kamino lost ten kilograms, his wife thirteen, and his eldest daughter eighteen. They awoke each morning to find their eyelids sealed shut by gluelike secretions. Once they opened their eyes, they could see clearly for only four or five meters. Wind or sunshine caused secretions to flow, sealing the eyelids again. Yet no doctor could diagnose the problem (pp. 15–16).

But the worst problem was the eruption of acnelike boils over their entire bodies. "After June, squeezing the boils became the biggest task of the day, and it always waited until evening. . . . The work and pain were unbearable. One of us would always start to cry, and by the end of the evening we were all in tears. We always went to bed at two-thirty or three in the morning. Any part of our bodies that touched the bed stung with pain. If we used pillows, then our necks hurt" (pp. 16–17).

The disease transformed the family's relationship with society. "People looked at us as if we were monsters, not human, like something other than human. If we went to buy things, the owner of the store hated it, because other people would not come" (Kamino interview). To visit the doctor, the family traveled by taxi, so that other people would not notice. They made every effort to stay inside their house and to hide from society, leaving only to buy food and visit different doctors (Kamino interview).

Kamino's daughters reacted acutely to the disease and its consequences. In Kamino's words, his younger daughter "could no longer bear the sight of her own face and feared to look in the mirror. After the boils appeared, she even avoided going out to throw away the garbage. On the train to the hospital, she hid behind a newspaper or a handkerchief, and my wife served as a screen. When my eldest daughter had to go out shopping, she always came home in tears, crying, 'Strangers stare at me, and I can't stand the way their stares cling to my body'" (Kamino, 1972, p. 17).

Throughout the summer of 1968, the Kamino family sought help from one doctor after another. "No one knew what the cause was. . . . All the doctors said, 'It's a strange disease.' We knew that the disease was something different, something strange, never before seen or experienced. . . . But we did not have any idea who was responsible for our disease." Kamino sensed that if the doctors could not name the disease, then the patient must somehow be at fault (Kamino interview).

In late August, at the urging of his sister-in-law, Kamino sent his younger daughter to a major medical school hospital in Osaka. But even in early October, after weeks in the hospital, the daughter's illness still baffled the medical specialists.

Then, on 10 October 1968, the western edition of the *Asahi* newspaper announced the spread of a "strange disease" (*kibyō*) around Fukuoka city on Kyūshū, due probably to contaminated cooking oil. The symptoms matched those afflicting the Kamino family: severe acnelike boils, extreme fatigue, excessive eye secretions, discolored fingernails. The article did not mention the brand of cooking oil, but it reported the type—an oil refined from rice bran. Doctors at the dermatology clinic of Kyūshū University Hospital had examined about thirty patients with similar symptoms. But one doctor suggested that a much larger number of patients might exist throughout Kyūshū.

The media announcement produced a huge impact, especially on the afflicted. As Kamino put it, "The strange disease was not just our family. Other identical patients existed. It was not hereditary and not contagious. Once the cause became known, I thought, Japan's advanced medicine would soon cure us. It was a great relief" (Kamino, 1972, p. 18). The family felt they now had companions, fellow sufferers. "We knew there was someone else with the same pain. Some light shone in the family— just that night" (Kamino interview).

The next day, on 11 October, Kamino's wife, daughter, and two grand-children visited the university hospital in Fukuoka city. The corridors of the dermatology outpatient clinic were jammed with patients, all suffering from similar symptoms, all hoping for some cure. Kamino and other patients soon learned that doctors at Kyūshū University had been studying the disease for several months.

There are several accounts of how the problem was first identified. The first patient, a three-year-old girl, visited the university's dermatology clinic in early June 1968. She suffered from severe acne. When the doctor called in her family, he found similar symptoms. The acne pustules covered not only the face, as normally occurs, but spread to unusual places: armpits, genital regions, breasts. The doctor recognized the disease from studies he had performed on workers at an agricultural chemical factory. Since the symptoms occurred in "males and females, babies and old men," the doctor reasoned the cause was not an occupational disease. "I knew it was chloracne, but did not know how the chlorine compounds entered the family" (Higuchi interview).

In early August, several other families visited the university dermatology clinic, among them the family of Kunitake Tadashi, an employee of the Kyūshū Electric Power Company. They had become ill in early 1968. Kunitake met others at the clinic with symptoms identical to his own. Moreover, two families lived in Kyūshū Electric Power Company housing in Fukuoka—Kunitake's own residence until six months earlier, when he was transferred to the nearby city of Omuta (Nishimura, p. 70; Kamino, 1972, p. 19).

In searching for a common cause of the disease, Kunitake remembered that in February he and his fellow employees had divided up a large drum of Kanemi-brand rice oil, produced from rice bran and therefore high in vitamin B. He knew that his own family had used this oil continuously since then, to prepare *tempura* and fried vegetables. When Kunitake returned to his former apartment house to visit the families still living there, he found that every household using the oil suffered the same horrible disease. Thoroughly convinced by late August that his theory was correct, Kunitake took a sample to doctors at Kyūshū University and asked them for an analysis. Weeks passed with no response. Finally, Kunitake gave up hope on the university doctors. On 4 October, he delivered a second sample to the Omuta health center and reported the suspected poisoning to the public authorities (Nishimura, p. 70).

As a result of Kunitake's request, a technician at the Omuta health center telephoned the Fukuoka prefectural health department that same day (4 October) to report a possible case of poisoning due to rice oil. Four days later, a health-center employee visited the prefectural office to provide details about Kunitake's illness, his examinations at the university's dermatology clinic, and the existence of other families suffering from the same suspected poisoning. The prefectural health department then instructed Fukuoka city officials to speak with the other families and directed Kitakyūshū city officials to inspect the Kanemi rice-oil factory. The next day, 9 October, a prefectural health official met with Dr. Gotō Masayasu, a dermatologist at the university's clinic, who had examined Kunitake and other similar patients (Akagi, p. 406). But the authorities issued no public warning. The *Asahi* scoop on 10 October came by word of mouth, through a student of Kunitake's mother, an instructor in the tea ceremony and flower arrangement (Nishimura, pp. 69–70).

Dr. Gotō provided a different perspective on who had identified the cause. In early August, he recalled, after several families had appeared at the clinic, he suspected some common contaminated food, first "margarine dermatitis," then poisoned instant noodles. When those two proved not to be common foods, the doctor questioned patients about cooking oil. Then, Gotō stated, he found the connection with Kanemi rice oil. He also denied that Higuchi made the chloracne diagnosis in June 1968 (Gotō interview).

In late August, Dr. Gotō received some Kanemi rice oil from a patient and asked a scientist in the hospital's central examination room to analyze the samples. The scientist replied he could do nothing during summer vacation and, moreover, had no available funds for such research. Consequently, the bottle of oil remained under the lab desk, untested (Gotō interview). According to another patient, doctors at the dermatology clinic began telling patients in early September that the cause of their disease might be Kanemi rice oil and that they should stop using the oil. "But the doctors asked us not to publicize this, because they did not have detailed proof" (Ujino interview).

On 7 September, however, three dermatologists from Kyūshū University reported to a regional academic meeting the outbreak of an unusual skin disease in five families. The doctors suggested a possible diagnosis of "keratosis follicularis," an abnormal thickening and scaling of skin growth around skin follicles. But they also noted distinct similarities to symptoms found in workers at a factory producing pentachlorophenol, a widely used agricultural chemical. The doctors stated that the disease's cause was uncertain, but all five families had used the same rice-bran oil for cooking. The ostensible purpose of the doctors' report was to search for other dermatologists with similar patients (Totokawa, Nakatani, and Kitamura). But according to another member of their department, the doctors primarily wanted to file the first academic report on the skin disease (Gotō interview). After the academic conference in September, this report did not receive any attention in newspapers and did not appear in a medical journal until almost one year later.

Dr. Gotō apparently tried to obtain help from public authorities to analyze the Kanemi rice oil. He took a patient with him to visit the Fukuoka prefectural health department and requested cooperation in examining the oil. But the prefectural officials replied that the problem was only one instance of an acne skin disease and that the prefecture could do nothing. The officials told him to try the city health department. He then visited city officials, who said they would consider the problem. "I went here and there, but nothing happened. No reaction" (Gotō interview).

Until the 10 October report in *Asahi*, consumers continued using Kanemi rice oil—except for the few families attending Kyūshū University's dermatology clinic. Doctors and public officials aware of the skin disease considered it a routine problem, limited and unimportant. The *Asahi*'s scoop changed all that.

With its headquarters located in Kitakyūshū city, the Kanemi Warehouse Company (hereafter, the Kanemi Company) in 1968 employed about 450 persons and was capitalized at ¥50 million or $140,000 (conversion to dollars, hereinafter, calculated at the prevailing exchange rate). The company ranked in the middle of cooking oil producers in Japan, but among the top manufacturers of rice-bran oil, and also operated warehouses. A public enterprise in Osaka, designed to promote investment in small and medium-sized industries, held 40 percent of the Kanemi Company's stock. In 1956, the Ministry of Agriculture and Forestry gave official certification to the company's warehouses, and in 1961 the ministry gave the company's president and founder, Katō Heitarō, an award for his work in developing the rice-bran oil industry in Japan. Three years later, in 1964, the Kitakyūshū facility received the ministry's recognition as a "model factory." The company also owned and operated factories in seven other cities in western Japan. This was not some simple or unprofessional small company; in the rice-oil business, they knew what they were doing (Katō Y., 1989, pp. 136–40).

Following the *Asahi*'s disclosure of human victims, public authorities

mounted an intensive effort to identify the poison in Kanemi rice oil. On
11 October, the prefectural health department formed a commission to
deal with the problem and reported the incident to the central Ministry of
Health and Welfare. The Kitakyūshū city health bureau entered the Kanemi Company's factory to collect samples for analysis and advised the
company to stop its sales of rice-bran oil. The company's president,
however, refused to comply, asserting that its oil had caused no harm.
Three days later, a doctor at Kurume University Medical School announced that he had found arsenic in the cooking oil—a charge the
Kanemi Company quickly denied. The arsenic theory was also rejected
by a special committee formed at Kyūshū University, including three
teams on clinical, analytical, and epidemiological aspects. On the fifteenth, Kitakyūshū officials legally ordered the Kanemi Company to
cease production of oil. This time it complied.

On the sixteenth, the central Ministry of Health and Welfare, in Tokyo,
declared the Kanemi Company in violation of the Food Hygiene Law. The
ministry instructed the chief health officers of all prefectures and designated cities to halt the sales and movement of Kanemi rice-bran oil, to
report the number and the symptoms of patients, and to analyze the
suspected oil (Akagi, p. 407). On the same day, Tokyo bureaucrats and
politicians began arriving in Kyūshū. Officials from the Ministry of
Health and Welfare came to inspect the factory and consult with local
health departments. Opposition politicians from the Socialist, Communist, and Komeitō (Clean Government) parties arrived in separate groups
to learn about the incident from the company, local officials, and the
victims. One week after the ministry's directive, and two weeks after the
Asahi breakthrough article, more than ten thousand potential sufferers of
the poisoning had registered at health centers in twenty-three prefectures.
By the end of October, that number exceeded twelve thousand.

On 22 October, the Health Research Institute of Kōchi prefecture
detected an organic chloride in the suspected rice oil. A national institute
substantiated the finding five days later. Then, on the twenty-eighth, a
food engineering professor of the Kyūshū University research group entered the Kanemi Company factory and returned with a sample of a
substance used as a heat-transfer medium, known by the brand name of
Kanechlor. On 1 November, the university committee announced that the
poison in the cooking oil was probably Kanechlor—or polychlorinated
biphenyls (PCBs), a highly toxic chemical. *Yushō*, meaning "oil disease,"
became both the medical and popular name for the illness.

The health problems in humans had been preceded by health problems
in chickens. In February 1968, flocks of chickens in western Japan began
dying, and veterinary officials first suspected a contagious illness. Soon,

however, several research institutes in Kyūshū traced the damage to two brands of chicken feed containing the Kanemi Company's "dark oil," a mudlike by-product from processing rice bran into cooking oil (Kōga interview). By mid-March, officials in the Ministry of Agriculture and Forestry knew that both feed companies used the same by-product and that the Kanemi Company's dark oil probably caused the chicken poisoning (Kohanawa, p. 117).

The chickens' symptoms closely resembled those of chick edema disease, which occurred in the United States in 1957 and was linked to contaminated feed (Schmittle, Edwards, and Morris; Kimbrough, 1972). In 1962, U.S. scientists produced chick edema symptoms with chlorinated biphenyls (McCune, Savage, and O'Dell), and in 1967 the disease's cause was identified as a chlorinated dibenzodioxin (Cantrell, Webb, and Mabis). In 1968, this information led a Japanese researcher in a feed company to suspect a similar substance in the Kanemi Company's dark oil. But apparently because of inadequate equipment, he could not find a chlorine compound in the oil until after the human illness of Yusho became public knowledge in the fall of 1968 (Kōga interview; Kōga et al., 1970, 1971).

Several newspaper articles had appeared in April and May 1968 on the chicken poisoning in western Japan. One article in April reported the probable cause as dark oil added to the feed (*Asahi*, 11 April 1968). The article stated that a fast response prevented any sick chickens from reaching the market, "but even if eaten by chance, [the chickens] would have no effect on people." How *Asahi* came to that conclusion is unclear, but the statement served to minimize the problem's importance for public health. Another article, in May, identified the source of the dark oil as the "K-Company" located in Kitakyūshū. But no reporter investigated the scope of the contamination. No one pursued the possibility that if the by-product (oil for animal feed) were contaminated, then the main product (cooking oil for human food) might also be contaminated. The media treated the problem as a routine matter, of concern to chicken farmers but not to general consumers. The press reported the chicken deaths, but did not understand or investigate the broader meanings.

In spring 1968, the two feed companies began paying settlement fees to chicken farmers, while seeking to recover damages from the Kanemi Company. Kanemi Company officials, however, stubbornly denied any problems with their dark oil. Hoping to increase pressure on the Kanemi Company, the feed firms convinced the Ministry of Agriculture and Forestry to test the dark oil (Kōga interview).

The ministry assigned the project to Dr. Kohanawa Makoto, a scientist at the National Institute for Livestock Hygiene, who was then director of the Isotope Research Office and had a personal interest and some research experience in toxicology. But the ministry did not provide him with additional staff, equipment, or funding to identify the toxic contaminant (Kohanawa interview). Because his staff members were unrelated to the poisoning incident, Kohanawa encountered a "very difficult situation" (Kohanawa, p. 118). He performed a simple experiment, showing that chickens fed Kanemi's dark oil died in a short time with symptoms similar to those found by chicken farmers. He presented the findings in a formal report in early June. Even then, the Kanemi Company's president refused to accept that his dark oil contained a poison (Kohanawa, p. 118). Throughout these tests, as the scientist later recalled, "we consciously avoided" thinking about the possible impact of contaminated feed on human health (p. 118).

Later that month, the ministry issued several directives to the feed companies and to the Kanemi Company, suggesting they improve their quality control procedures to prevent a similar occurrence (MAF, 29 Oct 1968). As one feed company scientist, Kōga Kiyomi, explained, "The ministry considered the dark oil incident a problem between private companies, to be resolved by the companies. . . . The ministry did not want to touch the problem" (interview).

Yet the ministry did take several more steps. In July, a ministry official wrote in a feed journal that several heavy metals were not detected in the dark oil and that the most likely cause of poisoning was a "deterioration" of the oil or its raw materials. The official noted the similarity between chick edema disease and the dark oil poisoning but stressed that the two incidents should not be considered identical (Fukuhara). Nonetheless, in August, the ministry quietly started a study group to examine the causes of the disease and the poisoning (MAF, 29 October 1968).

According to government records, the dark oil incident poisoned more than two million chickens, killing four hundred thousand of them immediately (MAF, 29 October 1968). In early spring 1968, the two feed companies stopped using the Kanemi Company's dark oil. By the end of April, the strange illness disappeared among chickens—while it continued unchecked among humans. One ministry official, who entered the Kanemi Company factory in mid-March to investigate the dark oil, apparently then asked the company president whether there were any problems with the cooking oil and was assured of its quality (Katō Y., 1989, p. 17). But no one else involved in the dark oil incident seemed to question the quality of the rice oil, despite connections in the production process (Figure 1). Months later, on 16 November 1968, the government finally identified the dark oil's poison: Kanechlor.

In 1968, eight rice-bran oil companies, including the Kanemi Com-

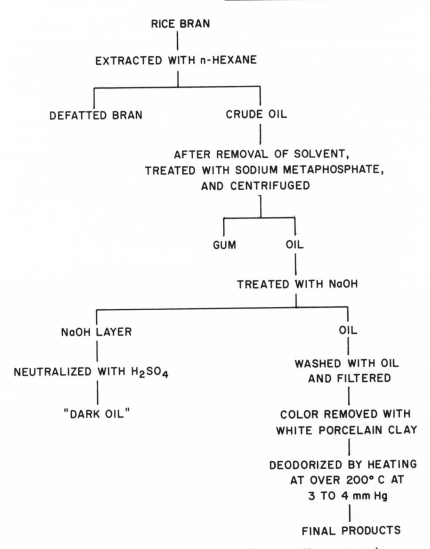

Figure 1. Production process of Kanemi rice oil. Source: Kuratsune et al., 1972, p. 125. Used by permission of the publisher.

pany, were using Kanechlor in the production of oil (MHW, 23 December 1968). Rice-bran oil had a foul smell, which could be removed only by heating it above 230 degrees centigrade. Kanechlor was used as a heat-transfer medium, and understanding the history of Kanechlor is essential to understanding how and why Yusho happened.

The industrial history of polychlorinated biphenyls stretches to the late nineteenth century, when two German scientists first synthesized the compound. It was seen as having potential for use in industry because of its chemical and physical properties: fluidity at room temperature, low flammability, extreme stability, low conductance of electricity. Industrial production of PCBs began in the United States in 1929. Initially, industry considered using PCBs in protective coatings (such as lacquers and varnishes), in die castings, and as heat-transfer agents (Penning). PCBs are actually a class of compounds, with varying number and arrangement of chlorine atoms around one biphenyl skeleton (two linked benzene rings), and commercial PCB preparations contain mixtures of different isomers. Kanechlor 400, the product used by the Kanemi Company, contained mostly tetrachlorobiphenyls plus some other isomers.

During the Second World War, use of PCBs as a coolant/dielectric in transformers and capacitors expanded rapidly, and production increased. In the postwar period, PCB uses continued to diversify, as industry added the chemical to paints, insecticides, lubricating oils, plastics, and inks in "carbonless" copy paper (Commoner). Japan first used PCBs soon after the Second World War, importing the chemical to replace insulating fluids previously used in the electric industry (Isono, p. 78). Then, in 1954, Kanegafuchi Chemical began manufacturing PCBs in Japan, producing 200 tons in 1954, 2,200 tons in 1961, and 5,000 tons in 1968. Although 90 percent of Kanegafuchi's PCBs in 1954 served as electrical insulation, by 1962, 400 tons (about 20 percent) went for heat-transfer media (p. 81).

Health problems associated with PCBs and similar compounds also began to appear in the late nineteenth century. A German scientist in 1899 identified a skin disease and named it chloracne, believing the cause to be free chlorine (Herxheimer). During and after the First World War, scientists linked chloracne with certain chlorinated hydrocarbons. Then, in 1936, an American scientist published a detailed report on chloracne and systemic poisoning due to PCBs (Schwartz). In Japan, several scientific papers on chloracne appeared in 1938–40; and in the late 1940s, Dr. Nomura Shigeru, a specialist in occupational medicine, studied PCB toxicity in animal experiments and reported on chloracne (Nomura; Nomura and Arimatsu). In the early postwar period, as Japanese companies began using PCBs for the first time, problems of chloracne spread. One condenser factory introduced PCBs in 1952, and five years later, 50 percent of its workers suffered from skin disease. The company, however, refused to allow a public report on the factory's PCB health hazard until after the Yusho incident became public. Even when the report appeared in 1969 in a scientific journal, it did not disclose the company's name (Hara).

Substantial production of rice-bran oil began in the mid-1950s, partly because of advances in oil-processing technology. One such advance was the use of PCBs to deodorize the oil by heating it. In the Kanemi Company's process, hot Kanechlor circulated through a coil of stainless steel pipes in the center of a large tank containing the oil mixture. Prior to late October 1968, few people knew that the Kanemi Company used a chlorinated hydrocarbon in pipes to heat its cooking oil. Even fewer imagined that the chemical might leak into the oil.

The announcement by the Kyūshū University research group on 1 November 1968 focused attention on Kanechlor as the most likely poison. During an on-site investigation, Kitakyūshū city officials found irregularities in the repair record for one deodorization tank. Kanemi Company officials denied any connection between the repairs and the pipes carrying Kanechlor (*Asahi*, 5 November 1968). Then, on the fourth, the university researchers, known as the Yusho Research Group, formally reported their conclusion that Kanechlor caused Yusho. Gas chromatograph analysis of the heat-transfer medium and of Kanemi cooking oil used by patients yielded identical patterns (Yushōhan).

The next morning, the Kanemi Company's president, Katō Sannosuke, announced at a news conference that the company would accept full responsibility if the university research group showed his rice oil to contain a poison. Katō stated that he had yet to receive the university scientists' report. He did not explicitly dispute the researchers' analytic findings, but he remarked that many ambiguities remained, especially regarding the path of contamination. "If a pipe inside the deodorization tank had leaked, then the company should have been able to check it with the pressure gauge or the tasting test." Katō stated that no one in his company knew that Kanechlor was so dangerous. The pamphlet describing the heat-transfer medium, he said, mentioned nothing about toxicity. Katō admitted that the company had not adequately investigated Kanechlor's toxic effects prior to using it in a food processing industry. "On that point, certainly, we can do nothing about criticism of the company's responsibility" (*Asahi*, 5 November 1968, evening).

On 16 November, scientists from the Yusho Research Group ran tests on the six suspected deodorization tanks, especially number six, which had been repaired in January 1968. The scientists expected to find holes in the pipes carrying the PCB fluid, thereby explaining the contamination (*Asahi*, 16 November 1968, evening). In the experiment, scientists filled the number six tank with water and pumped compressed air through the pipes that normally held Kanechlor at low pressure. Bubbles soon appeared on the water's surface, demonstrating the existence of three "pinholes," one releasing 114 milliliters in five minutes, one releasing 7 milliliters in five minutes, and a third of uncertain size (Akagi, p. 413).

Although the pinhole theory became the conventional explanation for the contamination, several points of ambiguity persisted. First, what caused the pinholes to appear? As a highly stable chemical, PCBs do not usually decompose or corrode metal, especially stainless steel. According to the explanation that evolved, the Kanemi Company did not maintain its heat-transfer pipes as a perfectly closed system and allowed water to contaminate the PCBs. The water reacted with the PCBs, producing hydrochloric acid, which corroded the stainless steel pipes (Isono, pp. 89–90).

But why did the pinholes appear on the pipes of just one out of six deodorization tanks and leak only for several days in February 1968? Analysis of Kanemi oil labeled 5 and 6 February showed PCB levels of 2,000–3,000 parts per million (ppm). On subsequent days, the level decreased suddenly, and by 20 February, the oil contained almost no PCBs. Samples from other months also showed almost no PCBs (Kuratsune, p. 47). Kyūshū University engineers explained that the pinholes resulted from repairs made on the deodorization tanks in January 1968. According to this theory, the pinholes then spontaneously closed, due to the contact of hot Kanechlor with cooking oil and pipe rust (Nishimura, p. 76).

Doubts arose over whether three pinholes could have leaked a large volume of Kanechlor in a short period of time (Takeshita). The Kanemi Company normally used forty kilograms of Kanechlor a month. But in February 1968, the company added fifty kilograms on the third, followed by several other fill-ups totaling two hundred kilograms by the month's end. Several commentators suggested that the Kanemi Company may not have noticed this sudden and significant overuse, because of inadequate equipment inspections and gross mismanagement (Isono, pp. 89–90; Nishimura, p. 78). Although someone in the factory must have noticed the giant leak, for ten years no Kanemi Company employee publicly disclosed what actually happened. But in the late 1970s, challenges increased to the validity of the pinhole theory (Katō K.; Katō Y., 1979).

While scientists worked on identifying the poison and the contamination pathway, the Yusho Research Group used the most severe cases to compile a set of symptoms, which the Ministry of Health and Welfare adopted as diagnostic criteria for Yusho. The list stressed eye and skin symptoms but also included neurological and other problems (Akagi, pp. 409–10). Meanwhile, public health centers continued to register potential Yusho patients. By mid-November, the number of suspected sufferers throughout western Japan reached fourteen thousand. Symptoms among these people ranged from severe to light. According to a ministry official, when doctors examined some registered cases of suspected poisoning in

Fukuoka prefecture, they found that "unmistakable" Yusho patients represented only about 10 percent of the total. The majority, the official said, suffered from "rice-bran oil neurosis" (*Asahi*, 29 October 1968).

The 14,000 registered cases spanned twenty-three prefectures of western Japan, but almost half of them (6,611) lived in Fukuoka prefecture, the home of the Kanemi Company. Other high-ranking prefectures were Nagasaki (1,265), Yamaguchi (1,178), and Saga (962) (MHW, 20 March 1969). The poisoning distribution roughly followed the sales pattern of Kanemi's cooking oil, cutting across class boundaries, geographic lines, occupational groups, and cultural spheres. Victims included: retired coal workers; union factory workers; poor, isolated, rural fishermen; and urban bus drivers. This distribution, skewed spatially and spiritually, set the stage for dealing with the difficult problems of treating Yusho as a physical, social, and political illness.

On 20 November, the Yusho Research Group presented an interim report that summed up five weeks of frenzied activity and provided an initial clinical picture of Yusho (Yushōhan). The report recognized birth defects, including stillbirths and dark pigmentation of the skin (called "cola babies" for the color of the beverage), as a serious problem for Yusho patients. The report also presented neurological symptoms, especially damage in sensory nerves (pp. 25–26). By correlating the distribution of oil and patients in an epidemiological survey, one group of scientists intended to demonstrate again that Kanemi rice oil caused the disease (pp. 54–55).

The day after the report appeared, Yusho patients met Kanemi Company officials for the first time. Patients had repeatedly requested a meeting, while company officials asked that it be delayed until research "results became clear" (*Asahi*, 22 November 1968). One patient in Fukuoka city finally arranged an appointment and contacted patients in two other areas, Kitakyūshū city and the Tagawa region (Figure 2). On the morning of 21 November, six patients arrived at the Kanemi Company factory in Kitakyūshū, expecting to meet with Katō, the company president. But company employees balked, saying that the president was out of town and could not be reached. After some argument, an employee telephoned Katō, apparently at his country home, and arranged an immediate evening meeting in Fukuoka city. During that meeting, the company initially denied entrance to reporters. But once the session began, the patients, angry and dissatisfied at what they considered Katō's unremorseful attitude, opened the doors to reporters in order to encourage the president's apology and to assure a public record (Kamino, 1973, pp. 34–39).

For two hours, the Kanemi Company president spoke with nine Yusho patients, including four women suffering from severe chloracne. He

Figure 2.   Places in Japan related to Kanemi Yusho.

stressed that the company wanted to settle out of court and avoid the costs of litigation. "I want to pay as much compensation as possible. But since my personal assets are only several hundred thousand yen [about $1,000], we must begin factory operations soon. Then, while working, we will indemnify you." A press account of the meeting described the Kanemi Company president as ashen-faced, speaking in almost a whisper throughout the evening, repeating quietly, "I am terribly sorry." The president and the patients agreed at the end of the meeting to start concrete negotiations by 1 December. But one patient objected to simple economic compensation: "Money is not our objective. We want you to return our original bodies" (*Asahi*, 22 November 1968).

Unbeknownst to the victims or to the public, on 16 November, the day that the pinhole theory was announced, Katō's father, who had founded the Kanemi Company, pioneered the industrial production of rice-bran oil in Japan, and was then Kanemi Company's chairman of the board, had suddenly died at the age of 88. According to Katō Yachiyo, the company president's elder sister and also a part-time member of the board of directors, the shock of the Yusho disaster had helped send her father to his grave (Katō Y., 1989, pp. 6, 243).

The patients at this meeting represented not only themselves and their families but three victims' groups in Fukuoka city, Kitakyūshū city, and the Tagawa area. The first group formed in Fukuoka city on 14 October, only four days after the *Asahi* disclosure. The founders were Kunitake and his fellow Kyūshū Electric Power Company employees, who had been instrumental in the discovery of the poisoned oil. An article in the next day's paper announced the establishment of their Rice-Bran Oil Victims Society and invited other sufferers to join (*Asahi*, 15 October 1968).

But when two victims from outside Fukuoka city asked to become members, they were told they lived too far away and should form their own groups. So they did. Ujino Kazuyuki, in Kitakyūshū city, obtained a list of victims from a journalist, and Kamino Ryūzō, in the Tagawa area, acquired a list from a friend working in the local health center. Both then made personal visits to victims' houses (Ujino interview; Kamino interview). As Ujino recalled, "I told them that if the victims remained quiet, the authorities would never find a cure for the disease." Ujino, a worker at a large steel company, feared he might lose his job if he went public and became involved in a movement. "But I felt if I did nothing, we'd die" (Ujino interview).

Yusho disrupted the lives of every family it struck. In addition to physical pain, victims suffered economic disaster, social pressure, and psychological conflict. Public disclosure of the poisoning brought the victims' problems to light but not to solution. Although patients no longer consumed PCB-tainted cooking oil, their health continued to deteriorate. Predictions by some doctors of early recovery proved overly optimistic. No effective treatment or cure for Yusho existed.

Economically, patients confronted mounting bills and declining incomes. Until early December 1968, two months after public discovery, victims paid their own medical expenses and transportation fees to hospitals. For families in which the main wage earner could no longer work, the effect was devastating. Even if a Yusho patient could work, companies were reluctant to employ a sick worker. One rural family, between July and December, spent more than ¥400,000 (about $1,100) for medical and transportation costs (Hazama, pp. 37–38). Although Japan had instituted universal health insurance in 1961, benefits varied among different insurance plans (Fujii and Reich). In the late 1960s, a retired worker would be covered under community health insurance and required to pay 50 percent of his and his family's medical expenses; someone employed in a large company would have 100 percent coverage under employees' health insurance, although dependents would pay 30 percent.

Socially, the victims' physical transformation and the visibility of their frightful symptoms separated them from others. One young woman, who had been engaged, had to call off her marriage. Another victim described a persistent psychological conflict associated with Yusho. "How do you tell your daughter, who is only ten years old, that she may never marry because she would probably give birth to a deformed, dark-spotted baby? And what do you say to your son, whose upper teeth have been completely ruined by the disease, when he explains his failure at school by saying, 'Why should I study? I'm not going to live long anyway.' These are the real agonies of being a Kanemi victim. Even more than the physical pain" (interview with author).

A great burden of guilt fell on mothers. According to conventional Japanese thinking, the mother is responsible for providing food and raising children. In many families, the mother selected and purchased the Kanemi rice-bran oil. One woman, in an interview in the late 1970s, painfully showed pictures of her children, then in their twenties, with faces still scarred by chloracne. She asked insistently, "Can you understand the pain of a mother who fed her children poison, who watched them suffer, who sees them scarred outside and inside for life?" (interview with author).

In late November 1968, Kanemi Company officials communicated to the victims' groups that the company president would not engage in direct negotiations. Instead, the company formed a mediation committee including several lawyers to work out compensation and other problems. The victims' groups refused to accept this proposed negotiating structure. Kamino and another representative from the Tagawa group decided to ask the company president why he rejected direct negotiations with patients. Kamino's two daughters, "who for a long time felt anger at the company's insincere attitude," accompanied them to the factory. But company officials said the patients could not meet the president, who again was absent. Confronted by the officials' cool reception, Kamino's twenty-four-year-old daughter exposed her back, completely covered by chloracne boils. Her father felt on the verge of tears. Even the company men, Kamino recalled, seemed to feel some pity (Kamino, 1973, p. 45).

In early December, the Kanemi Company announced it would pay for "all treatment costs" for Yusho patients (*Asahi*, 3 December 1968). Later in the month, however, patients received vouchers for dermatological visits, to pay the coinsurance fees not covered by national health insurance. The company also sent some patients ¥20,000 ($56), and others ¥10,000 ($28), to cover previous medical expenses (Hazama, pp. 87–88). For many families, however, the economic burden remained heavy (*Asahi*, 16 January 1969).

While the Kanemi Company consulted its lawyers, other groups considered legal action. The health department of Kitakyūshū city, in somewhat unusual circumstances, became the first party to initiate legal action. On 29 November, at ten in the evening, the director of the city health department hastily called a press conference and announced his intention that night to file an oral accusation against the Kanemi Company president for criminal violation of the Food Hygiene Act. Under the Japanese penal code, a criminal investigation can be initiated in several ways, including an accusation by a public official, a complaint by an injured person, or an accusation by anyone who believes a crime has been committed. Public officials are obligated to file an accusation about any crime encountered in the performance of their duties (Dando, pp. 322–28). After a brief statement, the health director drove to the police station, followed by more than forty startled reporters. The police chief, however, informed him that an oral, middle-of-the-night accusation would not suffice in such a complicated affair as Kanemi Yusho. "This is not a problem of yesterday or today. We want you to submit a written document. . . . This is not a simple incident like a robbery." The officials then argued over the legality of oral accusations. Eventually, at half past one the next morning, after extensive closed-door negotiations, the health director repeated his oral accusation, and the police chief accepted it (*Asahi*, 30 November 1968).

Political conditions in Kitakyūshū city shaped the administration's sensitivity on the poisoning case. The city health director's sudden desire for legal action against the Kanemi Company arose when city officials learned of plans by the Socialist party to file a similar accusation the next day. Local officials did not want to be one-upped by the Socialists and to open themselves to additional charges of incompetence on the Yusho incident (Nishimura, p. 82). Acute competition existed between elected city officials and the Socialist party, because in the mayoral election of February 1967, the previous mayor, Yoshida Hosei of the Socialist party, lost to Tani Gohei, supported by the Liberal Democratic and Democratic Socialist parties. The Socialist party lost the election largely because the Communist party refused to support the Socialist candidate, owing to conflict between the two parties within the city workers' labor union. The bitter dispute between Socialist and Communist parties persisted in other forms.

The city administration may also have felt some vulnerability on Yusho, because Katō Sannosuke, president of the Kanemi Company, played an important role in Mayor Tani's political support association and was a well-known rightist figure in Kitakyūshū politics. About a year after Mayor Tani's election, Katō set up a new company and received the catering contract for all city hospitals, whose kitchen staffs were fired to help balance the budget. Not unexpectedly, from April 1968 on, the hospitals all used large quantities of Kanemi rice-bran oil (Fukada, pp. 118–122).

On 2 December 1968, seven persons affiliated with the Socialist party and two Yusho patients filed a joint, written criminal accusation (*Asahi*, 2 December 1968). And on the eleventh, the Fukuoka city victims' group filed its complaint against the Kanemi Company president (*Asahi*, 12 December 1968). In response to these complaints, the prefectural police office formed an investigation headquarters for the Yusho case and began to collect evidence. Sixteen months later, in March 1970, after a full pretrial investigation, the Fukuoka district public prosecutor brought the Kanemi Company case to trial, charging company president Katō and factory director Mori with injury due to professional negligence. The decision to begin criminal proceedings was no idle act, for in Japan, public prosecutors are an elite professional service. They win convictions in 99 percent of the cases they take to trial (Nagashima, p. 299).

The other legal action in this period came from the Fukuoka city victims' group. When direct negotiations with the Kanemi Company foundered in November 1968, the victims turned to litigation. As one victim explained, "No one had any experience with trials. But everyone's body hurt terribly. We weren't able to move and wanted some representative to fight for us" (interview with author). One group member contacted a long-standing personal friend, a lawyer named Haraguchi Torio, who agreed to take the case. Haraguchi approached Yusho as a straightforward civil damage case, hoping that a conventional trial would reduce the time needed to reach a decision (Haraguchi interview). In early February 1969, he filed suit for eleven families, including forty-four persons, demanding total damages of ¥154 million ($428,000)—finally increased to ¥877 million by 1977 ($3.27 million)—against the Kanemi Company and its president, but also against the Kanegafuchi Chemical Company (*Asahi*, 1 February 1969; Katō I. et al., p. 18).

Haraguchi decided to sue Kanegafuchi Chemical, the producer of Kanechlor, primarily because he reasoned that even if he won the suit against the Kanemi Company, the firm would not have enough financial resources to compensate all Yusho patients. He considered two other possible defendants: the central government and the Kitakyūshū city administration. But Haraguchi rejected them because of difficulties in demonstrating a connection between administrative responsibility for damages and compensation under private law. Only Kanegafuchi Chemical remained. One patient brought Haraguchi several Kanegafuchi catalogs on PCBs, wondering if the material could be used in the trial. In those catalogs, Haraguchi found his evidence: statements that belittled Kanechlor's toxicity and corrosive potential. The lawyer believed the catalogs could be used to show the chemical company's negligence.

But Haraguchi also decided to sue Kanegafuchi Chemical to obtain proof about what happened in the Kanemi Company factory. Haraguchi thought that Kanemi and Kanegafuchi might fight and accuse each other of negligence. That conflict, the lawyer hoped, would produce the proof that the plaintiffs might not otherwise discover (Haraguchi interview). The decision in the suit came nearly nine years later, in October 1977. It supported all the plaintiffs' major arguments and awarded them about 80 percent of the damages requested. For forty-two of Haraguchi's plaintiffs, his strategy and hard work produced a total court victory (*Kubota v. Kanemi Sōko*). But, for two plaintiffs who died during the long trial, the decision came too late.

In contrast, the lawyer consulted by the victims' group in Tagawa advised them to start a social movement (Kamino, 1973, pp. 42–47). In late November 1968, this lawyer explained to Tagawa area victims that the law alone could not solve a new problem like pollution. The lawyer emphasized that the victims needed to begin a movement and then decide whether to go to court (p. 46). This approach, which defined Yusho as a pollution disease (*kōgai byō*) rather than as simple food poisoning, became the general strategy of the victims' groups in Tagawa and Kitakyūshū.

Support for these victims and their movement initially came from Socialist party activists. In October and again in early November, the party sent a team of parliamentarians from Tokyo to study the poisoning incident. The politicians used the materials they had gathered to interrogate government officials in parliamentary committees. One lower house Socialist representative from Kitakyūshū, Kōno Masao, criticized in especially sharp words the government's actions on Yusho. Two examples, from many, illustrate his accusations.

On 30 October, at the meeting of the Society and Labor Committee of the national Diet's upper house, Kōno criticized the director of the Food Agency for designating the Kanemi Company as a "model factory." The director explained that the Kanemi Company had received the designation from 1962 until 1967 because of data showing its modernization. Kōno then demanded to know why the company continued in 1968 to print "Food Agency designated model factory" on its oil labels, which some people interpreted as an assurance of safety. The Food Agency director had no answer, and Kōno insisted that the company be criminally prosecuted for mislabeling (Shakai Rōdō, 30 October 1968, p. 7).

At another meeting of the same committee, on 12 November, Kōno attacked the Ministry of Agriculture and Forestry for its mishandling of the dark oil incident. It did not contact officials in the Ministry of Health and Welfare, did not thoroughly investigate the poisoning's cause, and did not finally identify the toxic agent (which was accomplished by university scientists). Kōno charged that connections between the Kanemi Company and the Ministry of Agriculture and Forestry had slowed down the investigation, and named two high-ranking Kanemi Company officials who had previously worked as ministry bureaucrats. "When such a major problem occurs (such as the mass chicken deaths), isn't it the ministry's duty to identify the cause? And isn't it because calling Kanemi's attention to the issue was said to be sufficient that today's huge problem developed?" (Shakai Rōdō, 12 November 1968, p. 6).

In late November 1968, the Socialist party reported its findings at a meeting in Kitakyūshū city, followed by a discussion with Yusho patients. During the session, members of Parliament promised to continue raising the Yusho problem in the Diet and to criticize the government on issues of compensation, medical treatment, and administrative responsibility. The next day, the party filed its accusation of criminal negligence against the Kanemi Company president (Hazama, p. 34). At the end of December, Socialist party activists in labor and in women's groups formed the Kanemi Yusho Patient Defense Society, to support the victims' cause and to carry the word about Yusho to other citizens (p. 35).

A full-time union leader, Uejima Kazuyoshi, became director of the Patient Defense Society. He served as head of the Kyūshū Electric Power Company Union and as director of the Socialist party branch in Kokura (part of Kitakyūshū). He became involved in the Yusho movement because he personally knew one union member "who suffered miserably from the disease." Uejima believed that the union had an important role to play in helping Yusho patients, both inside and outside the union. In early 1969, the union's offices became headquarters for the defense group (Uejima interview).

In late December 1968, prefectural health authorities in Kyūshū reported 577 persons as "true" Yusho patients. Fukuoka prefecture counted 309; Kagoshima, 3; and Nagasaki, 265. In Nagasaki, 224 "true" patients, or 85 percent of the prefecture's total, lived in Tama no Ura, a small, out-of-the-way fishing village, located at the far tip of Fukue Island in the Gotō Archipelago, a string of islands off the west coast of Kyūshū. In 1968, the 110-kilometer trip from Nagasaki city to Tama no Ura required six hours of travel, by ferry, bus, then small fishing boat. But the town's seclusion, rather than serving as protection, delayed discovery. Although health officials in mid-October possessed a list of all buyers of Kanemi's contaminated cooking oil, doctors who first visited the town in late October found only seventeen Yusho patients. On four subsequent visits, they discovered more and more. By late December, the doctors recognized four percent of the town's 5,267 residents as Yusho sufferers. The town of Tama no Ura, with 224 patients, then possessed more than twice the number as Kitakyūshū city, at 105 patients, whose population exceeded one million (*Asahi*, 26 December 1968). The initial delay in finding Yusho patients in Tama no Ura, and the continuing increase each time the doctors looked, suggested that no one yet appreciated the full meaning of Yusho.

Conflict around Yusho increased markedly, when, in the early morning of 25 February 1969, a group of Yusho victims staged a sit-in at the Kanemi Company factory. Kamino explained what moved him and other victims to protest. "The company president at first said he would surrender his entire assets to compensate us, and we believed him. But after handing over a tiny solatium [*mimaikin*, a monetary expression of sympathy for the patients' suffering], he changed his attitude, telling us to wait for compensation until after the company had recovered. To put the company's reconstruction before the patient's compensation is the wrong order" (*Asahi*, 10 March 1969). The victims decided on the sit-in to force another direct meeting with company president Katō. "We wanted the company president to look us in the eye, to hear even one word about the patients' suffering from our own mouths" (Kamino, 1973, p. 76).

Yusho victims carried banners stating their demands: "Treat Patients as People," "Patients Have Not Received a Single Yen of Compensation," and "This Face, This Body, Make Them Whole Again" (Hazama, p. 39). Many newspapers and television stations covered the event. In appealing to the public, Yusho victims sensed that their strongest plea came not from the written banners but from their scarred faces (p. 39).

Patients at the sit-in represented a new organization, a victims' liaison council of the Tagawa and Kitakyūshū groups. Soon after its formation on 13 February, the council held street campaigns in three cities to seek donations for the organization. They also sought to collect signatures on a petition calling for revision of the Food Hygiene Act to prevent another Yusho incident. At the sit-in, Kamino, the council's head, went with

several others into the Kanemi Company offices to demand an appointment with the president. A company official told them the president was
out of town and in any case would not meet them. The employee then
requested that the patients leave the premises (Hazama, pp. 40–41).
While returning to the gate, a patient asked an employee what he would
do if he and his family were afflicted by the oil disease. He replied, "If I
were a victim, I'd burn the company down." Why then did he refuse to
listen to the patients' demands? "To put it plainly, my position is to
defend the company. Your interests and mine cannot agree" (p. 41).

This gap between Kanemi Company employees and Yusho victims
became public knowledge soon after the disease was disclosed in October
1968. The company reported that its employees in past years had regularly consumed an average of five liters of rice-bran oil each month. Yet,
in late October 1968, when Kitakyūshū doctors examined 231 Kanemi
employees, from company president Katō down, only one showed any
symptoms resembling Yusho. That person had several black acne spots
near his right ear, but complained of no subjective problems. Moreover,
no employee reported past health problems similar to Yusho (Mainichi
Shinbun, 30 October 1968). One commentator speculated that the absence of victims inside the factory suggested that the company knew the
oil was contaminated and kept employees from using it (Watanuki interview).

In early March 1969, the victims' liaison council began bringing its
demands to political bodies. On the fifth, the group met with the
Fukuoka prefectural vice-governor, who agreed to repeat a study of the
patients' status and to request cooperation from the Kanemi Company in
resolving the compensation dispute (Hazama, p. 44). The group also
asked the Tagawa city council to press prefectural and central governments on victims' problems concerning treatment and compensation and
on measures to prevent similar disasters (p. 44). Then, on the eleventh, a
group of six victims left Kyūshū by night train for Tokyo to appeal to the
politicians and bureaucrats of the central government.

In Tokyo, Socialist party politicians arranged the victims' appoint-
ments. Kamino called it a "dizzying schedule of petitions, moving ma-
chinelike from one person to the next" (1973, pp. 129–30). The delegation visited every Fukuoka prefecture member of Parliament, spent six
minutes with the health and welfare minister, about the same time with
the agriculture and forestry minister, and spoke briefly with the chief
cabinet secretary. The patients also appealed for their cause on Tokyo
streets, asking for contributions and for signatures on a petition. Their
petition to the Ministry of Health and Welfare urged the central government to act on treatment, compensation, and revisions in the legal sys-

tem. "This is no longer just a problem of individuals. It must be solved from a social point of view. We believe the only way is immediate action by national politics, which is supposed to protect the people's livelihood" (p. 136).

On the eighteenth, the Society and Labor Committee reviewed the Yusho problem. Two victims, Kamino and Ujino, appeared as witnesses, along with representatives from the Kanemi Company, the Yusho Research Group, and government agencies. Ujino explained to the politicians, "Patients go to the city and the prefecture, but no one even extends a helping hand. They just want to pass the buck. We are stuck in the uneasy situation of having no one to turn to. The city and prefectural administrations that are supposed to protect the citizens' health have a careless attitude and are totally undependable" (Hazama, p. 46). During the session, a Socialist member repeatedly attacked government and company representatives for the lack of compensation and inadequate provisions for treatment. Later, as the victims returned to Kyūshū, they realized that the trip had produced few concrete results (p. 52). The prefecture's subsequent decision to present each patient with a solatium of ¥3,000 (about $8) amounted to little real assistance.

In addition to difficulties with government officials, Yusho patients encountered conflict and frustration with medical institutions. By early 1969, relations between Yusho patients and Kyūshū University doctors had significantly deteriorated. Some patients began doubting the doctors' integrity in October 1968, when it became clear that several dermatologists had identified Kanemi rice oil as the likely cause but did not report the poisoning to health officials or obtain an analysis from competent scientists. Other patients felt that doctors irresponsibly experimented with drugs on Yusho sufferers (*Fukuoka Kanemi Yushō Genkoku*, p. a113).

Once the university Yusho Research Group formed its clinical division in mid-October, patients began visiting at least three different medical departments: dermatology, ophthalmology, and neurology. But these departments made little effort to coordinate treatment and each required a long wait before an appointment. Patients needed two or three days to make the rounds of the specialists, which imposed on the patients enormous time demands, physical strains, and economic costs. In addition, each specialist prescribed his or her own batch of medicines. The total volume of pills was "enough to feed a horse" (p. a114). Yusho patients, particularly sensitized about what they ingested, felt frustrated at not receiving adequate information about possible side effects of the drugs (p. a114). Finally, patients sensed and resented that the university doctors considered writing research papers more important than caring for sick people (*Asahi*, 16 October 1968, evening). "Treatment" thus tended to increase patients' anxiety, making them feel like guinea pigs in an experiment.

For the above reasons, the Fukuoka victims' group filed a complaint with the Fukuoka office of the Ministry of Justice on 24 February 1969, charging doctors at the Yusho Research Group with violating the patients' human rights (*Fukuoka Kanemi Yushō Genkoku*, p. a114). Several weeks later, the doctors in the Research Group started a single Yusho Clinic. The new clinic helped coordinate treatment schedules, but it could not remove the patients' mistrust. During 1969, the number of patients visiting the clinic dropped dramatically. By 1972, the clinic was largely inactive. One university doctor attributed the clinic's failure to two causes: the lack of an effective treatment for Yusho and the patients' "strong feelings of dissatisfaction, disbelief, and anger" toward the doctors (Kuratsune, 1972, pp. 55–56). Although the doctor called on his colleagues to "reflect" about the patients' estrangement, no significant changes resulted. Most patients continued to ignore or boycott the university clinic.

In late April, the victims' liaison council met with Kanemi Company president Katō. He assured victims that he had no desire to shirk his moral responsibility for the Yusho incident. He also told them, "At the present stage, with Yusho's treatment unrealized and the company's finances unsatisfactory, compensation negotiations cannot proceed. Above all, treatment is the first consideration. Don't misunderstand this and think that we are ignoring the patients and working only for the company's survival. But for a medium-sized company like ours, a halt in operations for six months is a major blow" (Hazama, p. 54). In mid-May, another meeting occurred. Katō agreed to consider transportation and livelihood fees for families with special financial problems (*Asahi*, 19 May 1969).

On 31 May 1969, ten months after public discovery of the problem, the Kanemi Company resumed its production of rice-bran oil. The company altered its deodorization process, replaced Kanechlor with an alkyl naphthalene heat-transfer agent, and installed automatic temperature and pressure recording gauges that would indicate problems at a single glance. In addition, the company purchased two gas chromatographs to test its oil products daily, to assure that no new heat-transfer chemical leaked into the product and that the Kanemi Company sold safe cooking oil. The company then received the necessary permits to resume production (Hazama, p. 55).

At the same time, the Kitakyūshū Victims' Defense Society and the victims' liaison council announced a boycott of Kanemi Company products. Both groups denounced the city's decision to allow the company to operate while under investigation for violating the Food Hygiene Act. The Socialist party branch in Kitakyūshū also sent a written protest to the city government. But a city official explained, "Legally, if the conditions are met, we have no reason not to issue the permits." A central government bureaucrat agreed that the operation permits based on new equipment were not connected to the criminal charges (pp. 56–57). The company president responded that different patients made different demands. While the Tagawa and Kitakyūshū groups demanded compensation, he said, patients in Fukuoka city and Nagasaki wanted the company first to resume production and then to provide help. He warned that if the boycott succeeded in restricting business, the company would not be able to deal with "remaining problems" (*Asahi*, 31 May 1969).

In June 1969, the Yusho Research Group issued its first annual report in a special issue of the *Fukuoka Acta Medica*. The journal contained nineteen articles of clinical and experimental research on Yusho and Kanechlor. The introduction proudly reported that the researchers had identified the cause within three weeks of the group's formation (Katsuki, p. 406). The first article stressed several times that since the disease's cause was "promptly" identified, serious liver damage did not occur (Gotō and Higuchi, pp. 425–27). The clinical assessment stressed the skin and eye symptoms, which formed the basis of a four-grade classification of the severity of Yusho patients (p. 419). Although the report noted the difficulties in finding effective treatment (p. 424), it did not include a separate critique of treatment methods. But earlier in the year, the group's head had expressed his opinion on the prognosis of the disease: "Since it is a poisoning, I think one must ultimately depend on time. It is important to restore the patient's entire body. It will be a long struggle, but they will certainly recover, probably with no lingering symptoms" (*Asahi*, 24 January 1969).

Then, in early July 1969, shortly after the university scientists issued their report, two young Yusho patients, a boy of 14 and a youth of 25, died suddenly. On autopsy, both patients showed atrophied adrenal glands. Within a week of the two deaths, three hundred victims and supporters held a memorial assembly in Kitakyūshū to protest the deaths. The assembly set forth these demands: the health and welfare minister increase research on treatment, the prefectural government mediate on compensation, the city mayor revoke the company's operating permit, and the company president show "good faith" (Hazama, p. 68).

Although the finding of atrophied adrenal glands could not be immediately related to Yusho, some connection seemed likely. At a meeting with patients, a doctor discussed the autopsy results and stated that

Yusho was not the direct cause of death but "made it easier to die." He told patients that the Research Group lacked sufficient funds to study Yusho and asked them to "make a movement" for more government grants. When a lawyer with the patients asked doctors to be kinder to Yusho sufferers, a doctor replied, "We are in a thoroughly neutral position" (Kamino, 1973, p. 165). One week later, a protest delegation of two patients and one supporter traveled to Tokyo to appeal to the health and welfare minister for increased research funds and for an integrated treatment policy.

The two deaths shocked both patients and doctors. For patients, the deaths represented a fear felt for months, that they might die suddenly from the poison they had unwittingly swallowed. For doctors, the deaths suggested that Yusho affected internal organs as well as the skin. Despite patients' complaints of internal disorders, doctors had restricted treatment primarily to skin problems. After the two deaths, doctors began slowly to expand their concept of Yusho (*Asahi*, 24 July 1969). One year later, in July 1970, the director of the research group's clinical division publicly reported its conclusion that in both cases, the cause of death was related to Yusho (*Mainichi*, 1 July 1970). A technical article, however, maintained more ambiguity on the deaths, concluding that more attention had to be paid to the relationship between Yusho and internal symptoms, especially circulatory problems (Kikuchi, p. 98).

In the early fall of 1969, an assortment of new groups formed around Yusho patients: victims' groups, victims' defense societies, and women's groups to support victims. The groups held rallies and demonstrations to stress the connection between the Yusho incident and modern consumer society. The Socialist party provided the main initiative to create these groups and the dominant political sphere for their activities (Hazama, pp. 99–105).

In March 1970, the public prosecutor's office instituted criminal proceedings against the Kanemi Company president and factory director, indicating that prosecutors believed they had collected sufficient evidence to obtain convictions. This step in the criminal trial encouraged the filing of additional civil damage suits against the company and raised hope among patients that eventually they would receive compensation. But both criminal and civil trials promised to be prolonged processes (*Asahi*, 25 March 1970).

In late May, at a meeting of the National Council of Kanemi Victims Defense Societies, a leading activist declared that the deadlock in settlement negotiations and the Kanemi Company's lack of good faith made legal action necessary. He called for a broad-based team of lawyers to fight the case for all Yusho victims. A lawyer then proposed they create a separate umbrella organization to coordinate the entire Kanemi Yusho movement: the victims' groups, the victims' defense societies, the proposed lawyers' team, and a doctors' team. During this meeting, arranged by both Socialist and Communist party members, Kamino began to feel a wall of separation growing between victims and supporters, and conflicts emerging within both groups (1973, pp. 194–201).

Supporters and victims met in July 1970 to discuss the umbrella organization, based on similar structures coordinating the movements for other major pollution incidents in Japan: the two cases of Minamata disease due to mercury poisoning and *itai-itai* (pain! pain!) disease due to cadmium poisoning (*Mainichi*, 6 July 1970). But plans for the Yusho umbrella organization eventually collapsed because of disagreement between Socialists and Communists over who would organize the group and who would control it (Kamino, 1973, pp. 208, 213).

The lawyers' team nevertheless began to take shape in August 1970. Some Tagawa area patients, however, felt ambivalent about the lawyers, because the core group came from a major Kitakyūshū law office closely related to the Communist party, even though the team included Socialist and Liberal Democratic lawyers as well. Moreover, patients felt uneasy about conflicts between the lawyers' team and the National Council of Kanemi Victims' Defense Societies, which reflected hostility between Communist and Socialist parties (Kamino, 1973, pp. 226–27). Tagawa victims consulted the lawyer who had helped them from the beginning, telling him that victims wanted a movement without a political party and that they feared that dependence on only one party would block help from other parties and citizens. The controversy seemed to overwhelm many patients in struggles they did not want and sometimes did not fully understand. Many hoped to resolve their problems without politics.

On 23 August 1970, the lawyers' team formally announced its plans to file suit. The team included nearly four hundred lawyers, mostly supporters in name only, with much of the actual work performed by the legal office in Kitakyūshū affiliated with the Communist party. Potential plaintiffs numbered five hundred persons, with an initial damage claim of more than ¥1 billion ($2.8 million).

The announcement of the trial did not eliminate controversy. On 13 September, a meeting of victims and supporters continued for hours, as personality clashes, party conflicts, and strategy differences arose one after the other. An added element of tension was a widely distributed series of letters containing false personal and political information about leaders of different groups (Kamino, 1973, pp. 244–47). Participants at the September meeting agreed on the need to restructure the relationships among victims' groups, support groups, and the lawyers' team. But they could not settle who should do it or how (pp. 249–69).

In early November 1970, the patients' lawyers filed suit in Kitakyūshū against the Kanemi Company, the company's president, and the city administration of Kitakyūshū. The next day, the lawyers added the central government as a defendant; much later, they added Kanegafuchi Chemical. Initially, the plaintiffs numbered 299 Yusho patients. But over the next few years, suits filed in other areas—except the one in Fukuoka city—became consolidated into a single mammoth trial in Kitakyūshū. By 1978, the suit included 729 plaintiffs and asked ¥20 million ($95,000) for each death and ¥15 million ($71,000) for each living victim, regardless of severity, for a total demand of ¥6.08 billion ($29 million) (*Asahi*, 10 March 1978).

For its defense in both civil and criminal trials, the Kanemi Company hired several lawyers serving in the local Human Rights Protection Commission. Yusho patients and their supporters objected to human rights commissioners defending the Kanemi Company and demanded to know the lawyers' idea of human rights. According to an *Asahi* column on the controversy, the company's lawyers responded that the constitution assures a right to trial and that even guilty parties have human rights. Therefore, they argued, the work of private lawyer and that of human rights commissioner are compatible. This argument, the column stated, might pass a strict legal test, but it would not be accepted so easily by society's common sense or by the people's image of a human rights commissioner. "The expectation is that [the commissioners] protect the weak—that they be the people's ally" (*Asahi*, 18 February 1971).

The Kanemi Company and its lawyers adopted one position in court and another outside. At the first session of the criminal trial, on 28 September 1970, the Kanemi Company president and factory director denied all charges, rejecting the accusation of professional negligence as well as the alleged causal relationship between the company's oil and the disease (*Asahi*, 28 September 1970). At the same time, the company neared completion of an out-of-court settlement with a victims' group in the far-off Gotō Archipelago, including patients in the town of Tama no Ura and on the neighboring island of Naru. The company initially offered ¥50,000 to ¥200,000 (about $140 to $560) for a Yusho patient plus an additional ¥70,000 ($195) for medical fees. When patients rejected those figures, the company raised its offer slightly. In October 1970, leaders of the victims' group accepted the new sums as adequate (Kaga, p. 212). The settlement contract, when made public, included several conditions: no admittance of legal responsibility for Kanemi, patients would accept the payment when treatment was "almost no longer necessary," and children under 12 and adults over 70 would receive reduced payments (pp. 198–200).

Not all patients in Tama no Ura agreed to the proposed settlement.

Some opposed the payments as ridiculously low. Others worried that Yusho symptoms would worsen and the Kanemi Company would not pay medical expenses. Many feared for their children's health and doubted that the settlement would protect their children's future. At a general meeting of the victims' group on 14 February 1971, 75 Yusho families (243 persons) approved the settlement—the first concluded anywhere by the Kanemi Company—and 25 families (49 persons) opposed it. The minority group immediately began preparations for a separate organization and for filing a civil damage suit. In Naru, the entire group of patients (21 families, 87 persons) decided to call off the settlement and to join the trial (Kaga, p. 215).

The split between settlement and litigation factions roughly matched geographical, religious, and other boundaries in Tama no Ura. The settlement group lived largely in one hamlet, Imochi no Ura, a Catholic community that retained a strong sense of separateness after surviving four centuries of relative isolation as "hidden" Christians (to escape persecution during the Tokugawa period, 1603–1868). Almost every family in this hamlet was poisoned by Kanemi rice oil. The trial faction was based mostly in the town's center, with a college-educated leader, who owned and managed a small lumber business (p. 213). Other factors exacerbated these tensions: money received by the settlement faction, personality conflicts among leaders, and political accusations about motives created a gap between the groups that evolved into a protracted and bitter feud (interviews with author).

Yusho patients in Tama no Ura and Naru suffered more than physical pain and community conflict. Because of their isolation, they lacked easy access to assistance from doctors, lawyers, or political activists. Mostly fishermen and farmers, they lost food supplies as well as incomes when Yusho stopped them from working (Kuroiwa interview). The area also had a high prevalence of congenital Yusho patients, poisoned while in the mother's womb and born with darkly pigmented skin. Some women gave birth to several children marked by Yusho, because the parents as devout Catholics refused to practice birth control. In some cases, children became poisoned solely by drinking the mother's breast milk, partly because women lacked the advice and the resources to buy infant formula (Takamatsu).

In early 1971, while Tama no Ura residents disputed whether to settle or to sue, scientists announced that PCBs contaminated Japan's general environment. PCB pollution of the environment was first discovered in 1966 by Soren Jensen, a Swedish scientist (Jensen, 1972). In the United States, scientists found PCBs in the environment in 1968 (Risebrough et al.). And the U.S. Food and Drug Administration in November 1969 began testing all raw agricultural products in a pesticide surveillance program for PCBs (Kolbye, 1972, p. 86). But in Japan, during 1969 and 1970, the central government did not sponsor research on PCB environmental contamination. And between 1968 and 1970, annual PCB production in Japan more than doubled (Isono, p. 102).

The Japanese public first learned the news about PCB environmental pollution in early February 1971, when an analytic chemist from Ehime University announced on television that his laboratory had detected PCBs in fish taken from the Seto Inland Sea. At about the same time, a Kyoto scientist independently found PCBs in fish from Lake Biwa, but the city government refused to allow public announcement until April (Isono, p. 116). In late February, scientists reported PCBs in carbonless copy paper and on the skin of people handling the paper. The media gave top coverage to this disclosure, and the Ministry of International Trade and Industry immediately issued an administrative guidance to halt that use of PCBs. The industry readily complied, recalled its supplies, and by mid-March started producing copy paper with a substitute for PCBs (p. 120).

Reports of PCB contamination continued throughout 1971, with high levels detected in water, air, land, and animals in Japan. In June, the central government began a comprehensive study of PCB contamination, and in December, it prohibited the use of PCBs in open systems, as had recently been done in the United States (pp. 118–19). In early 1972, researchers reported PCBs in mother's breast milk, with especially high levels detected in Osaka women, averaging 0.3 parts per million. Although the levels approximated the quantity of PCBs ingested by Yusho patients, the Ministry of Health and Welfare issued a "safety declaration," stating that such mother's milk, if ingested for less than six months, would not harm an infant. The Ministry of International Trade and Industry, however, placed more restrictions on PCB uses in Japan. By June, both of Japan's PCB manufacturers had halted production (pp. 136–41). In August, the Ministry of Health and Welfare set the world's first acceptable daily intake for PCBs at five micrograms per kilogram of body weight. The ministry applied this figure to different foodstuffs, calculating tolerance levels, such as 0.5 ppm for deep-sea fish and 0.3 ppm for coastal fish (pp. 146–53).

In 1971 and 1972, as awareness soared of PCB pollution, Yusho patients came to be seen as a "living experiment" with global implications. Spurred by recognition of PCBs as a worldwide contaminant, Yusho gained international attention as the only known instance of mass PCB poisoning outside a factory. In Japan, a sense developed that foreigners were watching the Yusho patients, their physical symptoms, and their court trial (*Asahi*, 28 July 1971). The National Liaison Council of Yusho

Victims assisted this process by sending two patients and a doctor to the Environmental Forum in Stockholm, held at the time of the United Nations Conference on the Environment. In June 1972, Jensen examined the two Yusho patients. "I was stunned by the severity of skin symptoms on the patients' faces and backs, even after four years. I was further stunned by the high level of PCBs remaining in their blood. If so many severe patients exist, why is there no information on PCB levels in mother's milk [of Yusho patients]? I wonder what Japanese scientists have been doing" (*Asahi*, 13 June 1972).

Yet Jensen viewed only a few of the lingering problems of Yusho patients. Even as the Yusho delegation visited Stockholm, the Tagawa victims' groups reported a survey showing reproductive abnormalities. The group found that many women had difficulties conceiving and then had problem pregnancies, often with spontaneous abortions. Almost all non-pregnant women in the study complained of menstrual disorders. But accurate data on these problems among all Yusho patients could not be obtained. The breakdown of relations between patients and most university doctors made such epidemiological studies impossible (*Asahi*, 14 June 1972).

In late June 1972, nearly four years after public discovery, Yusho patients sought more governmental assistance for their problems by appealing directly to the health and welfare minister and the environment agency director. Kamino criticized the government's passivity on Yusho. "The company gives us almost no livelihood compensation, and the government has ignored us for four years. A food approved by the government caused our illness. That government should think about assisting us." The ministers responded that they would try to encourage the company to pay compensation and would consider certifying Yusho as a pollution disease (*Asahi*, 20 June 1972). If designated as a pollution disease under the 1969 Law Concerning Special Measures for the Relief of Pollution-related Patients, patients could be compensated with company funds administered by the government.

Kodaira Yoshihei, an upper house Diet member of the opposition Komeitō party, actively supported the patients' demands and on 11 July 1972 submitted four formal questions to the government. One week later, the government provided its written response, which indicated only minor changes in policy. With regard to research, the government stated it would broaden the range of specialists and institutes studying Yusho, but would not create a special organization. On certification criteria, the government denied using the criteria to limit the number of patients, but agreed to review the criteria on the basis of the yearly follow-up examination. It refused, however, to form a new body to certify patients. On designation as a pollution disease, the government replied that the Basic Law for Environmental Pollution Control of 1967 did not apply to the Yusho incident, since the law's definition of pollution did not include health damage caused directly by processed foodstuffs, and therefore neither did the compensation law of 1969 apply. As for livelihood assistance, the government recommended the use of existing measures for needy patients and considered it difficult to create a special welfare act for Yusho victims (Fujiwara, pp. 75–77).

An *Asahi* editorial rebuked the government's narrow approach to the Yusho victims. "Even when considered a food hygiene problem, Yusho has some common aspects with pollution diseases. If the government takes the position not to intervene because this is a private affair, the victims will not be helped." The editorial urged the government to aid Yusho victims with measures used previously for pollution victims, such as administrative assistance or mediation between victims and the company (*Asahi*, 20 July 1972). Some Yusho patients expressed even greater dissatisfaction at the government's responses and refusal to provide more aid. Kamino felt the government unjustly discriminated against Yusho patients by not considering their illness as a pollution disease. He planned to protest (*Asahi*, 19 July 1972).

In late September 1972, Kamino and his family began a sit-in outside the Kanemi Company factory in Kitakyūshū, a protest that lasted much longer than anyone expected, nearly four years. Kamino stated that he had previously resigned as director of the victims' liaison council, because the organization stressed the trial too much, paid no attention to patients' individual problems, and neglected direct action against the company and the government (*Asahi*, 17 December 1972). He also criticized the intervention of political parties into the victims' movement, especially the competition between Socialists and Communists. "It developed into a political movement, separate from the victims themselves" (Kamino interview). Socialists and Communists, in turn, viewed Kamino's actions as destructive of the movement and as influenced by the new left (Uejima interview; Tsukauchi interview). Kamino denied both charges.

During the sit-in, which attracted supporters from all over Japan, Kamino verbally attacked the company. He demanded morality from the company president and from individual employees, sometimes reading the Bible through a megaphone, expressing his Christian beliefs. "I want the company and the government to have humanness. I want a conscience that believes a sin is a sin" (*Asahi*, 28 December 1972). Kamino stressed that rights contained in one's own flesh could not be given or sold to anyone. To make his point, he withdrew from the trial and refused any compensation (Kamino interview).

The criteria for certification remained a continual source of discontent for Yusho patients and came under increasing attack in 1972. That summer, patients in Tama no Ura boycotted the follow-up medical examination scheduled for "official" patients. According to a survey done by the local victims' group, more than two hundred uncertified patients lived in the town without any form of assistance (*Asahi*, 18 September 1972). Diet member Kodaira's questions to the government helped raise doubts about the validity of the criteria, which were established in October 1968, before doctors knew the disease's cause. By fall 1972, even some Kyūshū University doctors recognized inconsistencies and deficiencies in the criteria and their application. One doctor reported that in many families that had used Kanemi rice oil, only some family members became certified, while others were designated "doubtful patients" or said to show "no abnormalities." Among "cola babies" born to Yusho mothers, only some received official status. No accurate estimate existed of the number of uncertified patients (Kuratsune, pp. 56–57).

In late October 1972, the Yusho clinical research group revised the diagnostic criteria. The group's director explained that revisions had become necessary because of advances in technology, especially the ability to measure PCBs in patients' blood, and because of changes in patients' "disease image." The new criteria recognized Yusho not as simply a disease of the skin and eyes but as a disease of the entire body (Urabe, pp. 2–3). The Ministry of Health and Welfare then decided to use the revised criteria to examine uncertified family members in the next survey of Yusho patients in Fukuoka prefecture (p. 3).

In January 1973, doctors began the survey with a surprise: a sit-in by patients and supporters who demanded that persons outside Yusho families also be examined. Doctors initially resisted the demand from the protesters, a small group known as the Gathering House for Patients and Supporters, but the researchers finally relented (Jūta interview). Doctors examined what they called eight "healthy persons" and diagnosed six certain and two doubtful cases (Urabe, p. 4). The decision to examine uncertified people, who had eaten Kanemi rice oil but had no certified family members, represented a major change in policy and an important victory for the protesters (Jūta interview).

The ability to measure PCBs in blood became a key diagnostic criterion for Yusho. Researchers in environmental health at Kurume Medical School devised quantitative and qualitative criteria for diagnosing Yusho patients, using an improved version of the Jensen method (brought back from Sweden by a Japanese scientist who had attended the environmental conference in 1972) (Takamatsu et al.). In 1973, the method served as a major factor to certify 75 percent of 87 uncertified patients examined

(Takamatsu, p. 757). In 1974, a Kurume doctor went to Tama no Ura to collect 20 to 30 blood samples, but residents there insisted that he take more, finally from 230 people, because they continued to confront problems in getting certified. Twice previously, doctors had collected blood to test for PCB levels, but prefectural authorities had not used the results for Yusho diagnosis or certification (p. 757). In 1975, finally, 72 Tama no Ura residents became certified Yusho patients, largely as a result of blood analysis (Fujiwara interview).

Other scientific discoveries about the Yusho incident occurred in the 1970s. In 1970, scientists detected a contaminant, polychlorinated dibenzofurans (PCDFs), in European PCBs, and in 1974, American scientists found the same substance in Japanese Kanechlor, Yusho's cause (Roach and Pomerantz). In 1975, Japanese scientists detected high levels of PCDFs (17–18 ppm) in Kanechlor and about 5 ppm in Kanemi rice oil used by victims. The scientists suggested that the much higher toxicity of PCDFs (compared to that of PCBs) could account for the unexpectedly high toxicity of the cooking oil. They also revised the estimated level of PCBs in the cooking oil, reducing it to 1,000 ppm (Nagayama, Masuda, and Kuratsune). Three years later, scientists in America and Japan found another highly unusual contaminant, polychlorinated quaterphenyls (PCQs) in Kanemi rice oil (Kamps et al.; Miyata, Murakami, and Kashimoto), apparently caused by heating Kanechlor above 300 degrees centigrade (Yamaryō et al.). In 1979, Japanese scientists detected PCQs in the blood of Yusho patients (Kashimoto, Miyata, and Kunita). This provided the possibility of an unequivocal test of poisonous Kanemi rice-oil consumption because no one else in Japan was likely to have been exposed to PCQs (Kashimoto interview).

Yusho could no longer be considered simple PCB poisoning. The discovery of PCQs and PCDFs along with PCBs in Kanemi rice oil provided not only a marker for "true" Yusho patients (through PCQs) but also an explanation for the persistent symptoms of Yusho poisoning (through PCDFs). In June 1981, the official diagnostic criteria for Yusho patients were amended to include the blood level of PCQs (Katō Y., 1989, p. 110). The skin lesions of workers occupationally exposed to PCBs quickly decline once exposure ceases, whereas the symptoms of Yusho victims have proven to be more severe, more systemic, and more persistent (Kashimoto et al., 1981), despite the decrease in PCB blood levels of most Yusho patients to slightly higher than those of the general Japanese population (Kashimoto, Miyata, and Kunita). In 1979, another outbreak of Yusho, this time in Taiwan, also involved the contamination of rice oil by PCBs, along with PCQs and PCDFs, producing over two thousand victims with symptoms similar to those found in Japanese Yusho (Kuratsune and Shapiro).

While scientific knowledge advanced, social protest continued. In mid-1973, protests against PCB pollution of the environment spread rapidly through Japan in what came to be called the Great Fish Panic. Following the government report of widespread PCB contamination of coastal fish, at levels exceeding the standard of 3 ppm, and the announcement of new outbreaks of Minamata disease caused by mercury-contaminated fish, many Japanese consumers stopped buying fish. National sales dropped sharply, throwing thousands of fishermen out of work. During that summer, unemployed fishermen organized forceful and sometimes violent protests throughout Japan and in Tokyo. The government responded with a confused and ineffective attempt to indicate safe quantities of fish consumption and with a tolerance standard for mercury in fish (Huddle and Reich, pp. 169–81). Then, in October 1973, the Diet passed the Law Concerning Screening and Production Control of Chemicals, largely motivated by Japan's experiences with PCBs (Reich, 1981).

Protests related to the Yusho incident also continued, along various lines and with assorted targets. In January 1974, patients and the lawyers' team planned protests against Kitakyūshū city, the central government, and the Kanegafuchi Chemical Company. These demonstrations and their participants tended to be closely related to the mammoth trial in Kitakyushu (Fujiwara, pp. 35–38). They also occurred more or less in the sphere of influence of the Communist party. Another line of protest came from victims in Nagasaki, loosely affiliated with the Socialist party. In October 1974, the Nagasaki group held a sit-in at the Kanemi Company factory (taking a different gate from the one occupied by Kamino and his family). The protesters demanded a meeting with the company president and received it, the first in five years for any Yusho patient (*Asahi*, 19 October 1974). Yet another initiative emerged from patients and supporters aligned with Kamino, who demonstrated against the Ministry of Agriculture and Forestry, for the ministry's refusal to answer an open letter on the dark oil incident sent several months earlier (*Japan Times*, 10 September 1975).

At the same time, negotiations began between Yusho patients, their supporters, and central government bureaucrats. In December 1973, a national assembly of victims' groups approved a list of basic demands, which became the agenda for the first negotiating session with government officials in February 1974. Reverend Kaneda Hiroshi, a Baptist minister living in Fukuoka city, played a key organizing role. Kaneda in 1970 had formed a citizens' organization in Fukuoka to assist patients, "with the dream of everyone participating as individuals." But that group encountered internal conflicts and eventually disbanded. He then helped establish the National Assembly of Victims. The negotiations begun in 1974, however, confronted the government's standard positions: that the solution to the Yusho incident must depend on what was called the "Polluter Pays Principle" and that the government could not admit any administrative responsibility in view of its position as defendant in the Kitakyūshū civil suit (Kaneda interview).

In 1975, Kaneda helped start a new organization, in a renewed effort to unite all Yusho-related groups. Called the Kanemi Yusho Incident National Liaison Conference, it was located in the Fukuoka Prefecture Labor Federation, a group largely associated with the Socialist party. The Liaison Conference entered negotiations with the central government and the companies in the late 1970s (Kaneda interview).

About ten years after Yusho broke out, three long-awaited court decisions finally arrived. Two civil decisions (one handed down in Fukuoka city in October 1977, the other in Kitakyūshū city in early March 1978) found both the Kanemi Company and Kanegafuchi Chemical guilty of negligence and ordered them to pay damages to the plaintiffs: for 44 Fukuoka plaintiffs, ¥683 million ($2.54 million) (*Kubota v. Kanemi Sōko*) and for 729 Kitakyūshū plaintiffs, ¥6.08 billion ($28.9 million) (*Noguchi v. Kanemi Sōko*). The criminal decision, reached in late March 1978, found Kanemi Company factory director Mori guilty of injury due to professional negligence and found company president Katō innocent (*Japan v. Katō*). The Kitakyūshū trial in particular raised the issue of the government's liability for not preventing the Kanemi rice-oil disaster.

All three decisions accepted the pinhole theory to explain the immediate cause of the contamination. The two civil trials agreed that both the Kanemi Company and Kanegafuchi Chemical could have foreseen the accident (based on their knowledge or access to knowledge of PCB corrosiveness and toxicity) and that both companies were negligent. On negligence, the judges in effect accepted the Kanemi Company's argument that the chemical company had not provided sufficient information about the dangers of PCBs and also accepted Kanegafuchi Chemical's argument that the rice-oil company had been sloppy in its production procedures. The civil decisions thus extended liability from the corporate user of PCBs (the Kanemi Company) to the corporate producer of PCBs (Kanegafuchi Chemical), a precedent that seriously worried Japan's chemical industry. The Kitakyūshū judges refused, however, to find the local or central government legally at fault, arguing that these public authorities could not have foreseen the contamination and were not negligent in a strict legal sense.

These decisions did not end the litigation. Although the Kanemi Company accepted the verdicts, Kanegafuchi Chemical appealed both civil decisions. The chemical company publicly denied any responsibility for the Yusho incident in pamphlets published by the company, by the company labor union, and by an industry-related research group. Also, Kanegafuchi quickly sought and obtained an injunction to halt payments during the appeal process. But after the Kitakyūshū decision, victims and supporters launched direct protests at the chemical company's factory and its headquarters. Negotiations followed in which the company agreed to pay a large part of the compensation plus other expenses (Andō, pp. 589–90). The Kitakyūshū decision was also appealed by the plaintiffs, who hoped to demonstrate the liability of the city and central governments and the Kanemi Company president, and to raise the compensation payments at least to the level of those in the Fukuoka decision (p. 590).

Then, in March 1982, a third civil suit (filed in Kitakyūshū in 1976) was decided along the same lines as the two previous civil suits. The court awarded a total of about ¥2.5 billion ($10 million) to 342 plaintiffs, found the Kanemi Company, Kanegafuchi Chemical, and Kanemi Company president Katō guilty of negligence, and found the national government and Kitakyūshū city government not liable (*Asahi*, 28 March 1982). Both the plaintiffs and Kanegafuchi Chemical filed appeals.

Following the Fukuoka decision in October 1977, the Kanemi Yusho Incident National Liaison Conference began to seek assistance for patients not involved in any trial. After the second decision in March 1978, conference leaders provided a concrete proposal to the two companies through Ministry of Health and Welfare officials serving as mediators. Some conflict arose between the lawyers' team and Kanegafuchi Chemical, causing tension in the conference and in the negotiations. But in July 1978, the parties reached an agreement whereby the Kanemi Company paid each nontrial patient ¥220,000 (about $1,000) and Kanegafuchi Chemical paid ¥1.3 million (about $6,200). The agreement stipulated that if Kanegafuchi Chemical lost the final decision, payments to nontrial patients would be brought up to the same levels as awards to plaintiffs, and if the company won, no adjustments would occur ("Kanemi Yushō misoshō").

Evaluations of the agreement varied. Opponents of the agreement called it a compromise settlement engineered by Socialist party activists to compete with the Communist-oriented trial. Supporters called it an honest effort to provide fair assistance to all Yusho patients, including the approximately six hundred nontrial patients. Some of the one thousand patients who had struggled for years in trial proceedings, providing funds for lawyers and traveling to demonstrations and meetings, resented the agreement. They felt that the nontrial patients were taking a free ride, jumping on the bandwagon that they, the trial patients, had pushed and pulled for eight long years.

In October 1979, Kanemi Company president Katō's elder sister, a scientist, writer, researcher of "creative engineering," former member of the Kanemi Company's board of directors, and a Christian, reported in detail that company officials and others knew from the beginning that the oil was contaminated by Kanechlor. The article, titled "The Many Doubts I Have Harbored: A Time to Keep Silence and a Time to Speak (Ecclesiastes)," appeared in a trade journal of the fat and oil industry. The article stated that Kanechlor got into the cooking oil through faulty welding of a joint in the heat-transfer pipes, not through pinholes in the pipes. The company reprocessed the cooking oil in an effort to purify it and then knowingly sold its contaminated product. The sister supported her accusatory doubts by quoting letters, conversations, and personal opinions over a ten-year period. She believed that the story would provide important lessons to managers and technicians in the food industry, to consumers, and to administrators, doctors, lawyers, and others involved in preventing and dealing with cases of food pollution. She also published her doubts, she wrote, because of her sense of social responsibility as a scientist (Katō Y., 1979, p. 43).

The elder sister's disclosure generated new controversies and raised old doubts about how the contamination had occurred. Kanegafuchi Chemical adopted her article as a major piece of evidence in its appeal of the Fukuoka civil decision, to support its analysis that the Kanemi Company had falsified its production and repair records. Kanegafuchi Chemical sought to escape liability by disproving the pinhole theory and shifting all responsibility for compensation to the Kanemi Company (Kanegafuchi Chemical, 1979). The patients' lawyers, on the other hand, called the sister's claims "unscientific" and hearsay. Moreover, citing ties between the sister and people close to Kanegafuchi Chemical and referring to hostility between her and Kanemi's president (her brother), the lawyers suggested that the chemical company was using her to support its claims while she was using the company to get back at her brother. The patients' lawyers argued that the pinhole theory was based on concrete evidence while the accident theory proposed by Kanegafuchi and the sister was a product of "pure fantasy" (Hikōsojinra dairijin bengoshi).

About six months after the sister's article appeared, Japanese newspapers reported that Higuchi Hiroshi, a former Kanemi Company employee who had served as chief of the deodorization department in 1968, had met with lawyers for Kanegafuchi Chemical to disclose the true sequence of events behind the poisoning. Following these newspaper reports in 1980, the sister wrote Higuchi a letter and subsequently met with him; at his request, she wrote down his recollection of events for public disclosure and publication (Katō Y., 1989, pp. 113–22). Higuchi refused to testify in court, but told Kanegafuchi Chemical lawyers how PCBs had leaked into the number one deodorization tank after a faulty repair job, how some contaminated rice oil was reprocessed for distribution to humans and some was added to the company's dark oil for distribution to animal-feed producers, and how company officials had destroyed and doctored both repair and production records. According to Higuchi, "the Yusho incident from start to finish was covered with lies" (p. 116).

Some legal action in the mid-1980s continued to favor the Yusho victims. In March 1984, the Fukuoka High Court, in the appeal of the 1978 Kitakyūshū decision, upheld the ruling against Kanegafuchi Chemical and for the first time found the central government coresponsible and liable for damages. Both the company and the government appealed to Japan's Supreme Court. At the same time, the Fukuoka High Court decided on the appeal of the 1977 Fukuoka civil trial and upheld the ruling against Kanegafuchi Chemical, which also appealed to the Supreme Court. Both decisions, however, reduced damages to the plaintiffs.

In February 1985, the Kokura Branch of the Fukuoka District Court also found government officials at several levels negligent in the dark oil incident and held the government liable for damages to 73 plaintiffs (*Nihon Keizai Shinbun*, 13 February 1985). Both the central government and Kanegafuchi Chemical filed appeals to the Fukuoka High Court. The same year, two more groups of 17 and 75 Kanemi rice-oil disease victims filed damage suits against the two companies, the Kanemi Company president, and the national government, thereby becoming the fifth and sixth groups to go to court. Yet another group, finally reaching 577 plaintiffs, filed suit in January 1986, and in an effort to obtain a quick resolution excluded the central government and local authorities as defendants (Katō Y., 1989, pp. 247–48). By 1986, a total of 1,906 plaintiffs were involved in suits related to Yusho (pp. 247–48).

Then, in May 1986, the tide began to turn against the victims. That month the Fukuoka High Court announced a decision that reduced damages previously awarded to plaintiffs by a lower court in March 1982 and rejected the plaintiffs' demand for damages from the central government, the Kitakyūshū city government, or Kanegafuchi Chemical. This decision reversed trends in six previous Yusho trials. The court ruled that the poisoning was caused by an operational or repair error, not by the pinhole theory, and therefore neither Kanegafuchi Chemical nor the government could be held responsible. It marked a stunning setback for Kanemi victims, who appealed to the Supreme Court (*Asahi*, 15 May 1986).

In the late 1980s, the struggle for compensation finally ended. In July 1986, the Supreme Court announced that in October it would begin oral proceedings on the two initial civil suits (the Fukuoka trial and the first Kitakyūshū trial). The *Asahi* newspaper reported that the decision to hear oral arguments usually indicated an intention to change lower court rulings, in this case probably to review the interpretation of the central government's responsibility and Kanegafuchi Chemical's responsibility as manufacturer (5 July 1986). The oral arguments occurred as scheduled, and several months later, in March 1987, the Supreme Court proposed a settlement to the plaintiffs in the three cases under consideration, to the plaintiffs in the four outstanding cases, and to Kanegafuchi Chemical. Payments from Kanegafuchi to plaintiffs varied up to ¥3.0 million ($20,500), depending on how much individuals had previously received, resulting in total additional payments of about ¥2.0 billion ($13.7 million) from the company.

Yusho victims complained about the low payments and the lack of clear responsibility for Kanegafuchi Chemical. But on 16 March, the two main victims' associations announced their acceptance (*Asahi*, 16 March 1987). After nineteen years of struggle, the victims reluctantly agreed to end their battle with Kanegafuchi Chemical. Later in March, plaintiffs who had sued the central government announced the withdrawal of their suit still pending before the Supreme Court. Several months later, the government agreed, but still expressed a desire to recover the approximately ¥2.5 billion ($17 million) of public funds already paid as compensation to Yusho victims as a result of two court decisions (*Asahi*, 25 June 1987).

In the fall of 1987, court-mediated agreements were concluded between the Kanemi Company and plaintiffs in the remaining three trials, with payments of ¥5 million ($34,000) per person in compensation, plus "sincere efforts" to pay for medical expenses. In March 1989, the last three individual plaintiffs accepted the Supreme Court mediation plan with Kanegafuchi and withdrew their complaint against the government (*Asahi*, 23 March 1989). By that time, 1,860 persons had been certified as official patients, including 142 deceased. Over twenty years after the discovery of Kanemi Yusho, the legal battle had ended.

For many Yusho patients, however, the agony persisted. Their health problems continued to evade treatment. Physicians had failed to develop an effective method for accelerating the excretion of the toxic compounds from humans, as stated in the twenty-year report of the Yusho Research Group (Kuratsune, 1989). But researchers had found an unusually high rate of cancer deaths among Yusho patients (Urabe, Koda, and Asahi), especially liver cancer among one group of patients (Kuratsune, 1989). This finding heightened concern about the persistence of toxic compounds in patients and created understandable anxiety for Yusho patients. Victims also continued to worry about whether the onetime good-will payment from Kanegafuchi Chemical would provide for their living expenses and about efforts by the Kanemi Company to reduce its payments for medical expenses.

In an ironic and cruel twist of fate, victims had become dependent on their victimizers. Kanemi Company officials, who had lied to Yusho victims for years about the causes of the poisoning, would be continuing to provide victims with funds for treatment and livelihood. Yusho victims now depended on the continued economic viability of the Kanemi Company. For two decades, victims had fought the best they could, reaching Japan's highest court. But many felt that they had not achieved justice. The victims had received monetary compensation, but the Kanemi Company still existed, the government had escaped accountability, and Kanegafuchi Chemical had been exonerated. For many victims, their anger would never fade against the Kanemi Company and the other institutions, including the central government. Yusho had scarred the victims' lives as deeply and permanently as their bodies.

Chapter 3

Michigan PBB Contamination

Frederic "Rick" Halbert belonged to Michigan's new breed of dairy farmers called "progressive" farmers. After earning a master's degree in chemical engineering in 1968 and working several years for Dow Chemical, he returned home in 1971, at age twenty-six, to help his father and brother run the family dairy farm just north of Battle Creek. In 1973, their farm, with about four hundred milking cows, ranked among the largest in the state. The Halbert operation represented modern dairying in Michigan. It used the latest technology, including computers, depended on enormous capital investment, and aimed at achieving maximum milk production. Ironically, Halbert's modern dairying helped make his farm among the first affected by one of the worst chemical disasters in United States history.

In September 1973, cows at the Halbert farm fell victim to a strange illness. They developed teary eyes and runny noses and lost their appetite. As the cows stopped eating, milk production plummeted. And so did the farm's income. Like anyone confronting a problem, Rick Halbert first tried to fit the cows' symptoms into his past experience. He considered infectious diseases that commonly afflict herds, but the symptoms did not match. Halbert's veterinarian, Dr. Ted Jackson, also was baffled (Halbert and Halbert, p. 14).

With milk production dropping four hundred pounds a day, Halbert had to act fast. Suspecting that the problem might be in the feed, he telephoned the staff nutritionist at Farm Bureau Services, the company that provided the farm with a high-protein pellet feed called Ration 402. Farm Bureau Services was Michigan's largest feed distributor. As a corporation, it was a for-profit farmer cooperative operating only in the state of Michigan and a wholly-owned subsidiary of the state's most important

58

farmer organization, the Michigan Farm Bureau, a nonprofit corporation. The nutritionist, Dr. James McKean, was also a veterinarian. He listened to Halbert's description of the symptoms and, according to Halbert's recollection, concluded that the cows had an "inappetence problem." One possible cause, suggested McKean, was "moldy corn." He explained that since no other farmers had reported difficulties with Ration 402, the problem must be in Halbert's other feed rations (p. 15).

But Halbert persisted with questions to McKean, asking whether the company could have added manganese oxide instead of magnesium oxide to the feed, as once happened to a neighbor. McKean reportedly assured him that Farm Bureau Services had correctly added magnesium oxide, which "sweetened" the feed and pushed up milk and butterfat production, to help dairy farmers raise productivity and profits. Halbert then asked about other possible contaminants suggested to him in his review of veterinary texts. Halbert recalled that McKean admonished him: "You'd better leave diagnostic work to a trained veterinarian, Mr. Halbert" (p. 15).

Halbert asked a number of dairy specialists to inspect his cows, but none of them could say why the cows were so sick. However, Halbert sensed that something strange was happening at his farm. In three weeks time, daily milk production had dropped to half its previous level (Halbert, p. 37). One day he noticed that the rats and barn cats around his buildings had disappeared. The observation worried him. Was some "general disease" killing all the animals? (Halbert and Halbert, p. 34).

Halbert and his veterinarian, Dr. Jackson, decided to conduct their own feed experiments, and were soon convinced that the problem was in the pellets (Halbert, p. 54). But Halbert wanted more scientific proof. In November, he began feeding the pellets to a group of twelve recently weaned calves. Six weeks later, the calves had stopped eating the pellets and were dying. A year later, Halbert and Jackson published their report in the *Journal of the American Veterinary Association* (Jackson and Halbert).

If the Ration 402 pellets contained some poison, Halbert wondered, why was his farm the only one experiencing problems? McKean insisted to Halbert that Farm Bureau Services had received no other complaints. But to be safe, Halbert stopped ordering Ration 402 and changed to a different feed, Ration 412 pellets, without magnesium oxide, but sold by the same company. Ration 412 was the most commonly used dairy pellet in Michigan. Halbert reasoned that "if they have a serious problem with the [Ration] 412 feed, there should be a lot of complaints" (Halbert and Halbert, p. 41). Even after switching feeds, however, problems persisted at the Halbert farm.

Halbert also approached researchers at Michigan State University, bringing a sick calf from his feed experiment to the veterinary diagnostic laboratory for sacrifice and postmortem examination. Although the veterinarians found various internal lesions, their written report attributed the cause of illness to "malnutrition." Halbert commented, "The poor calf wouldn't eat for two weeks before it died. I told the vets that during the autopsy. . . . What I wanted to know is *why*." Halbert offered to donate animals and feed for an experiment, but the researchers explained they had no facilities for such an undertaking and that they had heard of no other farmers with similar problems. Halbert felt they also wanted to avoid getting involved in a private dispute between a dairy farmer and his feed supplier (p. 51).

Halbert urged a researcher at the state Department of Agriculture to perform its own tests of Ration 402 pellets. He hoped the official would verify his results, showing the feed contained some poison. The researcher agreed to do an experiment—but on mice, not cows. Ten days after beginning their diet of Ration 402 pellets, the mice weighed one-third less and all were dead. Halbert considered this "thrilling news, because someone had confirmed our tests" (Halbert interview). He asked the researcher to do the experiment again to be certain of the conclusion. The second round yielded the same result: dead mice.

A few days after Christmas 1973, Halbert called the executive vice-president of Farm Bureau Services, Donald R. Armstrong, to tell him the news. Armstrong reacted cautiously, suggesting that Halbert may have stored the pellets incorrectly, spoiling them in some way, and that the mice trial would have to be repeated. Armstrong insisted, "I can't believe those pellets could be causing any trouble" (Halbert and Halbert, p. 53).

In mid-January 1974, Halbert and his father met with officials of Farm Bureau Services. Armstrong asked the Halberts to sign a release giving the company access to the postmortem data at the university. Halbert agreed, hoping that "free and open exchange of information would eventually help solve the problem" (p. 57). Armstrong, however, still denied that the company's feed was at fault. He remarked that since no other farmers reported problems, the trouble must be limited to the Halbert farm. Yet Armstrong did promise to start a feed test on calves at a private research facility, to commission a laboratory investigation of the feed, and to notify Halbert of the results. What Armstrong did not say was that in October another farmer had complained to Farm Bureau Services of problems with Ration 402 pellets and that in December the company had decided to halt production of the feed (Armstrong, 1977, pp. 1622–23). The company reportedly notified the field inspector for the state Department of Agriculture of this decision to stop production (p. 1622).

After the meeting, the state researcher agreed to repeat the feeding experiment at the company's request. Again, the mice died. Halbert later recalled that company representatives continued to deny to him the existence of problems with the feed, suggesting the mice died because they had eaten "cattle feed" and not "mice feed." Halbert considered that response nonsense and was convinced that the experiments with mice proved his point (Halbert and Halbert, p. 58).

Within Farm Bureau Services, however, the third test on mice convinced some people that a serious problem existed in the high-protein feed pellets, for in mid-January, the company decided to recall Ration 402 pellets from farmers using it (Armstrong, 1977, p. 1623; Food and Drug Administration, 1974). The company offered to buy the pellets from Halbert, but he decided to keep them for his experiments. Only several months later did he learn about the recall from other farmers (Halbert, p. 63). The company also hired private research institutes to conduct chemical analyses of the feed and to perform a feeding trial on calves. The company apparently did not formally notify state agencies about the feed recall (Amstrong, 1977, p. 1629). According to the company's executive vice-president, there was "no requirement to notify the state government of a feed recall" (Armstrong interview).

Yet the health of Halbert's animals, even those on Ration 412, continued to deteriorate. Reproductive problems became increasingly common, with calf birth mortality rising to about five times normal. The appearance of many cows worsened noticeably. They had "ears dropping, coats dull and coarse, patches of hair missing from faces and necks to reveal a peculiar elephant-like skin, many stumbling over distorted hoofs" (Halbert and Halbert, p. 6). By the end of December 1973, Halbert's losses in milk production had reached about forty thousand dollars (Egginton, p. 37).

These problems took their toll on the Halbert family. The cows' sickness was more than a matter of economic survival. It threatened the family's integrity. Rick Halbert became obsessed with the problem, ignoring his wife and three young daughters and becoming more tense and withdrawn as time passed. Then Halbert's cousin, an experienced farm worker in his late thirties, died suddenly when he slipped from a tractor and was run over by its heavy wheels. The autopsy showed no reason for the accident, but Rick's father wondered whether the man's health had somehow been affected by excessive contact with the Ration 402 pellets. Rick's wife, Sandy, had also begun to worry about the health of her children and herself. She felt "so nervous and frightened" by the "invisible enemy" (Halbert and Halbert, p. 79). Within her grew a gnawing fear that the family had fallen victim to "the same unknown plague that had ravaged the dairy herd" (p. 75).

In late January, Halbert and Dr. Jackson began to look for assistance outside Michigan, approaching federal rather than state agencies. Jackson thought the problem might be a mold toxin, and sent a sample of the feed to a specialist at the National Animal Disease Laboratory (NADL) in Ames, Iowa. The researcher was unable to identify a known mold toxin, but he considered the result inconclusive, since ten unknown molds exist for every known one. In March 1974, an animal toxicologist at NADL tested for various chemicals with gas-liquid chromatography. None of the normal contaminants appeared. But by chance, the machine was left on for an unusually long time, producing a strange series of peaks and indicating the presence of some chemical substance. No one at NADL could identify what caused the peaks, and they became known in the laboratory as the "Rocky Mountain Range." This discovery marked a turning point in Halbert's search. He now had persuasive scientific evidence of a foreign substance in the feed.

But researchers at NADL could not continue their work on the project. The laboratory lacked the mass spectrograph necessary to identify the compound, and the scientists were unable to raise the money for testing at another facility. When the NADL toxicologist planned a trip to Michigan to perform postmortems on Halbert's cattle, his Washington supervisor forbade him to spend any more agency funds on the project (p. 83). Halbert desperately tried to revive the project by telephoning people in Washington and at NADL, but he repeatedly confronted the response that federal funds could not justifiably be spent to study the problems of a single farmer.

Confronted by these obstacles in public agencies, Halbert contracted with a private testing laboratory in Wisconsin to examine the feed. The laboratory confirmed the strange peaks on the gas-liquid chromatograph, as found at the NADL. The private laboratory also reported to Halbert that his feed sample contained no magnesium oxide, although this chemical was supposed to be a key ingredient in the Ration 402 pellets (Egginton, p. 51). Halbert met again with Farm Bureau Services in February. The company reported that tests on the feed for the standard pesticides were all negative. Armstrong concluded that the feed was "wholesome and pure." According to Halbert, the company official "refused to admit that there was even the slightest difficulty with the feed" (Halbert and Halbert, p. 92).

An internal memorandum, prepared by veterinarian McKean to brief Armstrong prior to this meeting on the "Halbert Dairy Problem," reviewed laboratory tests that had eliminated heavy metals and pesticides as possible contaminants in the feed. The memorandum included a comment from an NADL scientist that Halbert's sample "apparently contained an unpalatable compound of unknown origin." McKean provided this interpretation: "After perusal of all the tests run, history, and clinical signs, I am unable to conclusively indict the feed. I have a feeling that Rick's feeding and management programs contributed to the problem. If palatability is the only problem with the [Ration] 402 pellets, the serious effects seen in the herd since September cannot be totally laid at its door. I will continue to explore every available avenue to determine the problem source; but, to date, have found no feed-related problem which will explain the signs observed" (Egginton, p. 52).

In March, Halbert learned from the "risk manager" at Farm Bureau Services that three of the four calves in the company's experiment with Ration 402 pellets had died. Halbert immediately telephoned McKean to find out more details about the animals. McKean reportedly replied that the calves were "doing all right" and "had no problem." Halbert then said that he knew three had died. According to Halbert, McKean responded, "It's not my place to give out information" (Halbert and Halbert, p. 94).

Throughout these events, the agency responsible for regulating food quality, the Food and Drug Administration (FDA), was uninvolved. No one had contacted the FDA. Halbert, who along with his veterinarian had initiated most contacts, asked people along the way whether to inform the FDA. "The answer was a firm and uniform no. I was told that they would only create trouble" (Halbert, p. 40). Halbert feared that if the FDA discovered he had a health problem, the agency would shut down his farm, but without solving his problem (p. 57). When he finally called the agency, in March, an inspector came and took samples but provided no help. "Their attitude was that they were not a research agency but a regulatory agency, and unless we knew what was wrong there was little they could do to help us" (p. 40).

Later in March, Halbert argued again with McKean. Halbert had found two other farmers who had experienced similar problems with Ration 402 pellets. The local grain elevator had offered not only to buy back the remaining pellets of one farmer but also to compensate him for lost milk production. Halbert believed that the company was withholding information from him, was isolating affected farmers from one another, and was treating individual farmers differently (Halbert and Halbert, p. 110).

McKean and other officials at Farm Bureau Services believed they were doing their best. They had recalled feed Ration 402 in case it contained some contaminant. But health problems seemed to vary from herd to herd, making it difficult for the veterinarian to identify a common pattern or connection. Nonetheless, in February, McKean reportedly called the corporate offices of the supplier of magnesium oxide to check on the product's quality. He remembered that a salesman assured him that the chemical was a food-grade product, adequately protected in its packaging and warehousing (Egginton, p. 75).

In mid-April, Halbert received information from McKean that helped identify the poison. McKean told Halbert that a private research laboratory in Wisconsin had detected bromine in the feed. McKean then read off the molecular weights of peaks from analysis with low-resolution mass spectrometry. Halbert relayed that information to an analytical chemist he had contacted at the U.S. Department of Agriculture research laboratory in Beltsville, Maryland. The scientist identified the mystery compound as polybrominated biphenyls (PBBs), two linked benzene rings with varied numbers of bromine atoms attached, and produced by Michigan Chemical Corporation as a flame retardant. By chance, that scientist was one of the few people who had fed PBBs to animals in toxicological experiments. Halbert immediately recognized Michigan Chemical as the source of magnesium oxide that Farm Bureau Services supposedly added to Ration 402 dairy pellets. Halbert concluded there had been a mix-up. He informed both companies of his discovery. The report took Michigan Chemical completely by surprise. But the feed company, as Halbert put it, "felt that they had found somebody to take the crown of thorns from their head" (Halbert, p. 56).

Michigan Chemical first manufactured PBBs in 1970 as a fire retardant for molded plastic parts, such as the cases of televisions, typewriters, and business machines. By late 1974, when it stopped manufacturing PBBs, the company had made about 5.5 million kilograms of the substance for over 130 users. Michigan Chemical was not the sole PBB manufacturer, but the two other manufacturers of a similar product did not begin production until 1975 and stopped in 1977, after making only 680,000 kilograms for export (Wallen).

Michigan Chemical's PBB product, until the mix-up, represented a successful development for the company and its owners. Known as Firemaster BP-6, the product contained mostly hexabromobiphenyl (about 60 percent) and several additional isomers. Production of Firemaster BP-6 rose rapidly from 1,000 kilograms in 1970 to 1.0 million kilograms in 1972 and 2.0 million kilograms in 1974. Michigan Chemical had developed its PBB product in expectation of future federal government regulations on the flammability of molded plastic parts, as well as upholstery and other items (the regulations were promulgated in 1975). Michigan Chemical anticipated an expanding market in flame retardants and sought a competitive edge over chlorine-based products, since bromine-based products were considered to have a greater efficiency on a weight-cost basis (Northwest Industries).

Figure 3. Places in Michigan related to PBB contamination.

In 1935, Michigan Chemical began operations in the central Michigan town of St. Louis (Figure 3) because of the region's salt brines, which made Michigan the largest salt producer in the country. The brines provided numerous other chemicals, including magnesium and bromine, which contributed to the development of the chemical and pharmaceutical industries throughout the state. Michigan Chemical at first produced various bromine compounds and common salt, added magnesia in the 1940s for war use, and then shifted into production of DDT for the military and then consumer sales, with other chemical products to follow (Haynes, pp. 275–76).

In the 1970s, Michigan Chemical belonged to a complex web of corporations, under the control of Northwest Industries, a multibillion-dollar holding company based in Chicago. Michigan Chemical was purchased by Velsicol Chemical Corporation in 1970, and in 1973 and 1974 the company was jointly owned by Velsicol and Northwest Chemco, both owned by Northwest Industries. In 1976, Michigan Chemical and Velsicol Chemical merged; the resulting company of Velsicol was a wholly-owned subsidiary of Northwest Chemco, which was itself a nearly wholly-owned subsidiary of Northwest Industries. A variety of administrative and financial devices connected these companies, keeping them under the control of Northwest Industries.

Other private companies considered PBBs too risky for production because of the probable chronic toxicity. In the early 1970s, two of America's largest chemical companies, Dow and DuPont, decided not to manufacture a PBB product. Their decisions were based on separate animal tests that showed detrimental toxic and environmental effects—evidence of liver damage, bioaccumulation, and high probabilities of carcinogenicity and teratogenicity—and on knowledge about the high toxicity in humans of a related compound, polychlorinated biphenyls (PCBs), through occupational exposures and through the food poisoning case of Yusho in Japan. The companies publicly announced their decisions and published their research in 1972 and 1973 (Aftosmis et al.; Norris et al.).

Michigan Chemical, lacking an adequate internal research division, contracted a private firm to test the acute toxicity of its PBB product. The report, in May 1970, concluded that Firemaster BP-6 was nontoxic for ingestion or dermal application, not a skin or eye irritant, and not highly toxic when inhaled (Hilltop Research). Taking that report as a green light, Michigan Chemical rushed ahead with Firemaster BP-6 in the early 1970s.

Nonetheless, in late 1971, the company prepared a one-page health and safety statement on BP-6 to inform workers about possible health problems with PBBs. The statement recommended against prolonged exposure and noted that BP-6 probably accumulates in fatty tissue and the liver, "which certainly is undesirable and possibly could be dangerous." The statement warned against allowing BP-6 to contaminate any food or feed (Michigan Chemical, 1971).

In April 1974, when first informed of the mix-up, Michigan Chemical denied that its PBB product, Firemaster BP-6, could have been confused for its magnesium oxide product, Nutrimaster. Company officials explained to the U.S. Food and Drug Administration that the two products were stored and manufactured in separate buildings and were totally different in consistency and color. Firemaster was chunky and amber, and Nutrimaster was granular and whitish. In addition, the company reported that it packaged the chemicals in color-coded bags, one bright red

(Firemaster) and the other royal blue (Nutrimaster) (Kolbye, 1977b, p. 1022).

On 30 April, however, an inspector for the FDA discovered a half-used bag of PBBs in a Michigan feed mill, thereby decisively linking Michigan Chemical with the sick cows. The bag was Firemaster FF-1 from an experimental batch of PBBs in which the chemical was ground into powder and mixed with an anticaking agent (calcium polysilicate). The special processing transformed BP-6 into a substance remarkably similar to magnesium oxide in both consistency and color. Moreover, the discovered bag was not color coded, and its label did not list ingredients or manufacturer or a warning about toxicity. The plain brown bag showed only the trade name, Firemaster FF-1, stenciled across the top. Once the bag was opened, even those meager markings became nearly impossible to read.

Several factors at Michigan Chemical had caused the mix-up. In spring 1973, Michigan Chemical ran out of color-coded bags and used plain brown fifty-pound bags for both Firemaster and Nutrimaster. Neither product was clearly or adequately marked. Also, according to an internal company memorandum, the storage of the experimental FF-1 was "*very poor*," with broken bags in some warehouse areas (Michigan Chemical, 1973, italics in original). Additional confusion occurred because Michigan Chemical used three different commercial names for its magnesium oxide product.

Problems also existed at the feed company, Farm Bureau Services, as detailed in sworn court statements of several employees. The men employed to mix the feeds had little job training, and one employee could not read well enough to recognize the word *Nutrimaster*. Some employees, however, could read perfectly well and did report to a supervisor the appearance of a new trade name in the warehouse: "Firemaster." The supervisor told them it was just another name for magnesium oxide and to keep adding it as required (Courter and Lehnert).

FDA inspectors and state officials searched for other bags of misplaced PBBs, but reportedly found none. The chemical company initiated a recall of Nutrimaster and also checked for misplaced bags of flame retardant, but its efforts similarly revealed nothing. The half-filled bag was the only direct evidence of the mix-up found by the FDA. The other bags of PBBs involved in the misshipment, apparently, had already entered the animal feed and the food chains. Subsequently, Farm Bureau Services requested a private analysis of the "grab bags" of samples kept from different production runs of all the feeds produced at one feed mill (in Climax, Michigan). These tests showed that two direct additions of PBBs occurred, once in September 1973 and again in March 1974, along with a series of smaller pulses when returned feed was put through the machinery. Halbert's initial purchase of Ration 402 had contained a direct addition of PBBs; subsequent purchases of Ration 412 contained PBBs at lower levels probably from cross contamination.

The PBB mix-up in Michigan created a major case of contamination with a relatively unknown toxic chemical. In the spring of 1974, few people in Michigan or elsewhere had heard of PBBs. Federal as well as state agencies lacked information on PBBs, with little published in the scientific literature. The FDA found few scientific studies to help determine what level, if any, might be safe in human food. The main experimental study available compared the capacities of PBBs and PCBs to increase the activity of the liver's microsomal enzymes, which are involved in metabolizing toxins and drugs. It showed PBBs on a molar basis to be five times stronger than PCBs (Farber and Baker).

Nevertheless, on 10 May 1974, about two weeks after identification of the contaminant as PBBs, the Food and Drug Administration announced an "action level" of 1.0 part per million in the fat of milk and milk products. An "action" level is not a "tolerance," which requires formal and public procedures, but a temporary administrative guideline that can be informally and quickly set by the FDA. It requires no public participation and can be based on incomplete scientific data, yet can stand for years before a tolerance level is set. It provides for relatively quick regulatory action, despite scientific uncertainty. In late May and early June, the FDA set additional PBB action levels: 0.3 ppm for animal feeds, 1.0 ppm for the fat of meat, and 0.1 ppm for eggs.

FDA scientists chose the action level for PBBs solely on measurement capability, officials explained, since the Food, Drug, and Cosmetic Act requires in cases of "avoidable" contamination that the guideline be set at the lowest detectable level. The scientists regarded 1.0 ppm in the fat of milk and milk products as probably safe, since the PCB "temporary tolerance" in the United States then was 2.5 ppm, and the PBB molecule is heavier than the PCB molecule (Kolbye, 1977a).

On the same day as the announcement of the federal action level, the Michigan Department of Agriculture ordered seven farms to stop selling milk, because of PBB levels exceeding the FDA guideline, and forbade them to sell cows for slaughter and meat. Using the federal guideline for food safety, the state Department of Agriculture took steps to keep additional PBB-tainted products off the market and to contain the contamination; it imposed quarantines on dairy farms, exactly what Halbert had feared most. The state first quarantined the obviously contaminated farms, those that had purchased Ration 402 pellets from Farm Bureau Services. The Department of Agriculture also began testing milk in the tanks of milk dealers throughout the state to trace and identify other PBB-tainted herds. By the end of May, the department had quarantined thirty farms.

The heaviest burden of the quarantine fell on the farmers. They could not sell their cows or milk. Nor could they kill the animals, since no burial site existed to dispose of PBB-contaminated carcasses. Also, lawyers advised the farmers to keep sick animals alive for court evidence. As a result, farmers continued feeding the cows, dumping the milk, and losing hundreds of dollars a day. Yet farmers had no idea if or when the responsible companies would pay for losses. And farmers feared that the health problems in their cattle would appear soon in their families. They knew that for about nine months, from September 1973 on, while the problem was considered Halbert's private trouble, Michigan farmers and consumers had unwittingly eaten dairy and other farm products contaminated by PBBs.

On 13 May, the state Department of Agriculture made its first public announcement about the incident. The statement described the problem as limited to a "very few" farms and as posing "little need for concern about the public milk supplies." It noted that department officials had been working on the problem "since last October," but had not identified the cause earlier because the mix-up was a rare event and veterinarians had not experienced the problem before (MDA, 13 May 1974).

The state health director first learned of the contamination not from agriculture officials but from an article in the *Wall Street Journal* on 8 May (Chen, pp. 55–56). At that point, according to one health official, the department had "absolutely zero knowledge" about PBBs or the contamination (Wilcox interview). The first major assessment of the contamination occurred on 17 May, at a meeting called by the state Department of Agriculture. Participants included representatives from the U.S. Department of Agriculture, U.S. Food and Drug Administration, Michigan State University, Michigan Chemical Corporation, Farm Bureau Services, Michigan Department of Public Health, Michigan Milk Producers Association, and the state attorney general's office.

State Agriculture Director B. Dale Ball, a feisty, hardheaded, bureaucrat-farmer, set the tone of the meeting with an opening anecdote about how milk promotes health. Then, referring to the contamination problem, he stated that the public milk supply "was pretty much under control." One point Ball stressed clearly: "I think that milk is a pretty good product and I hope that out of this we don't do anything to injure the dairy industry" (MDA, 17 May 1974).

According to a tape transcript of the meeting, representatives from the two companies, Farm Bureau Services and Michigan Chemical, emphasized that the contamination was a limited problem and under control. James McKean confidently reported that Farm Bureau Services was tracing everyone who had received Ration 402 pellets. "We feel that the lid is on this problem" (MDA, 17 May 1974, tape 3, p. 4). Michigan Chemical's representative stressed that only a small amount of the chemical could have been involved, probably about thirteen 50-pound bags, or a total of 650 pounds. The mix-up, they said, was accidental, due to a mistake in the warehouse, and was not caused by cross-contamination or a processing error.

Potential human health disorders received scant attention at the meeting. The Michigan Department of Public Health sent only a lower-echelon employee to the meeting and reported that the department had yet to begin even a preliminary study of health effects. The meeting also avoided discussion of contaminated meat products. One unidentified participant made the only comment on meat in commercial markets, saying, "I think that's gone forever and I'd hate to have too much conversation develop on that, because it can't help anybody . . . " (MDA, 17 May 1974, tape 4, p. 1). In general, people at the meeting agreed that the contamination involved only one feed, affected mainly cows on a few farms, and contaminated milk but not meat.

The participants summed up the PBB problem in a news release. After a draft of the release was read, one person (unnamed in the transcript) suggested several revisions to minimize public "alarm." He proposed, among other points, elimination of references to plans by the Department of Public Health for a survey of families on PBB-contaminated farms and to possible effects of PBBs on human health. The final draft of the press release included all the recommended changes. In the last minutes of the meeting, questions were raised about whether the contaminated cattle would be condemned and how the farmers would be compensated, but the session closed without addressing those problems.

In late May 1974, soon after the meeting, the Department of Agriculture proposed emergency legislation in Michigan for authority to condemn animals and indemnify farmers. The bill allowed the state to compensate farmers and then recover costs through legal action against the companies. But the proposal never got off the ground. Governor William Milliken refused to support it, because he wanted the companies to pay from the start. Other state politicians accused Agriculture Director Ball of trying to bail out Farm Bureau Services. But Ball insisted that his primary purpose was to aid the farmers and allow them to survive the financial pressures of quarantine.

In the matter of disposal, the state initially had no idea what to do with PBB-contaminated animals. The Department of Agriculture instituted quarantines until it could decide on an appropriate policy. In late June, the state legislature passed a bill to deal with the disposal problem, and the governor signed it into law on 2 July, nearly two months after the first quarantine. Public Act 181 of 1974 permitted the Department of Agriculture to designate an appropriate site for killing and burying contaminated farm animals. The law allowed the state to condemn animals, to approve disposal facilities, and to use a civil suit to recover costs from responsible companies. But state officials decided not to condemn any animals and not to order disposal, because they did not want to open the possibility for farmers to file suit against the state or for the state to be held financially responsible. The state quarantined farms and monitored the disposal operation, while the farmers and the companies decided on their own whether to destroy the animals. As a private dispute, the matter had to be settled between the individual farmer and the private companies. The destruction of animals remained a "voluntary" decision of the farmer. Halbert later explained how this measure worked in practice: "Farmers were simply shut off from their markets and stuck with useless animals—after months of going backward in double time, most farmers who were faced with this impossible ruinous situation agreed to have their animals destroyed" (Halbert, p. 99).

Another problem related to disposal arose in July. The Department of Natural Resources selected a disposal site in a state forest located in sparsely populated Kalkaska County of north-central Michigan. Immediately after the governor signed the law, the killing and burying began at Kalkaska. In two days, Farm Bureau Services, under state supervision, disposed of 140 cattle at the site. However, the state had neglected to notify or consult local officials about the disposals (Brunn and Koons, p. 65). When the Kalkaska County Board of Commissioners heard about the disposal operation, it obtained a court injunction to stop the dumping of PBB cattle in the state forest. Five hundred local residents attended a public hearing, and many attacked the state's action (p. 66). For seven weeks, the conflict dragged on, as a county judge forced the state government to defend its proposal. A primary election in early August, for the offices of the commissioners and the judge, made Kalkaska county officials especially sensitive to protecting local pride, autonomy, and public health. Then, in late August, after the election, the judge ruled in favor of the state, and disposal operations began again.

Three initial studies were conducted to assess possible effects of PBBs on human health. The first study was done by the FDA in late May and early June 1974 as part of an inspection. This survey of sixty-five quarantined farms collected medical records from farmers along with medical release forms, and found various health problems. But the FDA did not analyze or publicize the data. The survey's results remained unknown to the subjects and to the public until 1977 when a congressional investigation committee confronted an FDA official with the unpublished report (Cordle, pp. 155–57). In response to aggressive questions from the committee's staff, the FDA official explained, "We did nothing with [the report] because that information is part of an inspection. . . . [The] FDA has no authority to follow up health problems which you are describing there and which the inspectors really collected on their own initiative. We had no mechanism" (Cordle, p. 156).

But the initial FDA survey became the basis of the second evaluation. The Michigan Department of Public Health performed a preliminary screening in June and July 1974 of 211 farm people, using the same group of quarantined farms as examined by the FDA. The department announced in late July that although PBBs could be detected in the farmers' blood and some people showed medical disorders, the results "have not revealed a medical syndrome, or group of symptoms, which can be related to PBBs" (MDPH, 25 July 1974).

The Department of Public Health next designed a "short-term" epidemiological study of 300 persons, divided into a PBB-exposed group and a nonexposed control group. The department used questionnaires to obtain medical histories and in the fall of 1974 performed physical examinations and laboratory tests. The Michigan health authorities drew conclusions from the short-term survey similar to those from the screening study, reporting no correlation between PBBs and "any symptom whatsoever." In a preliminary announcement in February 1975, the health director stated that he did not believe trace amounts of PBBs found in blood "have any clinical significance." Moreover: "So far we have found no real difference in the health status of those exposed and those not exposed, and we have no reason to believe any will be found" (MDPH, 14 February 1975).

Not everyone agreed with the department's approach or its conclusions. One person who found the department overly cautious from the beginning was Dr. Thomas Corbett, an anesthesiologist at the University of Michigan Medical Center in Ann Arbor. When the contamination became public knowledge in May 1974, Corbett was testing a commonly used anesthetic on mice for its ability to cause cancer or birth defects. He "felt obligated" to run similar tests on PBBs, because such information did not exist, and because he could find no one else—at the university or in government—planning or willing to do the work (Corbett, p. 127). As the experiment progressed, Corbett found mice dying of gastrointestinal bleeding and showing enlarged livers. His final results demonstrated several birth defects, including cleft palates and brain defects (Corbett et al.).

By the fall of 1974, the definition of PBB poisoning in cattle and animals had also emerged as a problem. Officials had reasoned that if humans could safely consume milk and meat contaminated up to 1.0 ppm of PBBs, then the contaminated animals that produced the milk and meat would be considered healthy. But many dairy farmers reported health and milk production problems in herds that showed PBB levels in fatty tissue below the official action level of 1.0 ppm.

To review animal PBB poisoning, the Department of Agriculture convened two meetings in the early fall. Corbett attended both and presented his experimental results, urging that the acceptable levels in animals be lowered to prevent serious illness in humans. In early October, at the second meeting, state officials expressed concern that animals with PBB levels registering below 1.0 ppm were suffering from PBB-related health problems. They proposed that the guidelines be reconsidered. In late October, two FDA veterinary toxicologists arrived in Michigan for a three-day review of PBB poisoning in dairy cattle.

The brief report by the FDA veterinarians presented the clinical effects of PBBs in cattle at both high and "low" levels. Based on their survey, they defined a chronic, long-term syndrome that differed slightly from the acute, short-term syndrome in Halbert's animals. The chronic effects included higher rates of mortality, reproductive problems, and congenital abnormalities. Cows under stress, particularly stress associated with giving birth, experienced various health disorders, including rapid weight loss, reduced milk production, weakness, and sometimes death. The veterinarians also showed that on a farm with chronic poisoning problems, the bulk milk tank did not always have detectable PBB residues and the fat tissue residues from individual cows ranged from 0.07 to 0.55 ppm— far below the FDA guideline of 1.0 ppm. "Animals exposed represent a long-term reservoir of PBBs and the economics of the situation appears to favor immediate and complete destruction of affected animals" (Teske and Wagstaff, p. 4).

Soon after the report, the FDA lowered the federal PBB levels acceptable for commerce. In milk and meat, the levels dropped from 1.0 to 0.3 ppm (in the fat of these products), and in eggs and feeds from 0.1 to 0.05 ppm. The Michigan Department of Agriculture also announced the new levels, noting that an additional fifty or sixty herds might be included in the quarantine. But its statement stressed that the department had not condemned the animals. Condemnation remained a private decision made by cattle owners and the responsible companies (MDA, 7 November 1974). Quarantine was a decision of public policy, and Agriculture Director Ball said his department acted "with great reluctance" to quarantine more herds (MDA, 3 December 1974).

By fall 1974, public officials recognized a broader contamination problem than the one initially understood. The contamination apparently resulted not just from the two direct additions of PBBs instead of magnesium oxide, which affected herds like Halbert's at a high level, but also from widespread secondary contamination (Figure 4). Feed-mixing machinery had passed PBBs to other animal feeds and feed additives that did not directly include magnesium oxide. And animals unsuited for human consumption were slaughtered, processed, and added to feeds. These two feedback cycles helped produce a low level of chronic contamination throughout Michigan's feeds and farm animals—not just cattle—and into large parts of Michigan's food chain.

The new action level, by defining a much larger group of cattle as unfit for production or consumption, expanded the problem of disposal. By November 1974, the state had quarantined about ten thousand cattle, and about nine thousand had been killed and buried at Kalkaska. Under the new action level, the number of cattle for disposal rose rapidly. The state Department of Natural Resources had approved the Kalkaska site for burial of thirteen thousand animals, but burial continued far beyond that number. The Department of Natural Resources agreed to continue disposal there because no other site was available and because it seemed safer than burying on individual farms.

The new action level also affected policies on compensation for damages. In mid-June, the insurers of both the Michigan Farm Bureau and Michigan Chemical had agreed to an insurance pool of about $10.5 million, with a 50/50 split and liability to be decided later (Chen, p. 132). In summer 1974, the two companies had begun to settle out of court with farmers whose herds were quarantined by the 1.0 ppm level. By October, seventy-four claims had been paid, nearly exhausting the available funds. The new action level of 0.3 ppm for milk, red meat, and poultry greatly expanded the number of potential claimants against the companies and raised the stakes for everyone involved. More money would be needed to meet the new claimants.

In February 1975, the Michigan Senate passed a resolution establishing a special committee to investigate the PBB contamination. In an unusual move, the Senate's majority leader, a Democrat, chose Republican John A. Welborn to chair the committee. Welborn's main concern about the PBB problem was to aid suffering farmers. He began his activities with a "grisly news conference," as one report described it, "to dramatize the plight" of the farmers. Welborn called the press conference on a constituent's PBB-tainted farm near Kalamazoo. Standing outside in subzero weather, amidst a pile of seventy frozen dead cattle, the state senator told newsmen about legislation he planned to introduce to help PBB farmers (*Detroit Free Press*, 11 February 1975).

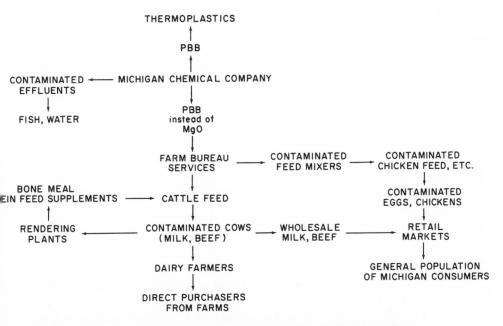

Figure 4. Pathway of PBB environmental contamination in Michigan. Source: Selikoff and Anderson, p. 10. Used by permission of the authors.

Also in early 1975, Governor Milliken intervened with the responsible companies, urging them to put together another insurance pool for compensation of damaged farmers. The first insurance pool (of about $10.5 million) was exhausted in the fall of 1974, and settlement payments stopped (Welborn et al., p. 11). Under threat of government action, in January 1975, the companies and their insurers agreed on a second pool of $15.5 million (Chen, pp. 133–34). This amount lasted until April 1976, providing settlements for about 350 of 650 outstanding claims from farmers. The process of creating these two insurance pools involved complicated negotiations among the various levels of insurers and reinsurers for the Michigan Farm Bureau and Michigan Chemical. Each time one level of insurance was exhausted, the next reinsurer had to be persuaded about the validity of the claim, thereby delaying the process of settlement (Welborn et al., pp. 12–14).

The state Department of Agriculture also became increasingly concerned in early 1975 about reports of cows with health problems in herds with low-level contamination (now under 0.3 ppm). The department counted 286 low-level herds, with about one-half reporting health problems similar to those in cattle with higher PBB levels. The problems included an increase in deaths and calf birth mortality, reductions in weight, and decreased milk production. But some agriculture officials were skeptical that low levels of PBBs caused the health complaints in all these herds. The skepticism was expressed in February 1975 by the director of the FDA office in Detroit in a background memorandum for Washington officials: "We do not have documented proof but have repeatedly heard that some farmers are anxiously trying to rid themselves of uneconomic dairy herds. Physical conditions of animals taken off high energy grain and protein supplements are identical to those found in animals suffering early symptoms of PBB toxicity" (Hoeting, p. 6). In late February, at a meeting in Washington, state and federal officials discussed the problem of low-level contamination of cattle and the adequacy of the PBB guidelines. While they discussed, concern about PBBs in Michigan began to spread.

In March 1975, the Michigan press published prominent reports of human illness allegedly due to PBBs. On 1 March, *Michigan Farmer*, a semimonthly farm magazine, published a special issue on PBBs, with a color photograph on the cover showing the burial of cows at Kalkaska. The issue included stories on PBBs in human blood, consumer fears about Michigan food products, and the disastrous consequences for farm families. The editorial's headline sounded an alarm: "Public Health Bomb Ticks away Unnoticed." The top manager at Farm Bureau Services responded with a sharply worded letter to the editor. He called the editorial an "inaccurate, over-simplified and downright libelous 'cheap shot' which was obviously the product of a minimum of research and a maximum of hindsight" (Armstrong, 1975).

Then, one week later, the *Grand Rapids Press*, a daily paper in western Michigan, located near many contaminated farms, began a thirteen-part series in daily installments called "The Poison Puzzle." The articles reviewed the discovery of PBBs in feed, described the agonies of farmers and consumers, and attacked bureaucrats for failing to contain the poison. Headlines included "The Effect on Human Health—An Unknown in PBB Contamination," "Victims May Not Know Cause," and "Official Tolerance Level Doesn't Mean It's Safe."

Partly in response to this press coverage, the Michigan legislature passed a resolution in March seeking to compel state officials to review the PBB action level and to assure the public about the safety of Michigan's food. The resolution urged the state Agriculture Commission to hold a hearing on the removal of all food products showing a detectable level of PBBs. The legislature directed the request to this body because it provided oversight for the Department of Agriculture and farm policy and appointed the director of agriculture. The commission's members were appointed by the governor to four-year staggered terms. This organizational set-up was intended to protect the department from undue political interference by the governor or the legislature. But this structure had limitations. The Michigan Farm Bureau maintained a high degree of influence in the department, and the commissioners "were invariably MFB activists" (Coyer and Schwerin, 1981, p. 717).

On 13 March, the FDA commissioner in Washington ordered a complete review of the PBB situation in Michigan. A major reason for his decision was newspaper stories reporting human illness due to PBBs (Schmidt, p. 1). That day, for example, the *Detroit News* had published a page-one story on the PBB experiments of Dr. Corbett, under a banner headline that declared "Human Birth Defects Seen in Tainted Feed." Corbett warned that the effects of PBB on people "may be far more serious than state officials believe." As part of the FDA review, the commissioner ordered a survey of herd health in Michigan, with special attention to herds with low-level contamination, defined as less than 0.3 ppm. Three days later, six veterinarians arrived in Michigan.

The PBB controversy kept heating up in March with continuing press reports of human health problems related to PBBs and with spreading commercial consequences. On 21 March, one of Michigan's largest supermarket chains, Meijer's, Inc., with twenty-five stores in the state, announced in public advertisements that it would sell no Michigan products contaminated by PBBs. That day, Governor Milliken telephoned the FDA commissioner to report that "the situation in Michigan was deteriorating." Worried about an outbreak of panic and a stampede away from Michigan farm products, Milliken asked the FDA for a public statement that food in Michigan was safe to eat. The commissioner obligingly agreed to help, and later that day, the FDA released a statement of reassurance to the people of Michigan (Schmidt). The agency announced that food products within the PBB guidelines were safe for human consumption, that the FDA knew of no commercial products not meeting the guidelines, that preliminary health studies had shown no case of "human disease" traceable to PBB contamination, and that the agency continued to study the health of animals and humans exposed to PBBs.

The governor's office next tried to rebut the articles printed by the *Detroit News* and *Grand Rapids Press*. In a brief report, a "government task force" claimed: "People in some parts of the state are being terrorized—for no justifiable reasons" by news stories "filled with rumor, speculation, half-truths, and errors of fact" (MEO, 27 March 1975, p. 1). It then reviewed each story of "The Poison Puzzle" and the "Brain Defects" article. A reporter for the *Detroit News*, however, contacted two task force members, who denied knowledge of the report and the task force (Chen, p. 76). The *Grand Rapids Press* responded publicly to the charges, attacking the official report's "baseless" conclusions. "The contents of the critique hardly match the boldness of the task force's cover statements" (2 April 1975).

In early May 1975, about one year after public discovery of the problem, the state Department of Public Health issued the final report of its short-term human health survey, and about the same time, the federal Food and Drug Administration released its final report on cattle-herd health. Both studies reached approximately the same conclusion. As one state health official wrote in an internal memorandum, "While there are some interesting findings, we cannot identify a significant adverse health effect associated with ingestion of PBB-contaminated dairy products" (Isbister). The cattle report concluded that no significant differences in health problems could be found between sixteen PBB-contaminated cattle herds and fifteen presumably non-exposed control herds (Mercer et al.). The two final reports appeared in time for a hearing scheduled by the state Department of Agriculture on 29 May to review the state action level for PBBs.

Both studies, however, suffered from a similar problem: contamination of the control group. The survey of cattle herds included only two control herds that were negative for PBBs. Data on other herds were reported as "not available" (Mercer et al.). The report showed that feed used on both contaminated and control farms was contaminated with PBBs, although at higher levels on contaminated farms compared to control farms. And one farm in the control group was later quarantined.

In an internal memorandum, one FDA scientist recommended a more competent epidemiological survey of cattle, with follow-up studies, to separate two groups: cattle with low levels of PBBs that were survivors of earlier high-level feed contamination and showed clinical symptoms, and cattle with low levels of PBBs that were later exposed to low-level feed contamination only and showed no clinical symptoms (Kolbye, 1975). The head of the veterinary team, in his memorandum response, rejected the suggestion and the epidemiological approach. "Further work of a survey nature, with the intent of further delineating dose exposure, would be non-productive" (Mercer).

The state's human health survey was criticized on similar grounds by Dr. Walter Meester, a respected clinical toxicologist and director of the Western Michigan Poison Center, who had begun to examine patients with PBB exposure. He pointed out that 70 percent of the subjects in the control group had detectable PBB blood levels and concluded that the study was "poorly planned, does not conform to the standards of adequate scientific medical and epidemiological evaluation, was incomplete, possibly biased, and does not support the conclusions reached and publicized in the lay press." In addition, the state's study had not been published in the scientific literature, and "it probably would not be accepted for publication in a bonafide scientific journal with a competent and unbiased editorial review board." Meester agreed with only one of the report's conclusions, that the detection of any long-term effects of PBBs must await a long-term study of the problem (Meester, p. 527).

At the hearing on 29 May, held by the state Department of Agriculture, scientists from state and federal agencies testified that a lower level was not necessary, while Dr. Meester and several farmers called for reducing the level to protect public health. In a notable shift of emphasis, the FDA representative no longer defended the new guideline of 0.3 ppm on the grounds of technical capability of measurement, which had been the basis for choosing the initial level of 1.0 ppm. Now that PBBs could be measured routinely at much lower levels, the FDA defended its guideline on the margin of safety indicated by toxicological data, mainly by comparing the level for PBBs with that for their chemical cousins PCBs.

Farmers at the hearing stood in two opposing camps on the question of health effects. One side complained of human health problems that they attributed to PBB contamination. One woman testified about various health problems in her family, as possibly related to PBBs, and argued that the level be lowered not out of concern for animals but to protect human health (Motz, 1975). Another farmer called attention to sickly PBB-contaminated cattle that he and others had trucked to the hearing and displayed outside: PBB cattle are "listless, stiff, emaciated or they just waste away. . . . How can we afford to take even the slightest chance of the same happening in humans?" (Zuiderveen). The other camp claimed psychological stress but no serious health problems in farmers, even those owning high-level PBB herds. A farm woman testified, "We feel there is a lot of health problems, but it's from nerves, the worrying to what is going to happen next. We feel the tolerance is safe as it is . . . and if this is lowered to [zero] there is going to be more health problems from the worrying" (Thomas).

Many farmers confronted PBB contamination of their animals as a deeply moral issue. "Just what do people do with these animals when the state says that below 0.3 you can sell them and when the Michigan Livestock Exchange says, 'Sign a paper, if you don't have PBB, you can sell them,' and we cannot in our own honesty sell animals that we know might possibly . . . have PBB, so consequently, we're stuck?" (Motz, 1975, p. 113). She spoke for those farmers who considered it a moral responsibility to provide safe products for the market and who refused to pass along to consumers the health risks of low-level PBB contamination. Agriculture Director Ball dismissed the question as irrelevant.

Not surprisingly, after the hearing, the Michigan Agriculture Commission rejected the proposed reduction and upheld the FDA action level. In July, the Department of Agriculture announced: "The decision must be based on either relying on present available scientific evidence and retaining the present action levels, or deciding in favor of a lower, unsupportable action level based on fear of unproven, improbable long range health hazards" (MDA, 10 July 1975, p. 17). Ball determined that the "evidence presented at the hearing overwhelmingly and conclusively supported the present guideline levels as safe" (p. 18). For a while, the department's decision stilled protest. But not for long.

Al and Hilda Green owned a farm in north-central Michigan with a herd of low-level PBB cattle. Throughout 1975 they tried to get state aid and action, but without success. The state would not recognize PBBs as the cause of illness in their animals, although tests showed low levels of the chemical. "One state inspector told me not to eat meat and milk from my own cows. But in the next breath he told me I could sell them for the public. That didn't make sense." Green decided not to sell the cows for slaughter. "I knew they had PBB in them, and if I deliberately sold those cattle, and someone died, then I would be an accomplice to murder." Through the summer of 1975, Green repeatedly contacted state officials to ask for permission to dispose of the animals at Kalkaska. But the attorney general's office rejected the low-level cattle. Finally, in early November 1975, in desperation, Green called neighbors and friends to help shoot his cattle and bury them on his own land (Green interview). The shooting attracted state and national media attention. It also brought together other farmers in the same situation, those faced with sick cows and decreasing milk production, low levels of PBBs in animal fat tissue, and the state's refusal to help. As Al Green recalled, "The state just did not want anything to do with low-level farmers" (interview).

Indeed, two months earlier, in August 1975, Governor Milliken had vetoed a bill passed unanimously by the state legislature to provide financial relief to all farmers with PBB herds. The bill would have provided low-interest loans through an emergency loan program with money borrowed from the state Veterans Trust Fund. According to the bill's sponsor, the recently elected State Representative Donald Albosta, the measure would have aided six hundred farmers waiting for insurance settlements, giving them financial support to remain in business. "This bill would have made it unnecessary for farmers to sell contaminated meat on the market. . . . Now they're going to have to do it just to keep their heads above water" (Albosta, 3 September 1975). Milliken vetoed the bill because he considered the trust fund not a "valid source of funding" and because the bill provided farmers with relief for all sick animals with a detectable PBB level. But the governor's main reason was his belief that the "responsible parties" had sufficient assets and ought to act on their own (MEO, 2 September 1975). Milliken later introduced his own loan proposal, but it never left the House Agriculture Committee. He also subsequently used funds from the Veterans Trust Fund to balance the state budget.

Albosta attacked Milliken for not acting. The state legislator said it "was apparent at that time that many animals were sick and dying although they had much less than 0.3 parts per million" of PBBs and that "many people who had eaten meat containing less than 0.3 parts per million of PBBs were showing signs of serious illness." Albosta accused the governor of "mismanagement, gambling and irresponsibility." "This meat is now for sale all over the state, reports of illness continue, and still the executive office turns a deaf ear" (Albosta, n.d.).

In October, Albosta held the first organizational meeting of a special House committee to review the PBB problem, focusing on the activities of the state departments of agriculture and public health. He announced that the committee would hold public hearings in areas severely affected by PBBs in order to hear firsthand accounts from individual farmers (Albosta, 16 October 1975). The Democratic Speaker of the House at first tried to control Albosta's efforts to politicize the PBB issue by denying funds for Albosta's hearings (Egginton, p. 237). But Albosta persisted in holding the hearings anyway, with important consequences.

In early March 1976, Albosta's committee held three hearings in farm districts. The first was at the Grand Rapids City Hall, in the southwestern part of the state; the next at McGuire's Restaurant, Old Mackinaw Trail, in Cadillac, in north-central Michigan; and the third at the Cisler Center, Lake Superior College, in Sault Sainte Marie, in Michigan's Upper Peninsula. Farmers with cattle contaminated at low levels gathered at these meetings. As they spoke publicly, they discovered common problems. They criticized bureaucrats who blamed herd health problems not on low levels of PBBs but on "poor management," that is, on the farmers themselves. They complained of their own PBB-related health problems, which resembled those of their animals and which the state and the companies refused to recognize.

At the Grand Rapids hearing, one farm woman told the special committee why it had taken so long to determine the cause of their problems. Farmers, she said, tend to be independent, and the state had spoken with farmers individually and not in a group. "Consequently, I think each one of us felt that we were all in this thing alone and, at the time they talked to you, that's just the way you felt, that you were just an individual and you were the oddball and that there wasn't anyone else having these problems" (Motz, 1976, p. 23).

At the second meeting, in Cadillac, an area with a large number of contaminated herds, Louis Trombley, a farmer, raised the call to protest. The bureaucrats, he charged, belonged in jail "for cheating the public" (Trombley, p. 9). Trombley's demands were clear: "We want zero toler-ance. We want it immediately. We won't attend one more meeting. . . . Governor Milliken can go to hell, we're not going to eat no more of this" (p. 10). While representatives from the state departments of agriculture and public health listened, the audience applauded Trombley's demands. Others echoed a call to march on the state capital. "All of you going to Lansing ought to take one of your animals along in your pickup and dump it down there" (Gleisen, p. 98). Resentment kept building: "Why the hell should I suffer, my cattle suffer, for these damn people in Lansing that don't do a damn thing for the farmer. I want something done. If that takes all the damn farmers to go down, we're going to go down" (Leeblow, p. 30).

While Albosta planned and ran what some critics called his "road show," the governor's staff initiated their own steps. In mid-January, Mil-liken sent letters to twenty-five scientists around the country, including several in Michigan, asking three questions: Is there evidence to indicate that the federal action guidelines are not safe? Is there evidence about long-term human health problems related to PBB exposure? Are cattle with low-level PBB exposure having health problems due to PBBs or could other factors be causing the problems? Milliken received a wide range of responses and decided that someone other than state bureaucrats must put the scientific information in order.

On 5 March, the day of Albosta's hearing in Cadillac, Milliken an-nounced a "five-point action plan" to examine the PBB problem. The plan's centerpiece was the creation of a PBB Scientific Advisory Panel to "help end the rumors and half-truths that often surround this issue." Milliken charged the panel to answer his three scientific questions and to review the scientific literature on PBBs (MEO, 5 March 1976). Milliken had concluded that he needed to go outside his state bureaucracy to independent scientists for clear answers and for public credibility, and also that the PBB problem had taken on a public and political life that demanded more prominent attention.

Farmers responded to Milliken's proposal with their own "nine-point action plan." They demanded, among other points, a zero tolerance level, immediate settlement of all claims, honest testing of livestock by private not state laboratories, a public hearing before the state legislature, and the closing and cleaning of grain elevators with inspection by a private firm. The farmers stressed that they, like Milliken, did not trust state bureaucrats for honest answers, but they also did not trust the governor.

During Michigan's Farmers' Week 1976, in late March, two weeks after the last of Albosta's hearings, farmers brought the PBB problem to Lansing. They also brought dead sheep and dead cattle, placing the carcasses on the steps of the state capitol, to confront the politicians with the PBB issue. The group had no formal organization, no special name. They came to Lansing, they said, to make the politicians act. Some had never before seen the state legislature in session. The demonstration attracted the attention of the national media. An article the next day in the *Washington Post*, for example, carried the headline "PBB: Little-known Poison Creates Nightmare in Michigan." The article ended with a farmer speaking about his dairy cows. "They fed our family, they supported us, and they sent 2,500 quarts of milk to town a day. How do you ever replace that? It's a hell of a way to end a dairy career—poisoning people" (22 March 1976).

U.S. Senator Philip A. Hart from Michigan had also been working on the PBB problem in Washington. He introduced a bill to provide farmers with emergency loans from federal funds, and the U.S. Senate passed the bill on 10 March. Next, in late April, a subcommittee in the U.S. House of Representatives held hearings on the bill. In 1975, the subcommittee had modified a program of emergency loan guarantees to aid Michigan's PBB farmers, but that program was not working. Private banks refused to provide loans to farmers, even when guaranteed by a federal agency. But the House subcommittee's chairman, Representative Bob Bergland, opposed Hart's proposal for direct federal loans to Michigan dairy farmers. The bill never left Bergland's committee.

In late May 1976, Milliken's PBB Scientific Advisory Panel presented its report. In a total surprise to the governor, the panel unanimously recommended a reduction in the PBB guideline to the minimum detectable limits, from 0.3 to 0.005 ppm in meat and from 0.3 to 0.001 in milk. The panel based its recommendation on "the potential acute and chronic health hazards, the uncertainty in regard to carcinogenicity and teratogenicity, the long-term tissue retention and possible accumulation of PBBs, and the present geographical containment of the contaminated stock" (PBB Scientific Advisory Panel, p. 3). The scientists proposed the lower standards even though they concluded that their review of the scientific literature and of studies by state and federal agencies showed that "no significant acute effects of PBBs have yet been documented in man" and "no specific disease or symptomology in animals or man can presently be associated with exposure to low levels of PBBs" (p. 2).

The panel's recommendations shocked Michigan's governor, bureaucrats, and public. Two years of government statements that no one should worry about the health consequences of PBBs suddenly appeared false. Reactions to the report varied sharply. Supporters of farmers with low-level contamination praised the report for vindicating those who decided to keep or kill contaminated animals rather than sell them for public consumption. On the other side, the president of the Michigan Farm Bureau announced that the report would wreak "economic havoc" on farmers and place the burden on consumers for greater government intervention and for possible bankruptcy of agricultural business. Governor Milliken turned the problem over to the Department of Agriculture, which was legally required to hold hearings on whether to maintain or lower the levels.

The Michigan Department of Agriculture scheduled its hearing for 10 June 1976. Although state officials knew that the chairman of the governor's scientific panel would be out of the country on that date, the officials made only halfhearted efforts to contact other panel members to attend the hearing. As a result, the panel had no official representative on 10 June to defend its report, its reasoning, or its recommendations. The panel also had no opportunity to respond to criticisms in the written record. The hearing was structured with a bias: to stress objections to the scientific panel and to emphasize support for the existing guidelines.

The hearing ran continuously for eleven hours, from nine in the morning until eight at night. Most of the scientists who spoke, including several who came specially from out of state, supported the existing guidelines. But two scientists supported the advisory report in its entirety, and two practicing veterinarians testified about herd-health problems probably due to low-level contamination. At the hearing, FDA official Dr. Albert C. Kolbye, Jr., testified that available toxicological data suggested that 0.3 ppm was probably safe and that the agency was no longer required by the Food, Drug, and Cosmetic Act to lower the FDA guidelines for PBBs because the contamination had changed from "avoidable" to "unavoidable."

The Department of Agriculture calculated that 125 people at the hearing called for keeping the guidelines at FDA levels and 38 called for lowering them. Farmers divided much as they had at the hearings in 1975: a group with herds contaminated at high levels, already compensated for their losses, opposed any changes; and a group with low-level contamination, still uncompensated, supported lowering the guideline. Predictably, Agriculture Director B. Dale Ball recommended to the Michigan Agriculture Commission, which was required to make the decision, that the guidelines not be changed.

At that point, Governor Milliken appealed personally to the Agriculture Commission to accept the conclusions of his scientific panel. But the commission, asserting its independence from the governor, rejected his request and adopted Ball's advice. Milliken then declared publicly that he disagreed with the commission's decision. "The issue is to try to find a way to resolve this problem in the public interest, and there is, frankly, a disagreement as to what the public interest is" (MEO, 22 June 1976). Regarding the division of expert opinion about long-term human health effects, Milliken commented, "I would rather err on the side of being overly cautious in this matter than the reverse" (p. 2). Unable to affect the commission's decision, the governor announced he would propose legislation to change the PBB levels—thereby shifting responsibility to the Michigan legislature. However, not everyone agreed that Milliken had erred on the cautious side; he came under fire for responding to the crisis only after it had become a political issue.

Among those actively pushing for a new PBB policy in the state legislature were farmers with herds contaminated at low levels but who had not received aid from the companies or the state. In August 1976, these farmers formed the PBB Action Committee to bring their complaints about state policy to the politicians, to the press, and to the public. The group, stressing that low levels of PBBs could damage animal and human health, helped redefine the PBB problem as a public health hazard to Michigan consumers. Many group members had participated in the public hearings organized by Albosta and also had filed civil suits against the companies responsible for the PBB contamination.

Others criticized Milliken for seeking to lower the PBB guideline. In the summer of 1976, the companies responsible for the PBB disaster initiated public efforts to influence state policy. The Michigan Farm Bureau became openly and actively involved in the dispute for the first time, organizing a telephone and telegram campaign to urge Milliken and state legislators not to lower the guidelines. Michigan Chemical hired a political consulting firm to lobby state legislators against lowering the guidelines. The chemical company hired the most powerful public relations firm in Lansing, with close ties to the Speaker of the House, and thereby tried to influence policy while still keeping the names of the company's owners (Velsicol Chemical and Northwest Industries) out of the public eye.

In early July 1976, the *Grand Rapids Press* reported that these lobbying efforts resulted in the legislature's decision to postpone consideration of a bill proposed by Albosta to lower the tolerance levels. The Speaker of the House, Bobby Crim, a powerful Democrat and possible challenger of Milliken in the 1978 election for governor, angrily denied that lobbying caused the delay (11 July 1976). On 13 July, Crim went "back on the offensive on the PBB issue" (*Pioneer*, 13 July 1976). He announced that Dr. Irving Selikoff of the Mt. Sinai School of Medicine in New York, internationally known for his research on industrial and environmental health, especially related to asbestos, would soon arrive in Michigan to begin the first comprehensive study of the human health effects of PBBs. Crim stated that Selikoff had offered to come to Michigan in 1974, but state officials had never responded with an invitation. Crim's energetic staff member assigned to the PBB issue, Edith Clark, had learned about Selikoff's earlier offer and arranged for the speaker to invite the scientist and his team to Michigan—outside the bureaucratic channels of the state Department of Public Health and the governor's office. The invitation represented both a scientific and a political challenge to the governor and the state bureaucracy.

In August 1976, the Department of Public Health made a startling discovery: twenty-two of twenty-six samples of human breast milk from the general population in Michigan showed the presence of PBBs, while none of ten samples from women living outside Michigan contained even a trace of the chemical. Moreover, PBBs appeared in all sixteen samples from Michigan's Lower Peninsula, even samples from highly urban areas, such as Wayne County in the southeast corner of the state. The highest sample measured 0.51 ppm—a level that would have destined a cow for slaughter and burial at Kalkaska (Reizen). The findings, produced from samples collected to test for pesticides, shocked health officials into realizing that PBBs had reached nearly everyone in the general population of Michigan.

On 19 August, the state health director announced the survey's results and tried to explain what it meant for nursing mothers. "There is absolutely no evidence to date that PBBs in mothers' milk causes babies to become ill. On the contrary, babies of such nursing mothers appear to be strong and healthy." But he continued, "The fact is that we don't really know what this means or what effect, if any, it may have on health. Although we are dealing with chemicals which may have some potential for harm in the long run, we have insufficient evidence at this time to recommend discontinuing breast feeding" (MDPH, 19 August 1976). In a separate memorandum to doctors, the health director explained that "insistent" women could pay to have their breast milk analyzed by a private laboratory, but that "laboratory tests on an individual woman will not provide information to help her reach a decision on whether or not to breast feed" (Reizen).

In October, the Department of Public Health completed a more scientific study that showed 96 percent of women living in the Lower Peninsula to have PBBs in their breast milk, with levels up to 1.2 ppm. Even then, the department said, "The levels of PBBs are not sufficient to discourage Michigan mothers from breast feeding if they so desire" (MDPH, 15 October 1976).

Dr. Selikoff disagreed with that position. When he arrived in Michigan in early November 1976 to begin his health study, he "warned that women should avoid breast feeding infants until it is proved their milk is free of the fire-retardant chemical" (*Grand Rapids Press*, 4 November 1976). Selikoff pointed out that PBBs are not metabolized in the body and that a baby would tend to store the chemical, gradually increasing the levels in its tissues. The Department of Public Health countered the next day with a press release repeating its original position that not enough scientific evidence existed to recommend against breast-feeding (MDPH, 4 November 1976).

Selikoff's thirty-five-member team planned a seven-day stay in Michigan to examine more than a thousand farmers and their families. The study gained front-page coverage throughout Michigan and in nationally prominent newspapers. Even after only one day of examinations, the doctors stated they had observed a pattern, which was widely reported in the press: "joint problems, fatigue, dizziness, memory problems, excessive sweating, wounds that won't heal, darkening of the skin, sensitivity to sunlight." Other common complaints included "lack of energy, muscular weakness, diarrhea, visual disturbances, sores and rashes on the skin." No one on the Selikoff team would say that PBBs caused these problems, but neither would they deny it (*New York Times*, 8 November 1976).

A few days into the new year of 1977, at a press conference in Michigan, Selikoff presented a preliminary report of his team's findings. The scientist summed up the team's main conclusion that "adverse health effects may occur in some people as the result of PBB exposure" (Selikoff, 1977). The report identified abnormalities of the nervous system, musculoskeletal changes, gastrointestinal problems, and other difficulties, including stress-related symptoms. Selikoff emphasized that the findings could not be extrapolated to the general population and that the data required more analysis before a final report could be made. The state health director thanked the Selikoff team for "expanding the frontiers of knowledge regarding PBBs. . . . This is the first time that I can recall we have knowledge of what [the] PBB effect is on people" (MEO, 4 January 1977, pp. 3–4).

Subsequent scientific reports from Selikoff's study supported the early evaluation. A comparison of the Michigan farmers with farmers in Wisconsin found significantly more musculoskeletal problems in the Michigan group, especially joint disorders (pain, swelling, and crepitation) and neurological symptoms (tiredness, fatigue, headaches, dizziness, and unusually long sleep hours) (Anderson, Lilis, et al.). Two indicators of liver function, SGOT and SGPT, were also significantly higher in Michigan men than in Michigan women and than in the entire Wisconsin group. These increases did not correlate with alcohol consumption, thereby supporting PBBs as probable cause (Anderson, Holstein, et al.). Also, Michigan farmers showed immunological abnormalities compared with Wisconsin farmers and New York City residents (Bekesi et al.).

Another study of the human health effects of PBBs, undertaken by state and federal health officials, reached different conclusions. The government study found no dose-response relationship. It reported the most symptoms among farmers with low-level PBB cattle contamination (contamination at levels not high enough to receive compensation). The report concluded that bias in the selection of the farmers examined and other non-PBB factors produced the appearance of a higher number of symptoms in the farmers with low-level PBB cattle when compared with the control group (Budd et al.).

February 1977 turned into a hot month for PBB politics in Michigan. The PBB Action Committee, the group of farmers with low-level PBB cattle, began to organize a recall of Governor Milliken. The Community Action Program of the powerful United Auto Workers criticized the governor for mishandling the PBB issue. On 10 February, Democratic State Representative Francis Spaniola introduced a bill to lower the PBB guidelines to 0.02 ppm in fat tissue of cattle. The *Detroit Free Press* ran a series of articles on PBBs, reporting that "PBB-laced cows" were legally sold in Michigan, because of the inadequate federal guidelines.

Milliken became increasingly defensive about his record on PBBs and increasingly aggressive about PBB legislation. On 14 February, he wrote a letter to Agriculture Director Ball asking the department and the Agriculture Commission to support the Spaniola bill. Milliken cited the report of his PBB Scientific Advisory Panel, the report prepared by Selikoff, the serious erosion of consumer confidence in Michigan, the refusal of Canada to buy Michigan beef and hogs because of PBB traces, and the growing support for Spaniola's bill in the public, the agriculture community, and the legislature. The chairman of the Agriculture Commission responded the next day. The commission and the department would now support lowering the state PBB level.

While the legislature debated what to do, the legal questions of animal health problems were being settled through both negotiation and litigation. Farmers with animals contaminated above 0.3 ppm were slowly receiving compensation from Michigan Chemical and Farm Bureau Services, backed up by their insurance companies. Farmers with animals contaminated below that level received no payment whatsoever.

On 28 February 1977, the first PBB case for low-level cattle contamination went to trial. Attorneys for Roy and Marilyn Tacoma charged in Wexford County Circuit Court that Michigan Chemical and Farm

Bureau Services had helped cause, through "willful, wanton neglect," the PBB contamination that poisoned the Tacomas' dairy herd. The attorneys argued that the companies contributed to the contamination and its spread and that low levels of PBBs damaged the herd. Many other farmers with herds contaminated at low levels had filed civil damage suits against the several companies involved (Farm Bureau Services, Michigan Chemical, and related companies). With their cases yet to go to court, they watched the Tacoma trial with great interest.

As the courtroom battle began, two subcommittees in the U.S. Congress held hearings on the PBB problem. Michigan's U.S. Senator Donald Riegle, a Democrat on the subcommittee responsible for toxic chemicals, organized field hearings in Michigan in late March "to lay the facts on the record, find out how best to help the farmers and consumers, and see how we can prevent disasters like this from happening again" (Riegle, p. 3). Michigan's senior Senator Robert Griffin, a Republican and good friend of Milliken's, worked on the same subcommittee and proposed two bills related to PBBs, one to provide farmers with low-interest loans, the other to provide financial assistance to states to protect and indemnify people harmed by toxic chemical disasters. The hearings produced a three-volume record of about eighteen hundred pages on Michigan's PBB contamination; the two bills went nowhere.

The second congressional group to examine PBB contamination was the House Subcommittee on Oversight and Investigations. This subcommittee, chaired by Congressman John Moss, was known for its hard-hitting reviews of government agencies. It held two days of hearings in Washington in early August 1977. The investigation focused on the actions of federal agencies and on the health dangers associated with PBBs. The first day of the hearings included testimony from medical and scientific experts and from one farmer who became ill after his farm animals were contaminated with PBBs and who was unable to obtain assistance from any state or federal agency. The second day presented testimony from representatives of the Environmental Protection Agency and the Department of Health, Education, and Welfare. One congressman concluded, "Since I have been a member of this subcommittee, I cannot remember receiving testimony which more clearly illustrated the Government's failure to prevent injury and assist the sick and needy. Federal action has been slow, uncoordinated, and ineffective" (Luken, pp. 49–50). The subcommittee grilled several FDA officials and attacked the agency's handling of the PBB contamination in Michigan for its failure to prevent the accident and its poor response afterwards. The subcommittee criticized the inadequacies of PBB-related research in several federal agencies (including the Environmental Protection Agency and the National

Institute of Environmental Health Sciences) and called for more intensive efforts.

These federal actions paralleled legislative efforts in Michigan. In March 1977, the state House of Representatives held hearings on Spaniola's bill to lower the state PBB tolerance levels. In a surprise move, Milliken appeared to testify—only the second time in nine years in office that the governor had come in person before a legislative committee. His chief aide explained, "There's no question that [the governor] has been getting clobbered [on the PBB issue]. . . . The governor and every other elected official could be hurt if the problem is not resolved" (*Detroit Free Press*, 7 March 1977). The House approved the bill in late March, and the Michigan Senate, after a new series of hearings, passed the bill in July. Milliken signed the law in August, as Public Act 77 of 1977. The law reduced the state's PBB tolerance level from 0.3 ppm to 0.02 ppm in cattle fat and required a test of each cow sent to slaughter. The Michigan Department of Agriculture and the Michigan Farm Bureau had desperately opposed the bill, but public concern and political action about the public health consequences of PBB contamination forced approval of the law—four years after the contamination of Michigan had begun.

The signing of Public Act 77 and a report of probable immunological damage in PBB-exposed humans touched off a new round of accusations. In a press conference in early August, House Speaker Crim attacked the state bureaucracy and the governor's office for "foot dragging almost from the outset of the PBB situation. They have reacted only when there has been great pressure from the scientific community and the public." Crim presented a chronology prepared by his office, showing points at which the governor and his department directors could have acted to contain PBB contamination and help the farmers but did not (Crim). Milliken responded the next day by calling the chronology "highly selective and misleading" and by listing positive actions of the executive branch not mentioned by Crim. Milliken accused the speaker of seeking to "politicize the very serious PBB problem" (MEO, 8 August 1977).

The fall of 1977, with the opening of campaigns for statewide primary elections, including for governor, saw several new political disputes over PBBs. A Democratic candidate for his party's nomination for governor purchased a copy of a documentary on PBBs, produced for British television and called "The Poisoning of Michigan," and began showing it around the state. Milliken publicly portrayed the film as inaccurate, outdated, and distorted. But the governor denied he had attempted to suppress the film in contacts with a Public Broadcasting Service producer in Washington who was planning to show the film on public television throughout the United States.

Problems of disposal also ended up in court. The lowered guidelines of Public Act 77 produced many more contaminated cattle carcasses and new controversies. State officials estimated that a minimum of five thousand additional cattle would need to be killed and buried, but people in Oscoda County, the sparsely populated area chosen for the disposal site, refused to allow it. A protest group affiliated with the PBB Action Committee obtained a court ruling that the Department of Natural Resources had not taken adequate precautions to protect the water table from contaminated runoff. The department's director lamented that no one wanted the cattle, even though they contained minute amounts of PBBs (estimated at a total of two ounces in thirty-five hundred animals), because of "public hysteria" (*Cadillac Evening News*, 14 October 1977). On 1 August 1978, Michigan's supreme court ruled that a plan using a clay-lined burial pit could continue, since the evidence presented in court was insufficient to show that disposal would result in environmental pollution. Under court order, some animals were buried in Oscoda County. Then in 1979, local opposition forced the state to ship fifteen hundred PBB-contaminated carcasses to a burial ground for radioactive wastes in Death Valley, Nevada (*State Journal*, 23 January 1980).

Another source of litigation was controversy over human health problems. In October 1977, lawyers filed a personal damage suit for 246 persons against Farm Bureau Services, Michigan Chemical, and their parent companies. They broadly defined their complaints as a "PBB syndrome," including fatigue, loss of balance, impotence, and aching joints and muscles. The suit consolidated cases from twenty-nine counties in Michigan and charged the companies with ignoring data on the toxicity of PBBs available in 1970, concealing information about the contamination once it occurred, and deliberately misleading the public and denying the toxic nature of PBBs. As explained below, however, this suit never reached the courtroom.

The federal government also sued the companies involved in the PBB disaster. In late November 1977, the U.S. district attorney in Grand Rapids filed the first federal criminal charges against Michigan Chemical and Farm Bureau Services. The suit charged the companies on four counts, each a misdemeanor with a fine up to $1,000, for contaminating animal feed with PBBs, a "poisonous and deleterious substance" under the federal Food, Drug, and Cosmetic Act. In May 1978, the two companies pleaded no contest, which the judge accepted, despite pleas from the district attorney that the companies had been "reckless and careless not only in their manufacturing process but in their statements to government investigations" (*New York Times*, 20 May 1978). The judge ordered the companies to pay $4,000 each in fines, ruling that a prolonged trial would not serve justice since the companies had already paid a total of $40 million in claims to farmers who lost animals because of the contamination.

By March 1977, according to Velsicol's chairman and chief executive officer, Michigan Chemical and Farm Bureau Services had made "available" about $39 million to settle 625 of the approximately 900 claims then pending (Hoffman, p. 1374). The chemical company executive did not specify, however, how much of these funds came from insurance companies and how much from assets.

PBB effects on the workers at Michigan Chemical's factory also attracted attention in the fall of 1977. Workers had reportedly experienced various health problems, had difficulty in obtaining life insurance because of suspected PBB effects, were denied workman's compensation, had difficulty in finding other employment because of PBB exposure, were exposed to other toxic chemicals while working for Michigan Chemical, and were threatened with no job if the factory closed. In October 1977, the Michigan House formed a special committee to examine these problems, and in December several state newspapers gave wide coverage to the workers' plight. The legislative committee published a report and made six legislative recommendations, two federal and four state (D. Wilson); two state acts passed. Then, in late 1978, the Michigan Chemical factory closed down rather than clean up, and the company ceased all operations in the state of Michigan. Worker efforts to buy the facilities failed, largely because of unresolved questions about future responsibility for past contamination. The local labor union subsequently filed two separate $250 million damage suits against the state and against the company. The suit against the state was dismissed for reasons of sovereign immunity.

In the fall of 1978, the civil damage suit for low-level PBB poisoning in cows, the Tacoma case, neared completion, with important political implications. Governor Milliken was running for reelection against Democratic candidate William B. Fitzgerald in a campaign dominated by the PBB issue. "PBB seemed to be in the news nearly everyday. Previously unknown landfills were being discovered; leaching at others was coming to light. And most of all, concern about human health—and the Milliken administration's handling of the PBB crisis—once more became a matter of public debate" (Chen, p. 272). Milliken came from a political family in northern Michigan and had held the governor's office since 1969 through a strategy he called "pragmatic republicanism" (p. 192). Relying on the staunch Republican base in rural farm communities, he had adopted liberal policies on urban problems, which appealed to Michigan's large urban population with its Democratic allegiance. That strategy had served him well in previous statewide elections. But in 1978 things promised to be different. Based on a survey of *Michigan Farmer* readers, two political scientists predicted that rural residents who were dissatisfied with the government's handling of the PBB issue intended to vote against Milliken, even though they identified themselves as Republicans and as Milliken supporters in the prior gubernatorial election of 1974 (Coyer and Schwerin, 1978).

In late October 1978, just before the statewide election, County Circuit Judge William R. Peterson handed down a decision in the Tacoma case, which had become the longest and most expensive in Michigan's history. After sixteen months in court, with sixty-three witnesses, 25,000 pages of court transcript, and about 1,500 exhibits of an additional 70,000 pages, the judge ruled against the farmer and for the companies. Simply stated, the judge decided that the Tacoma attorneys had not proven any damage to the health of the dairy cows or any reduction in milk production due to low-level PBB contamination. Their proofs, wrote Peterson, "fell shockingly short of their claims." The judge concluded that the evidence presented "proves, instead, that in small amounts PBB is not toxic." The real tragedy of the incident, in his opinion, "lies in the needless destruction of animals exposed to low levels of polybrominated biphenyl and even of animals who never received any PBB" (*Tacoma and Tacoma v. Michigan Chemical Corp. et al.*). Peterson's decision, written on tight legal principles and not on precedent, proved difficult to challenge. The case was not appealed, and other farmers whose herds showed low-level contamination subsequently settled out of court for small fractions of their losses.

Peterson announced his decision one week before the November election. Although the judge denied any political motivations, he must have been aware of the likely impact on voters of his decision, which indirectly supported the incumbent governor. In his decision, Peterson suggested that the state government had not mishandled the contamination problem, even though mishandling had not been an issue in the case. The judicial decision just before the election greatly boosted the incumbent governor's campaign. Milliken publicly welcomed the decision, claiming it helped restore what he called a rational perspective on an overly emotional issue.

In the election, Milliken trounced Fitzgerald, winning even Fitzgerald's home district in urban Wayne County and the Detroit area. Although the Democrat Fitzgerald won votes in traditionally Republican "outstate" farm districts, such as Oscoda County, there were not nearly enough votes to overthrow the popular Milliken. Fitzgerald's popularity in those rural districts reflected his position attacking Milliken's administration for mishandling PBBs. But Fitzgerald also may have lost some votes by stressing the PBB issue; his strident PBB television commercials, prepared by a Washington media consultant, were perceived by some people as overplaying PBBs and pushing the contamination charges too far.

In the same election, state representative Donald Albosta, who had organized and presided over the PBB hearings two years earlier, successfully won a seat in the U.S. Congress representing the tenth district in north-central Michigan. He defeated the incumbent Elford Cederberg, who had served in Congress for twenty-four years. Albosta's success in electoral politics depended in large part on his activism in PBB politics.

In 1979, PBBs faded away as a political issue in Michigan. Following the Tacoma decision, attorneys for the approximately one hundred outstanding cattle-damage suits began to settle with the companies. The attorneys for the Tacomas represented another 82 clients around the state; when the companies offered to settle, the attorneys felt they had no choice but to accept the offer. As one observer reported, "The superior resources of those responsible for the contamination had won out" (Egginton, p. 319). Farmers felt the offers were ridiculously low, but they had little hope of victory in the courtroom. Many wanted to put the problem behind them and return to dairying, if they could. As part of the settlement, the attorneys withdrew the suit of 246 persons for current human health damages allegedly caused by PBBs. The money from the settlement, $3.7 million in total, covered the legal expenses of the attorneys (28 percent of the funds) and the feed bills and bad debts that the farmers owed to Farm Bureau Services ($900,000), with the remainder for compensation. On average, the farmers received about $32,000, far below the losses most had experienced (pp. 318–19).

But the state did provide some assistance for those farmers who had not received compensation from the responsible companies. In 1978, the state established the Toxic Substance Loan Commission (Public Act 273), which was authorized to provide low-interest loans up to $75,000 for farmers with uncompensated damages related to chemical contamination. The agency was located in the Department of Public Health, not the Department of Agriculture, reflecting the deep abyss of distrust between state agriculture bureaucrats and low-level PBB farmers. By March 1980, the commission had disbursed almost $6 million in loans to about one hundred applicants (*PBB News*, March 1980).

In the early 1980s, despite subsidized interest rates on the loans (with no interest payments in the first five years), a number of farmers experienced financial difficulties and faced the prospect of defaulting on their loans. A former activist in the now-defunct PBB Action Committee contacted PBB farmers who had received loans and wrote a report on their financial and health problems in an effort to persuade state legislators to pass a law forgiving the loans. In the process, she discovered that the original loan law obligated the state to forgive the loans if the state recovered its administrative costs from the companies responsible, as occurred in 1982 (see below). That discovery persuaded both Democrats and Republicans to pass the law, which finally gave the farmers with loans some official redress (Miller interview).

Following the PBB disaster, the state of Michigan made various efforts to improve its capacity to handle environmental contamination by toxic substances. Individual state departments requested and received additional budgetary funds to expand their regulatory efforts for toxic chemicals. In 1979, the state legislature passed the Toxic Substance Control Commission Act, which created a special watchdog agency designed to prevent a recurrence of government inaction in the initial phases of a toxic exposure. Governor Milliken chose Rick Halbert as the commission's first chairman. One analysis of the commission, however, warned that the agency could become simply a declarer of toxic substance emergencies, constrained by the existing state departments, and might not develop into an effective watchdog organization. The analysis concluded that even if the commission had existed in 1973, it might not have made much difference in the PBB case, especially if it had adopted a passive rather than an active approach. For the future, "the choice is whether the Commission is reactive, narrow, and ad hoc; or anticipatory, broad, and policy-oriented" (Cardin and Brilliant, p. 1246).

Many questions about human health problems remained unresolved. Even five years after the contamination, about 97 percent of Michigan's residents showed measurable levels of the chemical, and the most highly exposed groups showed little significant decline in PBB levels (Wolff, Anderson, and Selikoff). A study of PBB levels in mother's milk between 1976 and 1978 confirmed the lack of a significant decline over time and showed a geographic gradient in PBB levels, with the highest levels in areas with the greatest number of contaminated farms (Miller, Brilliant, and Copeland). PBBs were shown to cause liver tumors in rats (Kimbrough et al.), indicating the need to follow PBB-exposed individuals indefinitely. Selikoff and state and federal researchers continued their long-term studies of the general population and of the highly exposed farm families. While not admitting that PBB exposure had any human health consequences, the state did open two medical clinics near rural areas for persons who believed they had PBB-related health problems (Chen, p. 297). But many farmers wondered whether the health questions about PBBs would ever be answered.

On 18 November 1982, the U.S. Environmental Protection Agency, the state of Michigan, and Velsicol Chemical Corporation announced a

consent judgment of $38.5 million to settle cleanup costs associated with
PBBs and other chemical contamination from Michigan Chemical. The
agreement included $13.5 million to the state government and $500,000
to the Environmental Protection Agency, to reimburse public expenses for
cleanup at the Gratiot County Landfill, which Michigan Chemical had
used, and to settle a $120 million suit filed in 1978 by the state of
Michigan for $50 million of exemplary damages and to recover costs of
investigating and managing the PBB disaster, estimated to exceed $59
million (*People v. Northwest Industries et al.*). According to Velsicol officials, the company had already set aside the other $24.5 million for
cleanup of Michigan Chemical's abandoned factory in St. Louis, Michigan, and toxic waste sites used by the company, and for materials and
services in the state's cleanup of the Gratiot County Landfill. Earlier, in
January 1976, Velsicol had settled a suit filed against it by Farm Bureau
Services for about $20 million (the original demand was for $270 million
in damages).

State officials considered the agreement a success for accelerating
cleanup at Michigan's worst toxic waste dump, for avoiding more years of
litigation, and for obtaining reimbursement for at least part of the costs of
PBBs to state government, thereby helping contain the conflicts in Michigan over the toxic consequences of PBBs. On the corporate side, Michigan Chemical no longer existed as a separate entity (after its merger with
Velsicol in 1976). Velsicol closed the responsible factory in 1978 and
finally sold its flame-retardant business entirely in 1981, as part of efforts
to clean up the company's image and act (*Chemical Week*, 1982). Then,
in 1985, leverage buy-out specialist William Farley purchased Northwest
Industries for about $1 billion and subsequently sold off Velsicol in two
pieces. Meanwhile, Farm Bureau Services continued to operate as an
integral part of the Michigan agricultural economy.

For some Michigan farmers, the PBB disaster permanently disrupted
their lives. Farm communities remained split by the controversy over low-
level contamination. Many farmers with high-level contamination had
received compensation early on, but then were struck by residual, low-
level contamination that compelled them to destroy their replacement
herds and begin again. Farmers with low-level contamination felt they
had been denied just redress from the responsible corporations, and felt
frustrated by the legal system and the state bureaucracy. In addition to
enormous financial losses, farmers confronted persistent residual PBB
contamination in their fields and barns and in their bodies. Some farmers
described lingering health problems they associated with PBBs. Michigan
health officials reported no cancer clusters among PBB farmers
(*PBB/PCB News*, Fall 1988), but some farmers remained anxious about

cases of cancer they observed among their families and neighbors who had been exposed to PBBs. Farmers wanted to put the PBB disaster behind them. But memories and fears persisted, as did environmental traces of the toxic chemical and the farmers' resolute disbelief in the private and public institutions that brought them the PBB disaster.

Chapter 4

Seveso Dioxin Contamination

Poverty and hunger ravaged parts of northern Italy at the beginning of this century. Peasants in Lombardy commonly died of pellagra, a devastating skin disease due to niacin deficiency, known locally as Lombardy leprosy. After the Second World War, Italy's rapid economic development transformed the area. Industry boomed and population soared. Brianza, an expansive semiurban district north of Milan, became one of Italy's most prosperous and most densely inhabited areas. The town of Seveso, less than twenty kilometers north of Milan's city limits, grew from about seven thousand residents in 1930 to seventeen thousand in 1978, expanding in all directions toward nearby neighboring towns. Growth benefited Brianza's family-run artisan industries, especially the furniture makers, as well as its large factories, such as SNIA, which produced artificial fibers, and ACNA, which produced coloring agents. But unbridled growth also brought social problems: lengthy traffic jams on highways; factories and houses jumbled tightly together; environmental degradation with air, water, soil, and noise pollution; rising crime and juvenile delinquency; and serious occupational diseases and deaths (Grigliè and Brera). Then, in 1976, Brianza suffered disaster.

The summer of 1976 will long be remembered in Seveso. June passed with little rain, an unusually arid spell, and the ground turned dry and dusty. The tenth of July, a Saturday, was a typically sweltering summer day in Seveso. Many families opened doors and windows to catch the light breeze and cool their houses. As noon approached, some families moved outdoors to eat the main meal of the day, preparing the table in the carefully tended family garden.

One woman of Seveso recalled the scene that day. "I was outside eating under a small trellis in front of my house. At half past twelve, I heard a loud whistle, then we saw a cloud rise in the sky. It looked like an ice-cream cone. Then this cone dropped some particles and the wind blew them toward our house. At that moment we escaped inside the house, because it seemed like a fog coming toward us. We thought immediately about some acid, about something that could harm us. After about ten minutes, we went outside and found the ground completely wet. Just after that, we saw the leaves of the flowers, the roses, all spotted, as if an acid had fallen from above. But at that time we didn't pay much attention to it, because there was always some cloud, some leak. And although there was a terrible smell, we didn't worry about it; we ate, we collected vegetables and flowers from the garden, as if nothing had happened" (Ferrara, pp. 60–61).

That huge reddish cloud rose from the ICMESA factory (Industrie Chimiche Meda Società Azionaria), which produced industrial scents and pharmaceutical chemicals. It stood in the town of Meda on the northern border with Seveso. The wind on 10 July, blowing mainly toward the southeast, carried the cloud away from Meda and into parts of Seveso. Smoke billowed from the factory roof for about twenty minutes, then thinned out.

Some families went outside to get a better view of the sudden cloud filling the sky. Other families, like the woman's family above, rushed inside as the cloud passed, but they soon returned outdoors to brush off residues on their food and finish the meal. Most neighbors of ICMESA found nothing unusual that day about the factory's cloud. According to popular local beliefs, ICMESA manufactured only perfumes, which explained why it often gave off strange smells, particularly late at night. No one imagined the cloud carried any great harm. Concern would mount in the days that followed.

On the day after the cloud, the mayor of Seveso received word that ICMESA officials wanted to speak with him immediately. Although surprised that anything could be so urgent on a Sunday afternoon, the mayor, Francesco Rocca, agreed to the meeting. According to Rocca, the factory managers explained in "general terms" that a chemical called trichlorophenol had escaped from the factory and that people should be warned not to eat fruit from nearby trees. The mayor recalled, "At that time it seemed a matter of no importance, a story of herbicides and nothing else" (*Corriere della Sera*, 15 August 1976).

On Monday, 12 July, the factory resumed production as usual. Workers returned to their jobs after the weekend to find all departments operating except Department B, which normally produced trichlorophenol. Management blocked off that facility with a new fence and signs saying "no entrance." Some workers demanded that management explain what had happened and whether it posed a danger to worker health. In a brief meeting the next day, company officials responded that no danger existed but that it would be advisable for workers to wash frequently and thoroughly. Management refused to provide additional information, and a group of workers decided to begin their own study of the situation (Chiappini interview).

Also on Monday, ICMESA's management sent a letter to the health officer who served the two towns of Meda and Seveso. The letter stated that management could not determine the cause of Saturday's incident but that it apparently resulted from an "inexplicable exothermic chemical reaction" in a reactor left in its cooling phase, according to normal procedure. The note said that the cloud of vapors was blown southeast and "dissolved in brief time." The company could not determine the substances in the vapors or their "exact effect." Management nonetheless took steps to advise people living near the factory not to eat their home-grown produce, "knowing that the final product [trichlorophenol] is also used in herbicides." The letter concluded with the statement that production of trichlorophenol had been halted and that the company's efforts now focused on explaining the event to prevent its recurrence (*Gente*).

The health officer of Seveso and Meda used the company's letter to notify the two mayors about the problem. As a result of the company's report to the police, one family near the factory was advised not to touch their fruits and vegetables because of a poison. Other families, however, heard nothing, even those living in Seveso's San Pietro district, which bordered the ICMESA factory. During the following week, reports spread in San Pietro of sick and dying animals, of children covered with strange skin rashes. On Thursday evening, 15 July, the mayors of Seveso and Meda met with ICMESA managers to review the situation. As a result of the meeting, the mayors declared the San Pietro district "polluted by toxic substances" and issued local ordinances that prohibited residents from consuming or touching homegrown produce. The substitute local health officer attended the meeting (since the regular health official was out of town) and proposed evacuating families near the factory, but his suggestion was not adopted (Galimberti, p. 20). He did, however, send additional letters describing the numerous cases of poisoning near ICMESA to various local, provincial, and regional officials. The letter to the health commissioner of the Lombardy Region arrived in Milan, about twenty kilometers away, five days later (RL, n.d., p. 2).

On Friday, ICMESA sent its workers to areas near the factory to post signs warning of a "danger zone" and announcing the ban on consuming home produce. While putting up the notices, workers spoke with the residents. One worker visited the Senno family, whose two small children, ages two and four, were covered with skin rashes. The sight of the suffering children, with their swollen, red faces, along with the sight of dead rabbits and chickens at surrounding houses, convinced the worker that something serious had happened. Back at the factory, which had continued to operate through the week, he called together his fellow workers and urged them to quit the factory until management gave them more information about the vague "danger." About 60 percent of the workers then in the factory left with him. Outside the gates, they declared their intention to strike until the company specifically told them no danger existed to workers (Chiappini interview). Their action shut the factory. Two days later, an ordinance from Meda's mayor closed the factory officially and for good.

Also in Seveso on Friday was Mario Galimberti, a local reporter for *Il Giorno*, a major Italian daily newspaper. Galimberti had heard of the strange happenings in Seveso from a friend the day before. The journalist arrived in Seveso on Friday, surprised to find once-healthy trees yellow-leaved and seeming somehow burned and the town looking "dead" compared to its former vitality. Galimberti also visited the Senno family, the sight of whose daughter shocked him; she appeared "deformed and with an inhuman face." Until the cause could be determined, the doctor had prescribed an ointment to smear on her face, but it did little to relieve the pain. The situation in Seveso and the lack of official action enraged Galimberti. He spoke with company and town officials, demanding an explanation. As a result of Galimberti's probes, Seveso's mayor entered the contaminated zone on Friday evening, for the first time. That night, ambulances arrived to take nineteen children to hospitals (Galimberti, pp. 15–16).

Galimberti's front-page article appeared the next morning, exactly one week after the ICMESA cloud of 10 July. It was the first public information about the event and its consequences. The article stimulated other journalists, who arrived en masse to report on the "mysterious gas that kills plants and animals" (*Corriere della Sera*, 18 July 1976). Seveso's mayor issued an ordinance to require burning all plants, vegetables, and animals in the polluted zone. Still no one in the public realm knew what toxic substance they confronted.

On the day Galimberti's article was published, the local health officer's letter arrived in the hands of Aldo Cavallaro, the director of the Milan Provincial Laboratory of Hygiene and Prophylaxis. Cavallaro called Seveso for additional information and learned of the animal deaths and skin diseases. He decided on an immediate visit to Seveso's town hall and then to the factory. That afternoon at ICMESA, however, Cavallaro found company technical staff reluctant to explain the incident's origins or the skin diseases that developed. They would only admit that the department responsible produced trichlorophenol. On returning to his laboratory, Cavallaro reviewed the scientific literature and soon discovered the normal contaminant of trichlorophenol production: TCDD, or dioxin, known fully as 2,3,7,8-tetrachlorodibenzodioxin, one of the most toxic chemicals made by man. The next day, Cavallaro pointedly asked company officials about dioxin contamination. Only then did they begin "with a certain reticence to admit the possibility" of dioxin pollution (Cavallaro, p. 225).

On Monday, 19 July, ICMESA workers organized a meeting in Seveso to discuss the cloud and its aftermath. A notice distributed by the group warned of "the serious behavior of the management that tended to minimize [the importance of] the event." The notice also criticized public and private authorities, who had waited "a full six days after the escape of the gas before taking the first countermeasures" (Corriere della Sera, 20 July 1976).

On the same day, ICMESA's owners, the Swiss multinational firms Givaudan and Hoffman-LaRoche, became openly involved, nine days after the toxic cloud. The director of Givaudan's chemical laboratory arrived in Milan and confirmed to regional officials the presence but not the quantity of dioxin outside the factory in Seveso. No one announced this information to the public. Givaudan's representative invited the director of the Milan provincial laboratory and the local Seveso health official to visit the Swiss firm's laboratory at Dubendorf, near the company's headquarters in Zurich, to obtain the latest data and a standard sample. The two specialists departed Monday evening and spent most of Tuesday observing measurement methods for dioxin on samples taken near the ICMESA factory. Tuesday evening, over dinner, Givaudan officials gave their Italian guests the bad news: the tests confirmed the presence of dioxin at high levels. The doctors called Seveso's mayor that night, telling him the poison in his town was "worse than the atomic bomb." It was dioxin (Galimberti, p. 22).

The confirmation from Switzerland that dioxin existed outside the ICMESA factory prompted the first administrative actions by regional officials. On Wednesday, 21 July, the health commissioner for the Lombardy region, Vittorio Rivolta, conferred with the Seveso mayor and various health officials. After the meeting, the mayor ordered several streets in the contaminated zone closed to all traffic, while the regional and provincial health commissioners began preparing reports for regional political leaders and for the central government's Ministry of Health in Rome. Rivolta, meanwhile, issued his first press release on plans for dealing with the contamination. He proposed to keep the ICMESA factory closed, to map the area contaminated by dioxin, to proceed with health-protection measures for the exposed population, to form an advisory commission of scientific experts, and to continue local measures to control polluted foods. Rivolta stated that children would be removed from the polluted zone (RL, 21 July 1976).

The judicial system moved on the same day. The magistrate at Desio, a town near Seveso, ordered the arrest of the factory's director, Herwig Zwehl, and production manager, Paolo Paoletti. After booking the two men on criminal charges of causing a disaster due to culpable negligence (*disastro colposo*), the magistrate released them under house arrest to return to the factory, since they were the most knowledgeable persons in the area regarding ICMESA and its dioxin.

In the 1920s, the ICMESA company started as a family enterprise in southern Italy near Naples. Bombardment during the Second World War destroyed the factory, and in 1945 the company began to build a new facility in Meda in northern Italy, with the full approval of the allied military command. From the early postwar days, the Swiss firm Givaudan maintained a majority ownership of ICMESA and purchased many pharmaceutical products from the Italian company. The "ties of friendship" between the families owning Givaudan and ICMESA were "so close," according to Givaudan Managing Director Guy Waldvogel, that when the ICMESA founder retired in 1969, the Swiss firm acquired the Italian family's share of the business (*Commissione Parlamentare*, p. 59). In fact, the company that purchased this part of ICMESA was Hoffman-LaRoche, one of the world's largest pharmaceutical firms and Givaudan's owner since 1963 (pp. 316–17). In 1982, Givaudan reported about $300 million in global sales; for Hoffman-LaRoche (hereinafter LaRoche) the figure exceeded $3.5 billion the same year. In 1976, at the time of the accident, ICMESA reported annual sales of about 4 billion lire or $4.8 million and employed 163 workers (Hill and Knowlton International).

Coincidentally, in 1969, Givaudan modified ICMESA's Department B, which had previously made vanillin for flavoring and scents, so that it could produce 2,4,5 trichlorophenol (TCP). This product has only two uses: as the precursor of hexachlorophene, an antibacterial agent often added to soaps; and as the precursor of 2,4,5 trichlorophenoxyacetate, better known as 2,4,5-T, a herbicide sprayed in agricultural fields around the world. Givaudan decided to produce TCP at ICMESA, and manufacture of TCP shot up from pilot production of 6,361 kilograms in 1970, to 38,400 kilograms in 1974, to 142,820 kilograms in the first six months of 1976. In 1975 and 1976, ICMESA reportedly sent its TCP to two buyers, Givaudan Vernier, of Switzerland, and Givaudan Corporation, of Clifton, New Jersey (*Commissione Parlamentare*, pp. 62, 63).

An inevitable contaminant of TCP production is dioxin, which results from the reaction of two TCP molecules (Figure 5). The higher the temperature of the TCP reaction, the more dioxin is created. On 10 July, the day of the Seveso explosion, ICMESA's final TCP production cycle for the week was halted at five o'clock Saturday morning as usual. But this time, the cycle had begun with a ten-hour delay (*Commissione Parlamentare*, p. 73). Although the reaction was cooled, the process had not finished the normal twenty-four-hour cycle, and the reactor contained solvent that had not been fully removed through distillation. Sometime after six o'clock, when workers left for the weekend, the chemical soup underwent an exothermic (heat-producing) reaction. As its temperature increased, pressure inside the reactor tank rose, as did the concentration of dioxin. When the pressure reached 3.5 atmospheres, it forced open a safety valve on a pipe leading from the tank. The vaporous mixture burst out the pipe, which led through the roof to the air outside, and emerged as ICMESA's toxic cloud. Had a worker named Galante not by chance been nearby and turned on the manual cooling system after twenty minutes, the cloud would have continued much longer, with even more serious consequences (pp. 71–72).

As officials and residents of Lombardy soon discovered, dioxin is one of the most toxic substances known. In guinea pigs, the minute quantity of 0.6 microgram per kilogram of body weight is enough to kill 50 percent of exposed animals (technically called the lethal dose-50 or LD-50). In humans, the LD-50 for dioxin was unknown, but incidents of dioxin contamination in factories had occurred around the world at least thirteen times between 1949 and 1976. The incidents all happened in factories producing trichlorophenol, in the United States, West Germany, France, Holland, Czechoslovakia, England, Austria, and Italy. These cases have demonstrated dioxin's effects on man: the severe skin disease chloracne, liver damage, emphysema, myocardial degeneration, hypertension, renal damage, depression, and disturbances of memory and concentration (May). The most well known incident of dioxin exposure to a nonfactory population occurred in Vietnam, where the U.S. military sprayed about 20 million kilograms of the herbicide Agent Orange, made up of equal parts of 2,4,5-T, and 2,4-D, and estimated to contain 167 kilograms of dioxin (Young et al.). The dioxin, according to a Vietnamese doctor, has increased incidences of liver cancer, skin disease, birth defects, miscarriages, and chromosome breaks in the general population (Wade). With that precedent, the Italian press soon began referring to Seveso as "Vietnamized."

Several ICMESA and Givaudan technical directors knew about a 1971 paper by M. H. Milnes in *Nature*, which described an exothermic reaction and explosion in 1968 at an English factory producing TCP (*Com-*

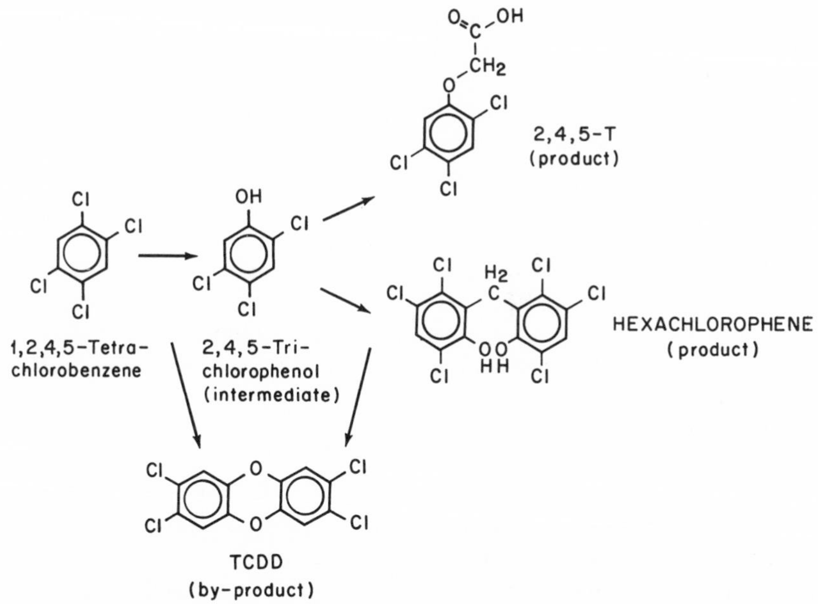

Figure 5. Production of hexachlorophene, 2, 4, 5-T, and the contaminant TCDD (dioxin). Source: Adapted from Reggiani, 1978, p. 163. Used by permission of the publisher.

missione Parlamentare, p. 72). Yet ICMESA and its Swiss owners took few steps to prevent overheating. Its TCP reactor had only a crude temperature gauge; the cooling system was manual and not automatic; the reactor's "safety" valve vented directly into the atmosphere rather than into a containment tank. The minimal investment for such a tank, or even a simple deflection shield over the mouth of the pipe, would have entirely prevented or greatly reduced the release of dioxin. In addition, ICMESA altered its TCP production process from the Givaudan patent in ways that significantly reduced costs but increased risks of forming dioxin and of causing an exothermic reaction (Gruppo P.I.A., Mazza, and Scatturin). Although the company monitored dioxin levels in the TCP product by sending samples to Switzerland for tests, ICMESA did not inform workers about dioxin or its dangers (Argiuolo interview). It should also be noted that after the explosion of the English TCP reactor in 1968, which caused persistent dioxin contamination inside the factory, the English company rebuilt the facility with a computer-controlled, five-step, "failsafe" design to prevent a similar recurrence—all described in a 1973 publication (May).

As the ICMESA incident became public knowledge in Italy, and as newspapers reported the dangers of dioxin, people in Milan began to worry. On 22 July, *Corriere della Sera*, Milan's major newspaper, with wide national and international distribution, announced in a large front-page headline, "Fears grow about the toxic cloud," adding in a subtitle: "The poisonous substances were deposited at the door of Milan." The next day, *L'Unità*, the daily paper of the Italian Communist party, countered the notion of a persistent moving cloud in its headline: "No poisonous cloud 'advances' toward Milan but the situation is grave around ICMESA." The same article noted that party members in Parliament and in the Lombardy regional assembly had already submitted interrogatives to the national and regional governments about their responses to the ICMESA disaster.

On the morning of 23 July, a large private conference took place in Milan. Participants included regional officials and the provincial health council, as well as experts from Milan University, the national Ministry of Health, the National Institute of Health, and the private pharmacological research institute Mario Negri of Milan—but no one from Seveso. The politicians and scientists reviewed the plan proposed by Rivolta and assigned various tasks to different institutions. In a press release afterwards, the experts affirmed the opinion that no further measures were necessary or urgent (Conti, p. 16). On the television news that day, Rivolta repeated the meeting's conclusion: "All is under control" (Cerruti, p. 10).

One person who believed otherwise and tried to attend the meeting was deliberately kept out. Dr. Giuseppe Reggiani, director of clinical research for LaRoche, was excluded because Italian officials did not believe he represented the company's position (Conti, p. 18). But he was able to speak at a second meeting later that day in Seveso. He stressed the necessity and urgency of evacuating the population with the highest risk of dioxin poisoning. He reportedly told local officials, "The situation is extremely serious, requiring drastic measures: removing the top twenty centimeters of earth, burying the factory, destroying the houses" (Cerruti, p. 10).

Lombardy's health commissioner reacted to Reggiani's unexpected intervention with incredulity and some animosity, expressed in an interview printed the next morning: "This character landed among us by parachute. No one expected him and, above all, no one expected declarations of that gravity. . . . I do not know whether he is a person officially accredited by the firm, and therefore I intend to ask him today on what authority he speaks. . . . I have the impression that this character is bluffing" (Conti, p. 18).

The next day, Saturday, Rivolta reconvened his collection of experts (from province, region, central government, public universities, and private research institutes). A technical representative from Givaudan, accompanied by Reggiani, presented two important documents. One was the first letter from Givaudan's managing director to any Italian public official about the accident; the other was a map showing dioxin levels around ICMESA. This meeting reversed the decision of the previous day—that nothing further needed to be done—and concluded that people living near ICMESA had to be evacuated.

The policy reversal occurred mainly because of the letter from Givaudan's managing director, Guy Waldvogel, who also served as chairman of the board at ICMESA. Dated 23 July 1976, the letter was carried by Reggiani to the meetings in Milan and was subsequently translated into English and published in an academic article by Givaudan and LaRoche scientists (Homberger et al., p. 368). On certain points, Waldvogel's letter was remarkably vague. Nowhere did he name the specific chemical, referring only to "contamination" and "the toxic substance." He referred to previous incidents in England and Germany similar to the ICMESA situation, but cited only "oral reports," although published articles, especially on the English case, were readily available. He stated that the levels of exposure at Meda and Seveso were lower than those in previous cases, again without mentioning the name or dose levels of the substance. But the letter did list five concrete recommendations, including the stunning proposals to evacuate the residents and to prohibit their carrying with them any personal belongings (Waldvogel).

After the meeting in Milan, Rivolta and the Milan prefect (a direct representative of the central government) traveled to Seveso to inform local officials about the decision to evacuate. At the Seveso town hall, officials used Givaudan's map to draw lines for evacuation. They planned to remove 171 residents of Seveso and 37 residents of Meda (RL, n.d.). The area for evacuation, with the highest contamination, they called zone A, as shown in Figure 6; a less contaminated area to the south became zone B, whose residents were advised to follow certain health precautions (Cavallaro, p. 228). The borderline between A and B was set at an average of fifty micrograms of TCDD per square meter of soil.

People in Seveso greeted the announcement of evacuation with anger and protest. According to one observer, the cars with the health commissioner and the prefect left Seveso that evening amid "kicks, spit, and insults." People demanded to know why the authorities had allowed them to live in the contaminated zone for fifteen days with no action at all. Although orders not to eat home produce had been issued, most residents continued to consume the goods as always, because no one gave them precise information about what had happened (Cerruti, p. 11).

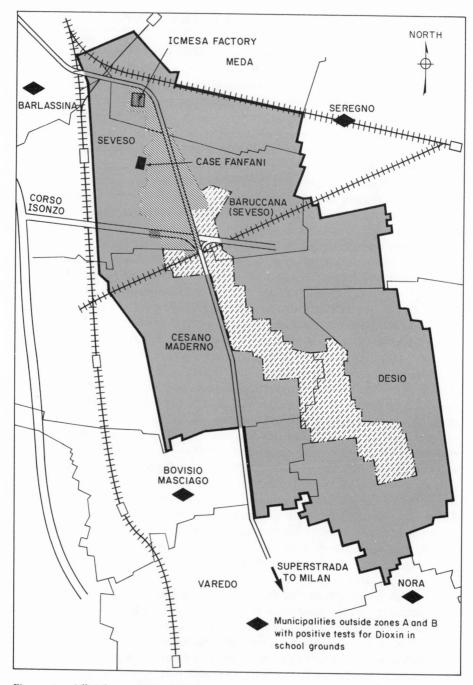

Figure 6. Official zones of dioxin contamination around Seveso: Zone A (narrow crosshatching), Zone B (wide crosshatching), and Zone R (shaded solid). Adapted from *Commissione Parlamentare*, p. 166.

On Sunday, 25 July, the military arrived in Seveso to erect a barbed wire fence around the San Pietro district, a residential area of fifteen hectares (thirty-seven acres) adjoining the factory. During a torrential rain, the soldiers worked in their uniforms, with no additional protection. One result of this unprotected approach was that some soldiers later filed suit against the government, claiming health damages due to dioxin exposure (Petrella). The sight of the soldiers also raised doubts in the minds of some local residents. If the poison were so dangerous, they reasoned, why did the government not protect its own soldiers (Brambilla interview)?

The barbed wire boundaries set by the soldiers also raised questions. One woman living in public apartments called Case Fanfani, about five hundred meters from the factory, watched soldiers unroll the wire in a line directly toward her building. "Then at some point they made a sharp turn, to go around the apartment house, excluding us from evacuation. We knew something was very strange." She continued: "We're lucky here in Seveso. We had an intelligent cloud. Look at how it made that right-angle turn and avoided our house." This woman explained the wandering lines of contamination not by the location of dioxin but by the location of powerful people (Brambilla interview).

The evacuation on Monday created additional trauma. That morning, all 179 inhabitants had blood samples taken. Children not hospitalized were sent for safekeeping to a "colony" or camp at nearby Lake Maggiore. The family heads received a sum of money for living expenses from the regional government, the amount based on the number of persons living in each dwelling. Finally, under guidance of the police, people closed their houses and left for the luxury hotel prepared for them, the Residence Leonardo da Vinci near Milan, comparable perhaps to an expensive Holiday Inn. People left their homes and, in the hotel's modern glitter, felt they were living in a "gilded cage" (Losa, p. 173).

ICMESA's dioxin also affected those who remained in Seveso. The larger public began to view all Seveso residents as carriers of dioxin, as if they were carriers of the plague. Newspaper articles reported that Seveso residents had been stopped at the Swiss border as well as refused reservations at hotels throughout Italy. People canceled orders for Seveso goods, especially furniture. The mayor of Seveso issued an appeal to outsiders to understand "that the citizens of this area, even those directly affected themselves, do not constitute any danger for others. I beg of you, do not add to our misfortune by isolating us from the rest of the world. It must be explained also that it is not dangerous to enter this region to buy our products. I am terrified by the thought that you may create a depressed area of unemployment due to an irrational fear of what has happened. A strong impression is arising that a sanitary cordon, psychological and physical, is being drawn around Seveso and environs" (*Stampa*, 28 July 1976, in Parks).

Health Commissioner Rivolta presented his first report to the regional assembly on 27 July. He reviewed the events in day-by-day detail, stressing ICMESA's responsibility for delaying actions by the Italian authorities. He also announced the formation of four committees of experts to advise on medical and epidemiological problems, analysis of the extent of contamination, veterinary matters, and cleanup. Rivolta added that science seemed impotent in confronting problems created by the cloud (*Corriere della Sera*, 28 July 1976). He rejected a proposal to deal with dioxin by burning the ground with flamethrowers, since a temperature less than one thousand degrees would only increase dioxin levels; he also rejected proposals to wash the area with water, since dioxin is insoluble in water, and to break down the contamination with ultraviolet lights, since it was considered an "impossible solution" (*Giorno*, 28 July 1976). Despite some critical questions, Rivolta and the regional executive (*giunta*) received the implicit support of other political forces. As one regional representative for the Communist party explained, the executive deserved support because of pressure on it from "ministerial and centrist forces" within the Christian Democratic party (Conti, p. 22).

While Rivolta reported in Milan, national political groups moved in Rome. Politicians from four political parties submitted separate interrogatives on the Seveso disaster, demanding to know what measures the central government had taken to prevent, investigate, and cope with the problems arising from ICMESA. The parties included most of the political spectrum from the far right neofascists (MIS-DN) and the conservative Christian Democrats to the secular Republicans and the Communists on the left.

In Milan, a major question was which level of government would take charge of the problem. Possibilities included: the central government, through a special commissioner; the regional government of Lombardy, with its executive and legislative branches; the provincial government of Milan, one of nine provinces in the Lombardy region; or the municipal governments of Meda and Seveso. The Lombardy regional executive, the group of governing commissioners, decided in a special session that the region would assume the major responsibility. The Lombardy executive followed the central government's pattern of an "open" coalition of the center-left parties, an arrangement that symbolized this special period of Italy's "historic compromise" between the Christian Democratic and Communist parties. The governing commissioners included the Christian Democratic, Socialist, Social Democratic, and Republican parties, with the external support of the Communist party.

On 27 July, the regional president traveled to Rome to explain Lombardy's decision to assume primary responsibility for the ICMESA disaster. The president also requested national action through an investigation of ICMESA, a special law to provide funds to the region, and scientific assistance on how to clean the contaminated area (RL, 27 July 1976). Yet disputes persisted between regional and central governments about whether regional authorities had notified the central government soon enough, about which administrative body was required to act by statute, and about who was responsible in the emergency (*Avvenire*, 28 July 1976).

Although to a large extent the region "won" and successfully limited the central government's direct role in the ICMESA disaster, the central government did take a number of steps in early August. On the fourth, the prime minister established a scientific and technical committee, under the Ministry of Health and headed by Aldo Cimmino, president of Italy's Superior Health Council. On the tenth, the central government appropriated 40 billion lire ($48 million) to support initial efforts at cleanup, health assistance, and welfare programs. Primary responsibility for managing the disaster's aftermath, however, would reside with the region.

The rapid transfer of funds and responsibility to the regional administration was facilitated by political circumstances, especially the good relations between Prime Minister Giulio Andreotti's new central government in Rome and Lombardy President Cesare Golfari's regional government in Milan, and their respective factions within the Christian Democratic party. The decision also reflected a sea change occurring in relations between the central government and the regions in Italy in the early 1970s, with the increasing transfer of administrative functions to the regional level. Regional government was finally emerging as a functioning administrative reality in Italy. In the area of environmental protection, as in other fields, however, the regions were constrained by responsibilities still legally assigned to the central government and by those assigned to local municipalities and provinces (Pototschnig, pp. 30–31). The ICMESA disaster offered the Lombardy region the opportunity to solidify and enhance its administrative capacities and also to demonstrate its superiority over the central government bureaucrats in Rome. The competition between the region and the central government over control of the disaster also reflected traditional tensions between northern and southern Italy.

The region made two decisions in late July, with consequences that would long reverberate. On Thursday, 29 July, the medical-epidemiological commission issued a statement suggesting that all pregnant women in the contaminated area come under strict observation at a prenatal clinic for preventive medicine and that it would be "prudent" for couples to avoid temporarily all new pregnancies. Further advice on these matters would soon be available at a clinic that was to open in Seveso (RL, 29 July 1976). On the next day, Rivolta announced that zone A, the most contaminated area, would be enlarged again. A second evacuation had already taken place on the twenty-eighth, involving a small number of families. This third evacuation would be the largest, including 36 hectares (89 acres) of land, ninety buildings, and 530 people, bringing the total to 108 hectares (267 acres) and 730 people. The commissioner also stated that pregnant women would receive medical examinations and that "every decision [on pregnancy] would be left to the free determination of those involved." This statement, Rivolta explained, included the possibility of "interrupting pregnancy," that is, abortion (*Corriere della Sera*, 31 July 1976).

The Seveso town council held an open meeting on Saturday night. About eighteen hundred people packed into the meeting place, a local cinema, in a tense and confused atmosphere. One resident later called it "an assembly of injustice. One professor said one thing; another professor said another" (S. Menaspà interview). The most often cited statement came from Professor Emilio Trabucchi, one of Italy's leading pharmacologists and a Christian Democratic member of Parliament in Rome. Trabucchi, according to a journalist present, "with the inappropriate aim of dissipating tension and minimizing the danger, declared his willingness to live in the evacuated zone and drink the contaminated milk." People responded with shouts, hisses, and insults (*Corriere della Sera*, 1 August 1976).

Sunday was a day of meetings, formal and informal, in preparation for the next day's evacuation. Family heads met with the mayor and requested a guarantee for compensation for lost goods. They told him they wanted their own houses returned or similar houses provided. The mayor responded, "Until now it was necessary to worry about health. . . . Now is the moment to evaluate the damages and study the best way to be compensated so that life can return to normal" (*Corriere della Sera*, 1 August 1976). In the streets people met spontaneously to discuss what to do. Some carried possessions to homes of relatives in order to escape the ban on removing objects. Others stayed home, crying about the lost house, the health threats, and the frightening future. Many thought about the aftermath of earthquakes in Friuli and Belice, where victims had lived for years in "temporary" dwellings, waiting for the government to provide suitable housing (S. Menaspà interview).

The evacuation created its own sense of disaster. "At nine in the morning, the soldiers arrived, took everyone outside, and put barbed wire in front of our houses. Then we waited, in empty cars, in the hot burning sun, for hours, from nine to twelve. Everyone was ready to go, crying, but no authority arrived for the evacuation—no mayor, no one from the region, no one from town hall" (S. Menaspà interview). Just after eleven, the vice-mayor arrived to announce that about sixty-five persons on one street would have to return to their closed and contaminated houses because, through a mistake in calculation, the motel chosen for the evacuees did not have enough rooms for everyone. This announcement only increased tension. The residents soon agreed on one point: either they all left or no one left. One group of evacuees organized a protest to block traffic on the street. Some people began a march to get the mayor at town hall. Soon Mayor Rocca appeared in the zone to lead the automobile caravan to the luxury motel of refuge. But many people "wondered if it all was serious, if it was possible to take decisions so casually at the cost of those subjected to a contaminant so dangerous. They began to doubt the truth of the pollution, suspecting some unknown political maneuver" (*Avvenire*, 3 August 1976). For one resident, at least, the confused evacuation was "the beginning of rebellion" (S. Menaspà interview).

Conditions improved little after the evacuees finally arrived at Motel Agip of Assago (about fifty kilometers from Seveso). They had to squeeze the extra sixty-five persons into a limited number of rooms. Among those evacuated was Giacomo Corna, an elected member of the Seveso town council. Corna's role evolved into that of direct representative of the evacuees in the town council. At the motel, Corna worked day and night to organize the new living quarters for people. He explained, "This was the tragedy of the motel. Usually only two people sleep in a room. But there, at the motel, five people stayed in a room, under extraordinarily cramped conditions." In the weeks that followed, Corna tried to make life in the motel more livable by giving families some autonomy, having them clean their own rooms, putting a refrigerator on each floor, providing hot plates for some cooking (Corna interview). But rearranging furniture could not remove stresses on individuals and families; families had to be separated, sometimes living on different floors, and they were unaccustomed to the confined apartment lifestyle (Ferrara, p. 63).

The hotel changed family relationships with the outside world and personal relationships within the family. One woman said, "We built our house working on Sundays, and we lived there only one year. . . . We never asked anything of anyone, we did everything by ourselves. But we have not lost just the house, the garden. Here in the hotel, life is impossible. In my room, my daughters sleep with me. The small moments of intimacy that I had at home with my husband, those I have no more. . . . It's not only impossible to have sexual intercourse; we cannot even speak of the economic side, we can't talk about anything anymore" (p. 64).

In early August, controversy over abortion became the most heated conflict in Seveso—and in Rome. The press had begun discussing on 24 July the strong toxic effects of dioxin on the fetus and its ability to cause birth defects, both shown in animal experiments. Over the next several days, the region's expert commission on medical and epidemiological problems considered the issue, leading up to Rivolta's statement about women having "free determination" to make all decisions. About the same time, women politicians in Rome presented a proposal for a law to allow abortions for women exposed to dioxin. The Radical party parliamentarian Emma Bonino made the proposal with the support of women members of Parliament from the Republican and Communist parties (*Stampa*, 30 July 1976).

In Italy in the mid and late 1970s, abortion was a major political issue. Under the Italian criminal code, article 546, abortion was punishable by two to five years in prison for the person performing the abortion as well as for the consenting woman undergoing it. In February 1975, however, a decision of Italy's Constitutional Court questioned for the first time the article on abortion. The court ruled that the law was unconstitutional and that pregnancy could be interrupted when the pregnancy implied medical damage or danger to the mother's health. Debate on an appropriate abortion law began soon after in Parliament. A national movement (supported by the Radical party) collected in 1975 more than the required seven hundred thousand signatures to hold a national referendum on the issue. In April 1976, a negative vote on the proposed bill in Parliament precipitated the government's downfall and subsequent parliamentary elections (Tognoni and Torri). The new parliament finally passed a law on abortion in May 1978, thereby avoiding a national referendum on the issue scheduled for the next month.

The ICMESA disaster thus occurred in the midst of a prolonged and bitter political debate over abortion. Groups advocating the legalization of abortion included Italy's growing women's movement, the Radical party, and other forces of the left. Those calling for a ban on all abortions included the Catholic church, the related new social movement Comunione e Liberazione, conservative elements of the Christian Democrats, and other forces of the right. Both sides of the debate viewed the issue of abortion in Seveso as portraying basic principles of the national controversy.

On 2 August, a "family counseling" facility opened in Seveso, designed especially to provide services to pregnant women and women of fertile age. On the first day, doctors explained that the clinic offered pregnancy tests and Pap smear tests (to detect cancer of the uterus and cervix) as well as instructions on contraceptives. On the sensitive subject of dioxin's effects on the unborn child, doctors would give pregnant women as much information as possible on the health of the fetus and on the risk of malformations, then allow the women to decide any actions.

On the same day, the Lombardy regional assembly voted its approval of steps taken by the regional executive to assist the local population and to define the contamination's limits. On the abortion issue, the assembly qualified Rivolta's proposal of an "autonomous and free decision" for women. The members voted to restrict the decision to "the possibilities offered within the existing judicial framework of therapeutic abortion." The result, as one assembly member later explained, was that women who wanted abortions had to prove psychological harm to a psychiatrist by demonstrating that they "were crazy or about to go crazy" because of the possibility of having a malformed baby. The assembly passed the motion with near unanimity (Conti, pp. 35–36).

Debate on abortion began almost immediately. A daily newspaper associated with the new left announced the next day that the counseling center had opened in Seveso, adding in the headline: "Now women can choose between the violence of a deformed child and that of abortion" (*Manifesto*, 3 August 1976). Another columnist wrote that "nearly all Italian and foreign scientists" agreed that dioxin could cause babies "malformed, mongoloid, phocomelic [thalidomide syndrome], also affected by hemophilia and leukemia." The columnist recommended removing the decision from the individual realm, thereby relieving individual anxiety, and making abortion at Seveso a social decision by legally requiring abortions for all pregnant women (*Stampa Sera*, 2 August 1976). The Catholic press responded, "When it becomes legal to kill, then it also becomes legal to oblige to kill. Euthanasia will become legal, because it has the same 'moral' principle of obligatory abortion. It will become legal to eliminate the sick, the old" (*Avvenire*, 4 August 1976). The Vatican also joined the fray with an article in its newspaper *Osservatore Romano*, calling the proposal for abortions at Seveso "disquieting" and susceptible to use for "instrumental purposes" (*strumentalizzazione*). As quoted elsewhere, the article in the Vatican newspaper stated the Christian moral position on abortion, repeating the church's opposition to abortion in any form or for any purpose (*Giorno*, 4 August 1976). Several days later, the Archbishop of Milan announced that he had families willing to adopt any malformed babies from ICMESA's dioxin (*Corriere della Sera*, 8 August 1976).

Rivolta supported abortion at Seveso because of scientific information and despite conservative political pressure. He noted that extremely low doses of dioxin in animals caused birth defects and chromosome damage.

"As a Catholic, I am against abortion; but as a public administrator, I cannot disregard situations that, objectively, can arise" (*Corriere della Sera*, 4 August 1976). Rivolta, who was affiliated with a secular faction of the Christian Democratic party, continued to be attacked by the church and by clerical factions of his party. Meanwhile, in Seveso, Christian Democrats requested that a "moralist (that is, a priest)" be added to the staff of the family counseling clinic. The Socialist and Communist parties opposed the proposal (*Repubblica*, 8–9 August 1976).

On 9 August, the region's expert commission on medical and epidemiological problems issued a position paper stating that dioxin was a teratogen in some animal species but that data were lacking for humans. The commission nonetheless concluded that an increased risk of malformed babies existed, depending on the period of pregnancy and the level of exposure. Implicit in this statement was the commission's support for therapeutic abortions. Several days later, the Ministry of Justice announced that abortions at Seveso did not require a special law. Therapeutic abortions could be performed, if the possibility of a malformed baby threatened the mother's psychological health, on the basis of Italy's Constitutional Court decision on abortion in 1975 (*Avanti*, 12 August 1976). One month had passed since the toxic cloud. For many pregnant women, the situation was becoming desperate.

In Seveso, and at the hotels for the evacuees, controversy over abortion raged. At the first meeting in one hotel, feminists who came to speak about abortion were "accepted by all," according to one participant. But the second time the feminists arrived, "one male leader sent them away, refusing to allow them to talk with us women. We had to meet them in the doorway. . . . But they had a strange way of living. They seemed like gypsies the way they dressed. That didn't create much faith in them" (B. Menaspà interview).

The evacuees also found themselves besieged by journalists. "Journalists wanted to speak with any pregnant woman, to ask her what she was going to do about abortion. Some journalists offered 50,000 lire ($60) to have the name of a pregnant woman. It was a terrifying atmosphere" (Corna interview).

The church in Seveso, a generally conservative Catholic area, took an active role in the controversy. "Public opinion in Seveso was traumatized by the problem of abortion, also because of the campaign waged by the church, by our parish, and by movements connected to the parish in some way, like Comunione e Liberazione and ACLI [Italian Christian Workers' Society]. They condemned everything, a priori, even gynecological visits, even explanations about contraception, even discussions done with women to help them live a life perhaps a little different from that which they had led before" (Ferrara, pp. 57–58).

Women requesting abortions confronted psychological and social obstacles. One woman, in her interview with the psychiatrist, was lectured about the other world and told that a baby could be happy without being beautiful. The doctor then had her listen to the heartbeat in her stomach with a stethoscope (Ferrara, p. 27). Another woman finally passed the psychiatric test and received permission for an abortion fifty-four days after she had made her initial request and in the fifteenth week of her pregnancy. The local hospitals at Desio and Seregno, both contaminated towns bordering Seveso, refused seven of the nine women who requested abortions. And in November, three women traveled to London to obtain abortions unavailable in Lombardy. The trip was organized and financed by the Sterilization and Abortion Information Center, an active group in the feminist movement for legalizing abortion and related to the Radical party (pp. 35–36). Twenty-six abortions were finally performed at the university clinic in Milan, beginning 13 August. These were the first legal abortions in the history of Italy. How many women ran the risk of clandestine abortion remains unknown.

In early August, the president of LaRoche announced that the company was insured for all damages and that everyone would be appropriately compensated (*Corriere della Sera*, 6 August 1976). Despite these public assurances from the company, representatives of Givaudan and the Lombardy government began criticizing each other in the press for not doing enough to control the dioxin disaster (*Corriere della Sera*, 7 August 1976). LaRoche then accused the region of refusing the company's offers of assistance, while Regional President Golfari denied the charges (*Corriere della Sera*, 12 August 1976).

Relations between central and regional governments also flared into conflict in the second week of August. On 11 August, the special commission created by the Italian prime minister reported on ways to cope with the dioxin contamination. Among its recommendations, the commission proposed burning all trees and vegetation in zone A in a special high-temperature incinerator to be built in the area. Interpreting the report, the commission chairman stated that although the eventual fate of buildings in zone A was unknown, they probably would have to be completely destroyed (*Giorno*, 14 August 1976).

The proposal of destruction touched a raw nerve in Lombardy, especially among evacuees from the southern portion of zone A who hoped to return to their houses. In a newspaper reply, Rivolta stressed that the commission in Rome was absolutely advisory and that all operational decisions would be taken by the region (*Giorno*, 14 August 1976). On the basis of contamination levels, Rivolta stated that perhaps four-fifths of the houses in zone A could be saved and reinhabited, which revived hope among evacuees. As Rivolta put it, "Seveso will not be destroyed" (*Re-*

pubblica, 14 August 1976). The mayor of Seveso agreed. "Seveso is not dying. We want to continue to live here, and the same for our children. And we want the evacuees, seven hundred and thirty persons who have been cut off from their homes and taken to the hotels in Bruzzano and Assago, to return quickly to our community, not to feel banished from us any longer" (*Giorno*, 13 August 1976).

In mid-August, Health Commissioner Rivolta declared that the emergency phase of the war against dioxin had ended. Rivolta distributed a map dividing the area into three zones: A (high contamination), B (low contamination), and R (a safety zone of "respect" that was not contaminated) (see Figure 6). Earlier, people in zone A had been evacuated, and residents of zone B had been told to follow certain rules of personal hygiene for safety. Now Rivolta informed people of the newly created zone R that they could not sell or eat any homegrown produce and must kill all animals, and that pregnant women and children would be allowed in their homes at night but must be evacuated during the daytime. Rivolta reported that thirteen people remained hospitalized, mainly for skin problems. At the special clinic in Seveso, 6,464 blood samples had been taken, 990 persons had received dermatological examinations, and sixty-one medical visits had been made; the women's clinic had carried out 117 discussions on contraceptives, 321 obstetric examinations, and fifty-five gynecological examinations (*Giorno*, 14 August 1976).

Also in mid-August, Professor Alan Poland, from the University of Rochester in New York, arrived in Italy at the invitation of the Lombardy regional government. Poland, who had researched the biochemical mechanisms of dioxin's toxic effects and had studied other incidents of dioxin contamination, was contacted initially by a private scientist in Milan. That scientist then arranged for a formal invitation, and Poland became the representative of the U.S. National Academy of Sciences to review the Seveso situation. Although hailed in the Italian press and television as "the world's greatest expert on dioxin," Poland stressed in press interviews that he was no expert on contamination incidents like that at Seveso, that his real home was in the laboratory, and that his expertise was in theoretical models of dioxin biochemistry (*Corriere d'Informazione*, 13 August 1976; *Repubblica*, 15 August 1976).

In an interview two years later, Poland explained that he was sent to Italy to provide the Italians with a list of American specialists on dioxin, to obtain information for the United States about what had happened at Seveso, and "to play a role for the Italian government." Poland initially resisted the request to go to Italy, because he felt he was being sent "as a politician." "The Italian [regional] government wanted an international expert to visit to make it look like they were doing something. They wanted someone to say, 'Look, we're doing what we can.'" By the time Poland left Milan he "would say whatever they wanted" (Poland interview).

Conflict continued in late August. On the twentieth, a press conference was called by the People's Scientific and Technical Committee. This group, named to contrast with the central government's Scientific and Technical Committee, included new left scientists and activists and was associated with the Workers' Movement for Socialism (Movimento dei Lavoratori per il Socialismo). The group criticized the region for its procedures on abortion—the boycotts by doctors who refused to participate, the "violence" of psychiatrists on women, the delays in decisions—and for the medical examinations, done on about nine thousand people and producing 250,000 bits of data. "If the examinations are useful . . . then it is ridiculous to have examined only 5 percent of the population at risk. If they are not useful, then it is criminal to use them as an instrument to tranquilize the population" (Petrella, p. 4).

On 24 August, in a special session of the Lombardy regional assembly, Rivolta presented his official report on Seveso. He read aloud the sixty-seven-page document, including a detailed proposal for decontamination. The report repeated the plan unveiled in previous days, but, according to Rivolta, it set a "firm point" on the boundaries and levels of contamination. The health commissioner admitted that the region had acted on two fronts, technical and political. He announced that damages totaled an approximate 20 billion lire ($24 million), including costs to private houses, personal belongings, agriculture, artisan businesses, commerce, industry, and infrastructure (RL, 23 August 1976).

Responses to Rivolta's report varied. Members of several political parties criticized aspects of the report and the actions taken by the regional executive. But after twelve hours of debate and discussion, the assembly voted to approve the proposed approaches for the regional, provincial, and local authorities. The assembly's measure was supported by Christian Democratic, Socialist, Republican, Social Democratic, and Communist parties, with abstentions by the rightist Liberal, far right MSI-DN, and new left Democratic Proletarian parties. No one voted against the measure. The assembly's motion also urged the national Parliament to form a commission to investigate the ICMESA incident.

One point of controversy that emerged in this session was the region's safety level for ground contamination. A member of the assembly from the Democratic Proletarian party questioned Rivolta about a report, allegedly from NATO, which stated that the tolerance level for dioxin was fifty micrograms per acre, a level far below the region's standard of 0.01 parts per million (set on the advice of the central government's commission). The Democratic Proletarian assemblyman demanded to know why the NATO report had not been made public and why such a large gap existed between levels. Rivolta replied that NATO's measurement method differed from that used by the regional commissions, and therefore the figures were not comparable. The discrepancy prompted the assembly to include in its motion for the day a requirement to remeasure the level of pollution in zone B (*Avanti*, 25 August 1976).

Over the next month, the region sought to find some way to decontaminate zones A and B. Experimental efforts to degrade dioxin with ultraviolet light, a suggestion from Givaudan, ran into difficulties due to a stretch of bad weather. Magical solutions surfaced one after another: a proposal from Vietnam to use a special white soap, a suggestion that NATO possessed a secret antidote, and reports that the U.S. government knew of antidotes. Barry Commoner, a prominent U.S. environmentalist, visited Seveso in mid-September and was rumored to know a dioxin antidote, a notion he strongly denied. The region's president even flew to America in search of a solution, while the health commissioner flew to Vietnam. In late September 1976, at an international meeting in Milan on the dioxin problem, suggestions from scientists abounded. But no one knew for certain what to do.

At Givaudan, however, management was beginning to understand what the company needed to do. In late August 1976, Givaudan hired a well-connected private Milanese management consulting firm, which designed a "social and political strategy" to cope with the contamination problems and the associated social conflict. The strategy's first step was to cease almost all public statements by company executives, such as those made by LaRoche's president about the company's willingness to pay damages. The consultant considered such statements as "crazy" for treating a social problem as an economic matter (Chiape interview). Those statements also made the company into a participant and a focus of social conflict.

The strategy's second step was to shift the management of contamination problems and the focus of social conflict from the private company to the public administration. The strategy sought to remove all private friction between the company and public authorities but also to pass the buck to public institutions. The consultant firm edited every statement prepared by Givaudan and then passed the statements to the Lombardy region to make public. In the consultant's opinion, the two-step strategy succeeded. By the end of 1977, "Seveso had become a problem of the state, the region, the Christian Democrats, the Communists, the local administration, the provincial health officer, etc. People had forgotten that Givaudan and LaRoche were responsible for the event" (Chiape interview).

But as the summer of 1976 drew to a close, the evacuees were still struggling to understand what had happened to them. They concentrated on resettling themselves, but did little in public, feeling "too frightened" (S. Menaspà interview). As another evacuee explained, "The people moved in the dark for the first two months. We knew nothing with certainty" (Corna interview). In mid-September, however, evacuees went to regional offices to meet with the regional president, demanding that the authorities speed up efforts to find them lodgings outside the hotels and that their houses be decontaminated. The president, "calming the representatives of the evacuees," explained that the region would find temporary apartments for the evacuees for one year. "Naturally . . . if it is not possible to return to your own houses or if some people anyhow prefer not to reenter their own houses, the region is obliged to provide compensation and construction of individual houses with gardens" (RL, 20 September 1976). That tranquilizer lasted only a short time.

On Sunday, 10 October 1976, Seveso evacuees staged a protest, encouraged by antiabortion groups who claimed that dioxin was just a pretext for promoting abortions. In a caravan of buses and private cars, the evacuees broke into the contaminated zone. The demonstrators carried signs declaring their demands: an immediate cleanup of the area, a plan for evacuees to reenter their homes as soon as possible, a reopening of Corso Isonzo to connect the isolated part of Seveso on the east side of zone A with the center of town on the west. Some demonstrators remained on the busy *superstrada* (highway) to erect a barricade and halt traffic.

The protest marked a culmination of pressures on the evacuees. Once in zone A, the evacuees streamed into their still-contaminated homes. Many people cooked the noonday meal and some slept in their own beds. They acted as if they had returned to normal life, to the predioxin days of Seveso. "It's the exasperation, the distrust, the fear," exclaimed one former resident of zone A. "It was supposed to be only a demonstration. A brief walk, to make a point. But when the evacuees saw their own homes, they lost their heads—an understandable reaction for people who for three months have tried to rebuild their families in the sterile environment of a motel" (*Corriere della Sera*, 12 October 1976).

Seventy *carabinieri*, the paramilitary police, soon appeared. An assistant priest of Seveso's main parish began to appeal by loudspeaker for "solidarity" among the protesters. Only the mayor's intervention, resulting in the police's withdrawal, averted open violence. All afternoon and evening, regional and provincial politicians arrived in Seveso, only to be greeted by shouts and insults. For fifteen hours, evacuees reoccupied their homes and resisted pleas to leave. Politicians repeatedly implored people to abandon their dioxin-tainted dwellings. Finally around midnight the last family agreed to reevacuation (*Corriere d'Informazione*, 11 October 1976).

The demonstration of 10 October brought together several local political forces: economic interests, which wanted Corso Isonzo opened to improve commerce; residents of Baruccana (the other side of zone A), who wanted the street reopened to connect them with the main part of Seveso; conservative clerical groups, which sought to counter any progressive steps on abortion; and evacuees, who wanted to be resettled in their own homes. Indeed, plans for the demonstration were made only the night before, when a group from Seveso arrived at the Assago motel. This group included the commercial and clerical interests. Many evacuees quickly accepted the protest proposal. As one victim explained, "For some time, there had circulated feelings of discontent, isolation, and abandonment" (*Corriere della Sera*, 12 October 1976).

On the morning of the eleventh, the disputing groups gathered in the Lombardy regional offices in Milan to discuss efforts to regain "normality." They reviewed problems with the cleanup operations, the resumption of commercial activities, the evacuees' temporary lodgings, and the construction of a street to connect the two parts of Seveso separated by zone A. To "avoid unjustified tensions," like the demonstration the day before, the authorities agreed to meet weekly with the affected population (RL, 11 October 1976).

By the end of the week, however, a new controversy erupted in Seveso. Local residents rejected the proposal by regional and provincial authorities to build a special high-temperature incinerator in Seveso to destroy dioxin-contaminated materials. On the fifteenth, officials from the region, the province, and Seveso held an open meeting about the incinerator, planned for the southern section of zone A. Local residents repeatedly interrupted the officials' presentations, shouting opposition to the *forno*. "Build it in Switzerland." "Meda polluted, make Meda depollute." "We don't want another ICMESA." "Why must Seveso pay again?" Residents proposed that the incinerator, if necessary, be built inside the ICMESA factory. Some people feared that the incinerator would continue to operate after the dioxin cleanup, making Seveso into the industrial garbage dump for all of Italy's toxic waste (*Corriere della Sera*, 17 October 1976). Some even came to believe that ICMESA's toxic cloud was just a *montatura*, an exaggeration created by the region and province in order to construct an incinerator in Seveso (Liberti interview).

On 8 November, about one hundred of the area's artisan workers blocked the superstrada in protest. They tore down some signs along the highway that declared the area a "contaminated zone" and plastered over other signs with their own posters that opposed the incinerator and the "mark" on Seveso (*Giorno*, 7 and 9 November 1976). That evening, when regional and provincial officials did not appear for a meeting to discuss the cleanup plan and the incinerator, local residents again took their complaints to the street. Once more they obstructed highway traffic, reopened Corso Isonzo, and marched through zone A to town hall. This time, however, the protesters also redirected traffic off the superstrada, onto Corso Isonzo, and through zone A (*Giornale*, 10 November 1976).

Rivolta responded to the protests, saying he could understand that Seveso residents did not want the incinerator. "But they must understand that dioxin does not melt like snow in the sun. And for now, incineration is the safest method" (*Giornale*, 10 November 1976). Rivolta's response convinced few people in Seveso. On 14 November, for example, Seveso's elected town assembly unanimously approved a resolution that expressed the citizens' concerns about the incinerator and requested the town administration to cancel its agreement to construct the incinerator (Giunta Municipale di Seveso). For Seveso residents, the enemy had become the region and the province, not ICMESA-Givaudan-LaRoche. The residents seemed to have forgotten who initially caused the dioxin problem (*L'Unità*, 19 November 1976).

The argument over the incinerator continued in December. The provincial, regional, and central governments insisted that incineration was the best method to dispose of contaminated materials. They proposed a facility with a daily capacity to burn 150 tons of earth, plant, animal, and other substances; it would require one year to build the incinerator and six years of operation to destroy the estimated 300,000 tons of dioxin refuse. Seveso residents adamantly opposed the construction of a single huge incinerator in their town. Artisans and businessmen argued that such a solution would ruin the town's economy (*Corriere della Sera*, 3 December 1976).

Again, Seveso residents demonstrated in public. In the early afternoon of Saturday, 4 December, about two thousand people of all ages converged on the convenient superstrada, blocking traffic in both directions and redirecting cars onto Corso Isonzo and through zone A. The protesters demanded that the economically important Corso Isonzo be reopened and that the economically ruinous plans for the incinerator be scrapped. The banner held over the highway barricade declared: "Open Corso Isonzo or the Superstrada Remains Closed." Indeed, the highway remained closed all that night and most of Sunday. After receiving a promise to reopen Corso Isonzo, the demonstrators removed their blockade

Sunday evening and proclaimed a truce. They also reached a written accord with the presidents of the region and the province on the priorities of cleanup (*Corriere della Sera*, 5 and 6 December 1976). The agreement made no mention of the incinerator, but it did specify decontamination techniques for the southern portion of zone A: dioxin-contaminated dirt would be carted into zone A's northern part, the area closest to the factory and with the highest dioxin levels (RL, 6 December 1976). The cleanup, it seemed, might begin soon.

On the ninth, the regional assembly met to consider and approve a law for measures to assist the area hit by the toxic cloud. (The decree originally issued by the regional executive on 10 August was converted into law on 8 October but still required approval by the assembly to continue in effect.) In the debate, assembly member Laura Conti (of the Communist party) criticized the regional executive for giving inadequate information to the local population and for delays in implementing the studies, experiments, and countermeasures required. She argued that both insufficient information and excessive delays contributed to undermining people's belief in the dioxin danger (*L'Unità*, 10 December 1976). The regional health commissioner responded that there were delays because the people of Seveso did not fully understand the proposals, "some people in good faith, some people in bad faith." The regional president explained that the region had tried to proceed with its measures by obtaining consensus from all groups but that it would proceed even if it did not obtain consensus. "I do not believe that—after having declared for three months that dioxin is deadly, an extremely potent poison, and has deleterious effects on the human body—public opinion can be allowed to expect dioxin to be left forever where and how it is" (Golfari).

In mid-December, the Lombardy region and Givaudan reached an agreement on the first cleanup efforts. The operation would be financed and supervised by Givaudan, with the region and other Italian technical bodies monitoring the results, especially final dioxin levels. The houses scheduled for first cleanup were in the southern portion of the evacuated zone A (*Corriere della Sera*, 18 December 1976), not in zone B, where people continued to live. One prominent national scientist called this a choice of "political nature; indeed, from the technical-scientific point of view, it would have been more appropriate to start decontamination in zone B" (Ministero della Sanità). From a public health perspective, the longer the delay in cleanup for zone B, the greater the possibility of continued exposure and of chronic health consequences.

A British newspaper published one of the first reports of widespread chronic health problems resulting from ICMESA's dioxin. On 17 October, the *Sunday Times* of London reported that tests of ten thousand blood samples showed about one thousand persons, mostly children, with a decreased number of lymphocytes, the white blood cells that fight disease. That change could reduce the ability to fight infection and possibly lead to leukemia later in life. The article stated that the findings had not yet been released in Italy. The next day, the Italian press reported the news, accompanied by official denials. Regional authorities called the diminished immunological capacity "absolutely not significant" (*Stampa*, 19 October 1976).

But Lombardy officials did announce the presence of several cases of chloracne. In fact, Italian doctors had previously reported cases of chloracne, though not to the public, not to the press, not to the patients, but to an international conference in Milan on 30 September. The head of the medical-epidemiological commission had stated, "Until a few days ago, cases of typical or even atypical chloracne had not been diagnosed" (Berlin, Buratta, and Van der Venne, p. 40). Only in late October did that commission admit to the public the existence of chloracne. The commission's statement concluded, however, that the distribution of results did not indicate "important differences" among persons of zone A, zone B, and other areas (RL, 20 October 1976).

In late January, several elementary school students in Seveso showed skin problems that looked suspiciously like chloracne. Dermatologists from Milan examined about three hundred pupils and identified twenty-five cases needing further observation (RL, 1 February 1977). That news put Seveso once again on the front pages of newspapers throughout Italy. But as Seveso's mayor noted, the discovery did not emerge as a result of the region's plan to monitor the health of people possibly exposed to dioxin. That plan, the mayor said in mid-February, "has practically not started. The cases of chloracne emerged by chance, thanks to the school doctors" (*Corriere della Sera*, 17 February 1977). In Seveso, the discovery of chloracne revived fears about health problems. It also suppressed, at least temporarily, skepticism about dioxin's danger and persistence.

In February and March 1977, as dermatologists examined more pupils in the Seveso area, they found more cases of light chloracne, reaching 215 suspected cases in 923 children examined. In some cases, exposure occurred in July 1976; in other cases, exposure occurred much later. The wide distribution of chloracne cases, throughout Seveso, Meda, and neighboring towns, suggested dioxin contamination in areas outside zone B, areas that the region had defined as clean (*Corriere della Sera*, 3 February 1977). The regional health commissioner blamed the "expansion of dioxin" on people who returned to their houses or entered zone A and thereby carried the poison away from the initial area (*Repubblica*, 10 February 1977).

"To isolate, defend, and protect" the contaminated zone of Seveso and Meda, the regional executive called on Rome to send in the army (RL, 11 February 1977). At a meeting in Seveso the day after the announcement, the regional health commissioner again faced an angry crowd. He said that even local authorities had difficulty managing the contaminated area and that "with the intervention of the soldiers, it will be understood that the situation is serious." As for cases of chloracne outside the contaminated zone, he said, "We gave precise advice on hygiene. Have [the instructions] been observed? To each his own responsibility" (*Repubblica*, 13 February 1977).

Several days later, in a session constantly interrupted by shouts, the regional council approved a three-part plan for the cleanup of Seveso. The initial phase would concentrate on houses in the southern part of zone A and on zone B. The second phase would prepare materials for final destruction. The last phase would destroy the contaminated materials with the infamous incinerator. According to one newspaper article, "This program is identical to that approved in August, then reapproved in October, and then never carried out." The article was titled "Seveso: The region, after seven months, presents the 'new' plan for cleanup" (*Manifesto*, 16 February 1977).

In early March, the Lombardy region began publication of *Il Punto* (The Point), "the official bulletin for the population struck by the toxic cloud in the territory of Seveso and surrounding towns, comprising the area of respect." The lead article in the bulletin declared: "Dioxin is also a problem of information. . . . The contradictions of the news and opinions expressed sometimes by single individuals, by social groups and forces, by the same newspapers could have created confusion and uneasiness." The region sought to resolve or at least reduce these problems by providing its opinions and its decisions directly to the people affected by the ICMESA disaster (*Il Punto*, 4 March 1977).

Il Punto also provided information on childbearing and child rearing. The first issue included an article that urged women in zones A, B, and R not to breast feed but to use infant formula and to continue to observe the proposal of August 1976 to "avoid procreation" until the area achieved the condition of "health safety" (p. 3). The next issue reported that pathologists at the University of Lubecca in West Germany had examined thirty-four embryos (from four "spontaneous abortions" or miscarriages and thirty induced abortions) and had found only one abnormal development. But, as the article noted, only limited conclusions could be drawn because of the small number of cases, the different ages and development stages of the embryos, and the damage done to the embryos during abortion (11 March 1977). Four of the induced abortions came from women living in zone A, three of them with at least fourteen days of exposure. Four other women in zone A gave birth to children in 1977, none showing any birth defects. Despite the limitations of the data, one regional report concluded that "at least eight pregnancies in zone A did not involve any" birth defects (Bisanti et al., p. 5). Dioxin seemed a less potent teratogen in humans than initially expected. But these early reports did not resolve the question.

In mid-March, the provincial authorities at last began cleanup efforts in zone B. Trucks started carrying contaminated dirt from zone B of Cesano Maderno into zone A of Seveso. Workers claimed that they washed the trucks before each load and that dirt from zone B was only slightly contaminated. But women living along the streets traveled by the trucks believed otherwise. They saw dirt and dust spread by the trucks, often with uncovered loads.

After several days of this "cleanup," a group of women blocked the street and stopped the trucks. They also distributed a leaflet: "We Sevesini, in addition to the crime of ICMESA, must suffer another pollution. All of this is absurd. Reclaiming Cesano to pollute Seveso. What are the local health officer and the town administration doing to protect our health?" The street blockade continued for hours. Politicians and residents finally reached a temporary accord, which included a promise by the authorities to consider new truck routes (*Giorno*, 18 March 1977), but several days later, when Rivolta did not appear at an appointment in Seveso to discuss the truck problem, women occupied the town hall (*Corriere della Sera*, 22 March 1977). The region, seeking to "eliminate every possible perplexity" about the decontamination, printed an article on the cleanup in its next issue of *Il Punto*. The article stressed that the hauling would contribute to the area's recovery and its return to normality (25 March 1977, p. 3).

Yet cleanup efforts continued to breed uneasiness. People living in zone B and in zone R watched men clad in white suits walking and working in fields adjoining houses. Electricians and plumbers working in the area, as well as soldiers riding around in jeeps, all wore special uniforms. Sometimes children played in the same fields that were later stripped of vegetation. It seemed to people that the government protected the workers but not the residents. These actions did not help persuade residents that the areas they lived in were uncontaminated or contaminated at a safe level (Blake interview). Some workers also wore their white suits into local bars and restaurants, creating additional confusion about danger and safety.

In many ways, the decontamination program merely moved contaminated materials from one point to another, so that disposal was postponed but not implemented. In the defoliation effort, workers collected presumably dioxin-contaminated leaves in preparation for final destruction, but, as one elementary school teacher recalled, children occasionally played in the piles of leaves, because no one removed the leaves or took adequate protection measures (Liberti interview). Authorities also collected thousands of dead animals (including 3,300 that died and 77,000 that were slaughtered). Some carcasses were burned in ordinary urban trash incinerators, probably dispersing into the air dioxin contained in the carcasses and creating new dioxin by heating residual TCP in the waste. The rest of the carcasses remained in barrels in zone A, awaiting some solution. As spring and warm weather approached, putrefaction of these carcasses began to pose an additional public health problem.

In early April, Seveso women again blocked the trucks. This time, some zone A evacuees objected to the protest, since every delay in cleanup postponed the return to their houses (*Giorno*, 3 April 1977). But in the Seveso town council, the head of health affairs (a Socialist and member of the town's center-left coalition) supported the protest and requested the suspension of trucking contaminated dirt. The cleanup, he said, continued to contaminate. A journalist agreed with this assessment, reporting stories told by townspeople of uncovered trucks spilling dirt, which children used in their play (*Repubblica*, 16 April 1977).

Several days later, French television planned a live broadcast from the Seveso town hall. Speakers included the mayor, the town's councilman for health affairs, a private scientist, and the Senno family with their two daughters, Alice and Stefania, the two children most severely harmed by the toxic cloud. As show time approached, about three hundred local residents gathered outside the town hall, honking their car horns to disrupt the broadcast. In a sudden surge, the people stormed the town hall, cut the television cables, and interrupted the program. They refused to allow any more "negative publicity" about Seveso and opposed the notion that Seveso's children all resembled Alice Senno, the child most disfigured by chloracne. *L'Unità* commented that the protest was almost asking "for a 'silent press,' as if that were sufficient to cancel the problem" (20 April 1977).

On 23 April, a group of five hundred new left activists marched in Seveso against the multinational company LaRoche and against the Italian political authorities in a protest that once again shut the superstrada. The protesters called the move the first moment of "real opposition." They criticized the people who disrupted the French broadcast for preventing the participation of scientists in France who were waiting to speak on the program. The group also accused the region of hiding data that showed high levels of contamination (*Giorno*, 24 April 1977).

In late April, the People's Scientific and Technical Committee released what the press called "the secret map of dioxin," based on official regional data. The map showed the distribution of chloracne cases throughout zones B and R, suggesting widespread and persistent dioxin contamination beyond the boundaries of zone A. The committee also published its own epidemiological study of symptoms in eight apartment houses near ICMESA, showing an increase in many skin and respiratory problems (*Corriere d'Informazione*, 28 April 1977).

Another protest erupted in mid-May, when about two thousand Seveso residents demonstrated in Milan against the still-planned incinerator. The protesters proposed a "natural degradation" of dioxin using physical and chemical methods other than incineration. The motorcade of 250 automobiles was led by motorcyclists dressed in the white and yellow suits of the decontamination workers. "We are the guinea pigs of dioxin; we don't want to be the guinea pigs of the incinerator as well." "The regional cleanup plan was created by technicians selected not for their scientific experience but for their political faith" (*Giornale*, 15 May 1977).

Two incidents of planned violence also occurred in this period, one against a public official, the other against a private executive. On 19 May, three young men entered the office of the local health officer for Seveso and Meda and demanded some documents. The health officer was publicly considered one of several persons responsible for not controlling ICMESA. Indeed, the court magistrate on the case in August 1976 had officially charged this doctor with criminal "omission of acts" with regard to ICMESA. The doctor refused to comply with the demands of the three men. They shot him three times in the legs and once in the arm. Later, one of Italy's extreme left terrorist groups, Prima Linea (Front Line), claimed responsibility for the shooting (*Giornale*, 20 May 1977). The second act of violence occurred in July, when a bomb exploded in the Swiss villa of a LaRoche executive coordinating corporate responses to the ICMESA disaster. The group claiming responsibility for the bombing called itself "The 10 July Commando." According to a notice left at the scene and sent to newspapers, the bombing revenged the dioxin at Seveso. The group accused not only LaRoche but also the Italian government, the Christian Democrats, and the Communists (for their historical compromise to support the government) (*Giorno*, 8 July 1977).

In late May, the region's medical-epidemiological commission held its first official meeting with the people affected by ICMESA's dioxin. For the meeting, the commission prepared a preliminary report on the state of health in the dioxin-polluted area (Fara). The commission registered six malformations in 1977 (from March to May), but the cases could not be compared with past data because of prior inadequate reporting of malformations, that is, lack of an adequate baseline. The report did find health differences between residents of zone A and those of zones B and R, but no differences between zones B and R. The report stated that the incidence of hepatomegaly and enzyme alterations "clearly demonstrates the involvement of the liver in dioxin poisoning" (p. 3). The document also compared the geographical distributions of three dioxin indicators: levels of soil contamination, animal deaths, and human skin symptoms. This comparison showed that dioxin had contaminated an area much larger than initially expected, and at levels in zone A much higher than previously announced. The report concluded that the dermatological situation appeared "more worrisome" than it had several months earlier, because symptoms showed signs of becoming chronic (p. 24). With regard to preventive and curative measures, the report noted delays due to inadequate health services and to problems in creating structures capable of handling the additional demands (p. 4).

A critical review of the commission's report faulted the document for not fulfilling one of its stated objectives: to propose preventive or curative measures to deal with the effects of dioxin. More basically, the review criticized the study for not meeting scientific standards (not providing criteria of representativeness, definitions of normal, definitions of suspected chloracne and nonclassifiable skin lesions, or complete data) and for not assisting affected people in understanding their problems (not providing criteria to identify symptoms as due to dioxin or as possibly compensable) (Terracini).

In early June, the regional assembly approved the plan for Seveso proposed by the Lombardy executive. The plan covered five programs—decontamination, health services and surveillance, social and educational assistance, reconstruction, and commercial assistance—in more than two hundred pages of detailed prescriptions. It soon became known as the "dioxin bible." The assembly also approved the creation of a special office for Seveso, an agency headed by a regional, not a national commissioner. The Lombardy executive proposed the special office in order to end further discussion and to quickly implement the regional plan.

Through its plan and new institution, the regional executive sought to avoid additional negotiations "with all the social realities and institutions, already convinced at this point of the plan of action." By concentrating executive power "in a few highly responsible hands," regional officials intended to facilitate "intervention and coordination." The regional executive deplored the prior pattern in which the incident had been "minimized by one side or instrumentalized by the other, in any case made more incomprehensible to the population." The executive wanted to start over, to "begin the path again united, because united we can conquer the fearful and invisible enemy, fruit of a contradictory development, but above all, product of a serious and irresponsible negligence on the part of the polluting company. Its name is ICMESA, and its owners are called Givaudan and Hoffman-LaRoche. They are names not to be forgotten, even if for some it has become easier and more convenient to attack the inevitable complexity of the intervention" (RL, 2 June 1977).

The man chosen as special commissioner for Seveso, called the "man of dioxin" (*l'uomo della diossina*), was lawyer Antonio Spallino, mayor of Como, a major city north of Seveso. Spallino was chosen as a politician and not as a technician. In an early interview, he declared, "I don't know what dioxin is, but we will succeed in conquering it" (*Corriere d'Informazione*, 27 June 1977). But Spallino was not just any politician. He belonged to the secular faction of the Christian Democrats, the same faction as the regional president and the regional party secretary (a faction known as "Base"). Moreover, his city government at Como had a center-left coalition executive, with the outside support of the Communist party. Thus, Spallino's appointment met no opposition from Lombardy's major political parties, though one problem arose. Spallino intended to continue as mayor of Como, thereby defeating the objective of having a full-time director of operations at Seveso, which was an important reason for creating the institution (*Giornale*, 27 June 1977).

Seveso's special office soon encountered a host of difficulties. Spallino had not anticipated "the enormous efforts" necessary to set up the institution. The region located the office in zone R of Seveso, making it accessible to local people but also to demonstrators. "For the first months," recalled Spallino, "the offices were occupied every two to three days." Spallino had not expected that high level of tension. In responding to the protests and the tension, Spallino asserted, the special office lost energies that should have been directed toward solving the dioxin problems. Another unanticipated difficulty was recruitment of personnel, especially technical and professional staff: chemists, health workers, and epidemiologists. The commissioner attributed this persistent problem to social and political tensions, job insecurity associated with the special office, and the low pay of public employees. Yet in 1979, Spallino believed that despite these problems, the office eventually met its objectives (Spallino interview).

In mid-June 1977 in Rome, another institution, the special parliamentary commission of investigation on Seveso, finally began its activities. After almost a year of debate, both houses of Italy's Seventh Legislature passed a bill establishing the special commission to investigate: the causes of the ICMESA disaster and the responsibilities of the private company and public authorities; the initial responses of local and central administrations and of the judiciary; the consequences for health, environment, and economic activity; the calculation of damages and the provision of compensation; and indications for a more effective regulatory system. In July 1978, the commission issued its unanimous 412-page final report, signed by all fifteen senators and fifteen deputies of the committee. The report was based on interviews with more than a hundred persons and a large collection of documents from the involved institutions. The commission stressed ICMESA's fault but also the inadequacies of public authorities and regulatory bodies due to overlapping responsibilities and inadequate resources, especially for the control of complex technologies of advanced industries. As one reviewer put it, "In sum, is no one responsible because everyone is more or less responsible?" (*Cittadino della Domenica*, 3 February 1979).

During the summer and fall of 1977, the cleaning company hired by Givaudan continued working on the houses in the southern part of zone A. In mid-October, the first 24 of 139 families returned to their homes. The resettled evacuees expressed joy on the occasion. One woman reflected, "They were fourteen months of prison" (*L'Unità*, 16 October 1977). Some accusations of improper cleanup procedures arose, but many people returned. In 1979, Corna, leader of the evacuees/returnees, emphasized, "We're not here to live as guinea pigs, to die. We came back because they did everything possible so that we could come back. We're not reckless people. . . . I believe, honestly, that it was not possible to do more than was done. Everything was thrown out. Everything is new. I challenge anyone to find another thing to do or to change, that something was not done" (interview).

But some families continued to doubt the safety of the cleaned houses. From late August 1976, some families demanded that they not be forced to return and that they be provided with equal, uncontaminated houses elsewhere (*Giorno*, 20 August 1976). In 1979, Corna said, "I know of no one who for health reasons does not want to return. Five families do not want to return, but each has personal or economic reasons" (Corna interview). According to Givaudan, 500 people who lived in the southernmost parts of zone A (A6 and A7) finally returned to their cleaned homes at the end of 1977. Forty families (236 people) received compensation payments from the company and moved elsewhere (Hill and Knowlton International, p. 12).

One woman insisted that she did not want her family to return because of health problems. They had lived in the zone for twenty days before evacuation, and she had diarrhea for fifteen of those days. Skin problems, initially severe, later improved but still remained in 1979. Her family and several others had not participated in the 10 October 1976 invasion of zone A and occupation of the houses. "After that protest there grew a distance between those who wanted to return and those who didn't. At assemblies, the leaders refused to allow us to speak. They kept saying, 'Why don't you want to go back? There's nothing there anymore.'" In conclusion, and in a tense voice, she stressed, "We want to go back and grow plants and have a garden. We want what we had before. We repeat that many times but they accuse us of wanting more than what we had. That's not true. We believe in dioxin and we want health" (Motta interview).

While some families resisted returning to cleaned houses, others struggled to be evacuated from uncleaned houses. Two areas of protest arose, one on each side of zone A. On the western side, not far from the ICMESA factory, stood eight public apartment buildings, known as Case Fanfani. This area bordered the most contaminated section of zone A, but was designated zone R—not even zone B. As part of zone R, the region excluded these residents from the health-control program. After much public criticism, and following medical examinations by the People's Scientific and Technical Committee, the region agreed to include the Case Fanfani in the official medical examinations. The region then found high levels of dioxin, skin lesions, and hepatomegaly (Fara, pp. 10–11, 17), but regional authorities continued to deny health problems, attributing any symptoms to anxiety and psychosomatic causes (Spallino). Finally, after two years of struggle, the regional assembly approved a "voluntary" exit of some families—but for reasons of mental health rather than physical health (*Repubblica*, 30 March 1979).

On the east side of zone A, in Baruccana, residents of another set of apartment buildings also protested. The Seveso health official had approved these new buildings for occupancy on 8 July 1976, two days before the ICMESA accident (Maffii). People began moving into the apartments in August 1976, because the region designated the area as zone R and safe. The wave of chloracne in early 1977 eroded the residents' faith in the authorities' assurances. Throughout the winter of 1976–77, the apartments lacked gas and therefore heating and hot water, making it extremely difficult to follow the rigorous hygiene standards recommended by the region. Moreover, while regional officials claimed that no health problems existed, the residents had yet to be medically examined. Some residents concluded they had been allowed to enter the new apartments, which had not been cleaned inside or outside, in order to "tranquilize" other families in the area and encourage the reopening of factories (Blake interview; Quartiere Case IACP). In 1983, some residents of the apartments continued to protest against persistent health problems and inadequate medical care (Maffii).

In early 1979, two old problems resurfaced in Seveso: dioxin mapping and birth malformations. Both developed into political conflicts. On 9 January 1979, the People's Scientific and Technical Committee presented the regional assembly's health committee with a map showing wider contamination than was ever admitted by the Lombardy executive. The map came from a published scientific article by LaRoche's director of clinical research, Giuseppe Reggiani. The article noted the map had been "handed over to the Italian authorities of the Lombardy region Saturday July 24" (Reggiani, p. 169). Several assembly members submitted formal questions demanding to know why the executive never passed the "Reggiani map" to the health committee for review. The regional president, in a written response of forty-four pages plus eleven appendixes, replied brusquely that the region never received the Reggiani map. But the president's document showed that the region did receive three other maps from Givaudan, none of which were provided to the regional assembly. Nonetheless, the president rejected the accusation of hiding or manipulating data. Differences in dioxin levels shown on maps, the president asserted, arose from the region's measurement of ground contamination and Givaudan's initial measurement of vegetation contamination (RL, 1 February 1979). After the president's reply, the map controversy subsided.

Controversy over birth defects exploded in early February 1979. On 1 February, at the special office's weekly press conference, a health official announced that the number of recorded birth defects in all eleven towns of Brianza had risen from thirty-eight cases in 1977 to fifty-three cases in 1978. Zone B reported no defects in 1977 but three in 1978. The health official also noted an increase in the number of serious congenital defects, but stated that no one knew if dioxin caused the changes (*Repubblica*, 3 February 1979). In a press release, the special office sought to calm a possible "alarmist" phenomenon by reporting that the number of malformations in zones B and R remained the same (nine) for 1977 and 1978 and that the overall increase resulted from a better reporting system (RL, 3 February 1979). The Catholic newspaper titled its article "Unjustified alarmism on the malformations" (*Avvenire*, 4 February 1979). *La Repubblica*, on the other hand, interviewed a gynecologist who accused political and health officials of ineptitude and said the fight against dioxin "is in the hands of dilettantes" (6 February 1979).

As scientists, politicians, and journalists argued in press conferences and in newsprint over the meaning of the increased number of birth defects, the People's Scientific and Technical Committee filed criminal suit against Spallino, the commissioner at the special office. The suit charged Spallino and his medical officer with "omission of official acts" and "omission of aid" for reporting only one-third of the officially recorded birth defects. The committee, moreover, possessed proof of an additional ninety-three cases of birth defects in copies of the medical reports. After Spallino denied knowing about any additional birth defects, the committee published a letter, dated 18 December 1978 and addressed to Spallino, from the doctor responsible for the team studying birth defects. The letter, which stated the number of birth defects as 101 in 1978, put Spallino in an awkward position (*Corriere della Sera*, 22 February 1979).

The emerging conflict over the number of birth defects threatened the coalition government of the Lombardy region. In late February, the Communist party began to take a more critical stance toward the regional executive. Then, in March, the People's Scientific and Technical Committee filed two additional charges of official improprieties: that ordinary urban incinerators in Milan had burned eighty-six hundred tons of waste material from Seveso and surrounding towns, probably releasing dioxin; and that regional officials found changes in certain biological indicators but did not perform follow-up examinations (*Stampa*, 11 March 1979). Several months later, Spallino revised his statistics for 1978, from the previous number of fifty-three cases to ninety certain cases of malformations and twenty-four suspected cases (*Repubblica*, 6 July 1979). This revision, however, occurred only after a magistrate officially charged Spallino with the allegations filed by the committee.

The regional plan for the ICMESA disaster had also established an International Steering Committee for Seveso. In Italian, the group was called the "Comitato Internazionale dei Garanti," clearly indicating its role as provider of legitimacy as well as oversight for the region's operations in Seveso. A working group of the committee met in Milan in the spring of 1979 and in full plenary session that fall.

The special office reported on these activities in its newly revamped newsletter for the affected population (the reporting was required by law under the regional plan). The old newsletter, *Il Punto*, had been "a bulletin of war," brief and dedicated to the battles underway; the new version, *Settimana 3*, received its name (Three Weeks) from the frequency of publication and was designed to be more friendly, to provide information on a broad range of local issues in addition to those related to the contamination, and to create an "open discourse" with the people (*Settimana 3*, no. 0). The newsletter explained the various activities of the special office; and as noted above, it provided brief news reports on the "guarantees" of international experts (*Settimana 3*, nos. 12 and 17).

The Italian government had also filed suit in 1978, more than two years after the incident, to recover 40.4 billion lire ($47.6 million) for administrative costs of the ICMESA disaster. Remarkably, the suit was against ten bureaucrats considered coresponsible for not preventing the dioxin escape, the charges against them based on a comprehensive report of the episode's causes by the Ministry of Labor. The Italian government occasionally files such suits against bureaucrats when faced with unpredicted financial burdens, but no one expected the ten bureaucrats to be capable of paying the enormous sum from their own pockets (*Corriere della Sera*, 1 August 1978).

National and regional governments also filed damage suits against the ICMESA-Givaudan-LaRoche conglomerate soon after 10 July 1976. Two and a half years later, the court's panel of experts issued a report of more than two hundred pages. Although a secret document, some journalists (and at least one foreign researcher) read the report. As expected, the eight technical experts attributed major responsibility to ICMESA-Givaudan-LaRoche. One newspaper headline summed up the report's conclusion: "At ICMESA they knew the dangers of dioxin; the firm's managers understood, the workers did not" (*Paese Sera*, 27 December 1978). This technical opinion contributed to LaRoche's decision to settle out of court with the national and regional governments in March 1980, nearly four years after the disaster started, for a total of 103.6 billion lire ($121 million). Signed that year in December, the settlement covered administrative costs for decontamination, reconstruction, and health surveys, with funds disbursed to the Lombardy region and the central government.

Just before that settlement was reached, terrorists struck once more. In February 1980, three men and a woman ambushed and assassinated Paolo Paoletti, aged thirty-nine, a director of the ICMESA factory (*Nature*). The guerrilla group Prima Linea claimed responsibility for the attack, as it had in 1977 for shooting the local health official of Seveso and Meda. Paoletti was one of the senior company officials at ICMESA who had been arrested and charged by the government with criminal violations related to operating a dangerous factory.

In June 1983, the criminal trial resumed against five remaining company executives from ICMESA and Givaudan, and a decision was reached later that year. The defendants were convicted of negligence, failing to take safety measures, and causing an environmental disaster, with prison terms ranging from two and a half to five years. The defendants appealed the convictions, and in 1985, an appeals court in Milan dropped charges against three executives and reduced charges against the other two (from "intentional omission of providing safety measures" to "neglect of providing safety measures"), with prison terms suspended (*Chemical & Engineering News*).

In the 1980s, Givaudan paid settlements to various parties involved in the ICMESA disaster. The towns of Meda, Cesano Maderno, and Desio received payments to compensate for administrative costs. The town of Seveso, in hope of reaching a rapid settlement, filed its case in Switzerland. Most private individuals negotiated with Givaudan for out-of-court settlements for property damage. At the end of 1982, the company had paid 8.3 billion lire ($6.1 million) for damage to agriculture, 4.7 billion lire ($3.5 million) for damage to industries located in zones A and B, and 12 billion lire ($8.9 million) for damage to individuals, including land, furniture, houses, decontamination, medical and other costs (Hill and Knowlton International, p. 16). Yet in 1983, about three hundred persons, including over one hundred exemployees of ICMESA, persisted in civil litigation to obtain compensation for damages from Givaudan (*Repubblica*, 21 June 1983).

In the 1980s, reclamation and decontamination—pleasantly called *bonifica* in Italian—continued for dioxin-tainted areas around ICMESA. Two huge "basins," with special impenetrable seals, were constructed in zone A to hold contaminated materials from other areas. In late August 1982, technicians from Givaudan entered the ICMESA factory and at last emptied the remaining wastes from the chemical reactor responsible for the explosion. The highly toxic mixture, about 2.2 metric tons, filled forty-one special barrels designed for safe transportation of dangerous substances (such as radioactive waste). The barrels left ICMESA in trucks on 10 September, passed the Italian borders, and then became lost in Europe. The customs documents did not indicate the final destination of the wastes, for the relevant regulation for the European Community would not go into effect until three months later, in mid-December (*Settimana 3*, no. 58). The Lombardy regional president argued his responsibilities ended at the border, since the agreement with Givaudan made the company responsible for the wastes left in the chemical reactor. In May 1983, French authorities found the forty-one barrels in the slaughterhouse of a butcher in the small village of Anguilcourt-le-Sart (three hundred inhabitants); they had been stored there because the subcontractor for the wastes could not locate anyone willing to dispose of them. Givaudan and LaRoche agreed to repossess the barrels for proper high-temperature incineration in Switzerland (*Le Monde*, 21 May 1983). Back in Seveso, the huge basins in zone A continued to be filled. When completed, according to scientists from the Superior Institute of Health in Rome, "a natural park [sic] will be established" (Pocchiari, Silano, and Zapponi, p. 235).

Questions about health effects of the disaster persisted in the 1980s. The skin disease of chloracne appeared as the most striking health consequence of the ICMESA disaster, with most identified cases in the age range of five to nine years. Some biochemical indicators showed abnormalities; but as in other chloracne cases, the investigators concluded: "No clinically definable systemic disease has been diagnosed to date in children of the Seveso area" (Caramaschi et al., p. 142). Follow-up studies of birth defects showed that some increases occurred throughout the area in 1978 at rates greater than expected according to both national and international standards. Although no single striking malformation appeared and not all pregnancies were affected, even among dioxin-exposed women, some clustering was noted in place and time. The Central Scientific and Technical Committee, in its health review of 1980, concluded: "At the moment, it is not possible to establish whether the excess of malformations found is due to TCDD pollution or represents a random fluctuation" (cited in Silano, p. 193). One neurological report did demonstrate subclinical changes in the conduction of nerve impulses, which showed a correlation to dioxin exposure (Boeri et al.). On the other hand, an analysis of the health status of the cleanup workers for zone A (sections A1–A5) after the first two years (1980–82) "showed no case of overt clinical disease which could be attributed to TCDD (i.e. chloracne, porphyria cutanea tarda, peripheral neuropathy, liver disease)" (Assennato, p. 66). This analysis concluded that the preventive measures taken had effectively protected the workers from dioxin exposure and its consequences, although long-term effects remained in question.

In 1988, a team of Italian and United States scientists analyzed blood samples that had been taken from exposed persons between July and October 1976 and kept frozen since then in nearby Desio Hospital. The results showed the highest dioxin levels in blood ever reported in humans, with the top levels in children who showed severe chloracne (Mocarelli et al., 1988). The tests demonstrated a gradient in dioxin blood levels, with no detection in samples from the zone of respect (zone R), detectable levels in samples from zone A without chloracne, and with the highest levels in samples from zone A with chloracne. Some persons without chloracne nevertheless showed levels above those with chloracne. The small number of total samples (fifteen) and the age differences between those with chloracne (sixteen years and under) and those without (fifteen years and up) make it difficult to draw firm conclusions, but the preliminary report noted that none of the exposed persons showed significant abnormalities in the laboratory examinations carried out between 1976 and 1985 (Mocarelli et al., 1986).

In 1989, a group of epidemiologists published a preliminary analysis of mortality among people in the Seveso area, comparing populations in zones A, B, and R with those in surrounding towns (Bertazzi et al.). The scientists reported "some peculiarities" in the mortality of the dioxin-exposed groups (p. 1194), especially higher-than-expected cardiovascular deaths and deaths from cancer of the bile ducts (in females), cancer of the brain and leukemia (particularly in males), and lower-than-expected mortality from breast cancer. The study suggested that the elevated cardiovascular deaths might be attributed to increased stress of the population due to events around the ICMESA disaster. The study did not conclude that the elevated cancer risks were associated with dioxin. A major problem remained: residence in zone A, B, or R did not necessarily correlate with actual levels of dioxin exposure. Indeed, "data exist in the literature suggesting heavier involvement of some parts of zone R than previously thought, at least immediately after the accident" (p. 1198). The final official figures for residents of the zones were: 735 people in zone A, 4,700 people in zone B, and about 31,800 in zone R. The total area covered six municipalities, with about 220,000 residents.

ICMESA's dioxin did not create the health disaster feared at first, but long-term effects, such as increased incidence of cancer, may now be emerging. After years of debate and poor epidemiological work, a cancer registration system was finally set up, along with efforts to identify and follow up the individuals present around the ICMESA factory in July 1976. Problems in determining individual exposure levels to dioxin remain confused by the administrative boundaries that did not correspond to actual probabilities of exposure. The persistent confusion may obscure the detection of dioxin-related health consequences. Other consequences, however, have not passed unnoticed. The contamination wrought devastating disorder in the fabric of society, the stability of community and family, the psyche of individuals. Seveso will never be the same.

Chapter 5

Nonissue: Private Trouble

Each of the three chemical disasters began as a nonissue. This phase lasted until the toxic agent and the contamination became publicly identified. In Seveso, the phase as a nonissue was short because of the visible nature of the contamination and its source. In Michigan and Japan, the phase as a nonissue persisted because of the invisibility of the contamination and the responses of both social institutions and individual sufferers. In such cases, the social causes of a private trouble can remain masked for a long time. This chapter examines both individual and institutional responses to chemical disasters to present the phase of nonissue and the transition to public issue. It explores the kinds of power and knowledge possessed by social institutions and by victims to explain how a major social problem could be perceived as a nonissue and to identify the consequences of that perception for the victims' redress. A special concern is how social institutions exercised power to maintain the definition of nonissue.

The Concept of a Nonissue

Not all private troubles have social causes or are public issues waiting to happen. But in the area of toxic contamination, most problems have important political and social components; their solution raises questions about, or demands changes in, the structure of society. The existence of nonissues can be demonstrated by proving that objective problems are kept off a community's agenda, as Matthew A. Crenson showed for urban air pollution. The empirical validity of nonissues as a concept is also shown by the many political activist groups that seek out potential issues for projection into the public domain.

The study of nonissues raises some methodological questions. How can something that doesn't exist (a *non*issue) be examined? My answer has been to choose objective problems that developed into issues in both the public and political spheres and to trace the problems back historically to their origins, before they became issues. Not all nonissues evolve into the public and political spheres as issues, but it is reasonably safe to assume that all issues at some point were nonissues. (To use a biological analogy, not all eggs become chickens, but all chickens began as eggs.) Historical reconstruction of the origins of chemical disasters shows not only that nonissues exist but that the persistence of nonissues can increase the scale of both individual and social harms.

Do nonissues result from limits in the administrative capacity of political systems, which can only put a certain number of items on the agenda? Or do nonissues result from conscious and intentional efforts to shape the agenda to support vested interests? The debate over capacity versus intentionality has a long history in political science. In brief, some political scientists argue that nonissues exist because "the number of potential public issues far exceeds the capabilities of decision-making institutions to process them" (Cobb, Ross, and Ross, p. 126), requiring the development of filters or gatekeeping mechanisms to determine which troubles become issues (Easton, pp. 87–96). Others argue that those mechanisms are not accidental but are consciously designed to inhibit the creation of issues that would threaten the existing social order (Bachrach and Baratz, 1975). Social institutions have power in their ability to shape the agenda and keep some issues from finding a place on the agenda. This is what I called the second dimension of power in chapter 1. Crenson's empirical study of air pollution argued that the gap between objective reality and political agenda reflected "evidence of bias in the makeup of local agendas" and "politically imposed limitations upon the scope of decisionmaking" (Crenson, pp. 176, 178). But Crenson stressed "that undemocratic restrictions on the scope of local political activity are the products of indirect influence. They are not the result of suppressive acts or directly applied pressure but are responses to the power reputations of various local groups, organizations, and individuals" (p. 181).

The analysis of nonissues thus leads to basic questions about the unequal distribution of resources and power in society. Which issues are kept private—as nonissues—and with what consequences? (Crenson, p. 184). Which groups benefit when a social problem is perceived only as a private trouble? And which groups suffer? What changes might be adopted so that problems are not systematically kept off the agenda as nonissues? I return to these broader questions and their implications for democracy in the last chapter of this book.

The Nature of Toxic Contamination

A chemical disaster can begin with a bang, for example, a factory explosion or chemical spill, as happened in Seveso and in Bhopal, India. But toxic contamination can also follow an invisible and slow process, not immediately obvious to its victims or to social institutions, as in the silent poisoning of the food chain. The circumstances of contamination have important implications for the initial responses of social institutions and victims. Four key factors determine the nature of contamination: the invisibility of the toxic agent, the nonspecificity of toxic symptoms, the geographical distribution of victims, and the difficulties of identifying the causative substances.

Invisibility of the Toxic Agent

The less visible the toxic agent, the greater the difficulty in defining a problem as toxic contamination and in identifying the source of the chemical.

One cannot detect contamination by some chemicals in samples of food or air by looking with the naked eye or by tasting. High concentrations of a chemical may make something taste strange, while lower levels may be impossible to sense. A victim can be poisoned without knowing it. The fact and the path of chemical contamination remain hidden. This insensibility makes toxic contamination disquieting: it could happen to any one of us, without our suspecting it until too late.

This kind of "environmental invisibility" creates various forms of ambiguity and uncertainty for the persons exposed. They are unable to determine "(1) if they are in an environment in which the invisible contaminant is present, (2) whether a contaminant that is known to be present is actually being absorbed by the tissues of their bodies, (3) how large a dose of the contaminant they are absorbing, and (4) whether the absorbed dose is dangerous" (Vyner, 1988a, p. 14).

Environmental invisibility significantly delays the discovery of chemical contamination and of its source. In the Yusho case, over a period of months, consumers unknowingly used many liters of rice oil contaminated by PCBs. And in Michigan, Halbert ate feed pellets in an effort to find out what was wrong, but he tasted nothing out of the ordinary, just a mild bitterness (Halbert and Halbert, p. 27).

In sharp contrast, in Seveso, the white cloud from ICMESA provided something visible with which damage could be associated. The cloud created a short-lived but tangible connection between factory and home that could be followed by eye, logic, and past experience. The cloud's visibility facilitated a relatively rapid discovery of the contamination's source. Nevertheless, local residents still encountered other problems of environmental invisibility in assessing whether they were poisoned and if so, what it meant. Many residents knew from past experience that the factory could contaminate the environment in harmful ways. But they did not know the names of the chemicals, and they never imagined how harmful it could be.

The degree of visibility does not necessarily correlate with the degree of toxicity. The specific chemicals involved and the exposure dosage can dramatically alter the toxic effects. The toxic cloud at Bhopal, for example, turned out to be much more acutely lethal than the toxic cloud at Seveso. Visibility simplifies the ability of victims to identify the problem as toxic contamination and to locate the source of chemical exposure, but still leaves unanswered the many other questions of environmental invisibility.

Nonspecificity of Toxic Symptoms

The more nonspecific the toxic symptoms, the greater the difficulty in diagnosing toxic contamination.

Physical symptoms related to toxic chemicals frequently are nonspecific and mimic ordinary disease (Hardy, p. 1191). Such symptoms in humans include headaches, gastrointestinal disorders, acnelike skin problems, dizziness, wheezing, and reduced immunological resistance. A similar problem of diagnosing symptoms can occur with infectious diseases, representing a general difficulty of attributing meaning to particular images of illness (Mechanic). In general, multiple gaps arise between the individual's subjective experience of illness and the physician's efforts to identify a specific biomedical disease. This distinction between illness and disease is increasingly recognized as a key factor in explaining how individuals and social systems respond to ill health (Cott; Kleinman). The difficulties in diagnosing toxic contamination help delay the transition to public issue.

Some exceptions to the general rule of nonspecific toxic symptoms are worth noting. One infamous case is that of thalidomide, the drug that caused the specific and rare birth defect syndrome of phocomelia, an extreme shortening or complete lack of long bones in the limbs (Insight Team). But even with thalidomide, the dangers of the drug did not become a public issue until November 1961, about four years after sales began (p. 104). The problem became a public issue one month after a Hamburg pediatrician determined that the incidence of phocomelia in his city had increased from one case in twenty-five years to fifty cases in thirteen months (pp. 97–98). Other examples of toxic substances associated with specific symptoms (specific cancer sites, in these cases) are asbestos and mesothelioma, petroleum oils and scrotal cancer, and vinyl chloride and angiosarcoma of the liver. Such "signal" neoplasms, however, are rare (Selikoff, 1980, p. 581); and even with specific toxic symptoms, the transition to public issue can be delayed by other factors. Specificity of cause and effect is helpful in shortening the phase as nonissue, but it is not a sufficient condition to create a public issue.

In our three cases, the toxic symptoms resembled various known diseases, thereby frustrating individual efforts by victims to associate their symptoms with poisoning by a specific toxic chemical. Moreover, the symptoms did not respond to usually effective treatment and therapy. The failure of normal medical care made the symptoms more incomprehensible to the afflicted, more strange and extraordinary. These obstacles to understanding exacerbated the human tragedy of families and individuals and raised the psychological toll exacted by the illness. The nonspecificity of toxic symptoms confounded medical and scientific specialists consulted by the victims. The experts generally assumed normal medical disorders and rarely diagnosed toxic contamination. Many Yusho sufferers, for example, went from doctor to doctor, visiting internists, gynecologists, dermatologists. Only at Kyūshū University did dermatologists tentatively diagnose the symptoms as representing a specific disease, chloracne due to chemical contamination.

The failure of diagnosis in Japan reflected a blind spot in medical training: inadequate instruction on recognizing chemical disease, even for occupational settings (Kuratsune interview). In the United States, the great occupational health physicians Alice Hamilton and Harriet L. Hardy wrote in 1974, "Most medical schools still leave industrial medicine out of the curriculum and students graduate with more information about endocrine diseases than about lead poisoning" (p. 3). Inadequate training in occupational medicine continues as a significant gap in medical education (Peters, p. 277). Michigan dairy farmers, like Halbert, ran into similar problems with veterinarians, who knew much more about infectious disease than chemical disease.

The lack of a public health perspective also contributed to problems in diagnosing nonspecific symptoms as toxic contamination. Physicians tended to focus on individual cases, viewing the patient's disease as an isolated phenomenon, as a private trouble. Physicians generally lacked an epidemiological or system-oriented perspective. They tended not to consider individual cases as part of a population or to explore social causes for disease.

A public health perspective can lead to a correct diagnosis, even for nonspecific symptoms. For example, the poisoning of workers and their families by a pesticide called Kepone, in Hopewell, Virginia, was discovered by a young Taiwanese physician, Dr. Yi-nan Chou, who suspected that his patient's neurological symptoms were due to chemical exposure at work. He sent a blood sample to federal authorities for testing and notified them of his suspicions, which initiated a full investigation (Reich and Spong, p. 233). This case illustrates that the actions of an individual physician can make a difference in relating a patient's perceived illness of nonspecific symptoms to actual sources of toxic exposure. In the Yusho case, Dr. Gotō reportedly sought to alert local authorities about the cooking oil contamination but was unable to elicit a response from the officials. The diagnosis of chemical disease thus can be a critical step in moving a nonissue to the public realm, but it also depends on the responses of social institutions. A correct diagnosis alone is not sufficient.

Geographical Distribution of Victims

The greater the density and proximity of toxic sufferers, the greater the likelihood of discovering a common problem.

The geographical distribution of victims and of the toxic agent depends directly on the pathway of exposure, which in turn depends on technology and its use. Modern commercial systems, which can rapidly distribute defective products over wide geographical areas, create a substantial barrier to discovering a common problem and to tracing the problem to the source. On the other hand, when contaminants are distributed over a relatively confined geographical area, the concentration of victims can facilitate the processes of discovery and definition of the problem.

Defective products were spread over wide geographical areas in the Yusho and Michigan cases. Physical distances separating victims from each other produced social isolation for most sufferers until the problem became a public issue. The Yusho case represents one extreme of distribution, with people poisoned throughout western Japan, a lateral spread of about seven hundred kilometers (the distance from Boston to Washington, D.C.). Similarly in Michigan, the poisoning affected farmers throughout the state (21,937 square kilometers), although most lived in the Lower Peninsula (about 10,000 square kilometers). In both cases, most victims did not already know each other, which impeded their ability to recognize and define their common problem. The case of Seveso presents a contrasting example. Toxic contaminants were spread over a relatively limited area by the ICMESA factory explosion. Physical proximity—along with the cloud's visibility—hastened recognition of a common plight. The cloud settled on several contiguous communities. People did not have to travel for hours to find fellow sufferers. They needed only to look around, to walk next door, to speak to their neighbors.

Defective products can also be distributed in ways that create clusters of victims. In the Yusho case, for example, Kunitake found fellow sufferers in his previous apartment complex, where other workers from his company continued to live. As at Seveso, the overlap between existing communities and the new community of sufferers hastened the discovery of toxic contamination and the transition from private trouble to public issue.

Difficulties of Identification

The less information available about the contamination, the more difficult the search for the identity of the contaminant.

Even when a victim suspects chemical contamination and locates a possible source, identification of the specific chemical can be difficult for technical, financial, and organizational reasons. The victim needs assistance from scientific experts. Yet, as shown in our cases, experts often have no simple answers. A scientist from the U.S. Food and Drug Administration spelled out the problem: "You can't just put your contaminated substance in a machine and have it come out with the answer. You must search. Too many [toxic] chemicals are found by the process of elimination. It depends on serendipity and human error" (Cordle interview).

In considering an estimated sixty-three thousand chemicals in common use (Maugh) as possible contaminants, the chemical detective must use sensitive and selective methods to identify unknown contaminants. One report in 1978 estimated that it would cost about $2 million to purchase the equipment necessary to identify and measure unknown organic, metallic, and radioactive contaminants (U.S. Office of Technology Assessment, p. 104). "The identification process is often accomplished by comparison of the physical and chemical properties of the unknown compound against the same properties of an authentic standard compound. For complete unknowns, the identification process can be very difficult" (Laseter, p. 187). The process can also become extremely expensive, with tests costing from ten thousand to one hundred thousand dollars, and sometimes more, to identify a single uncharacterized peak in gas chromatography (U.S. Office of Technology Assessment, p. 97).

The search to identify an unknown contaminant thus does not depend simply on "serendipity and human error." In two of our cases (Yusho and Michigan), both governments and companies lacked the resources—equipment, finances, and personnel—and the strategies necessary for an effective and timely search. The result was a prolonged phase as a nonissue. In Seveso, however, after a public official took the initiative to enter the factory and inquire about the production process, he was able to identify the probable contaminant with remarkable speed. This contributed to shortening the phase as a nonissue. In all cases, the scientific process of search depended not only on the nature of contamination but also on social and institutional factors.

Social Institutions and Nonissues

Private institutions have the greatest potential to discover, understand, and resolve the problems of toxic contamination. They have much greater resources than individual victims and often possess or have easy access to information about the causes and effects of the toxic contaminants. Yet, they often respond initially with denials, stonewalling, and evasions, thereby impeding public recognition of a social problem. These responses reflect the general tendency of corporations to react to claims about problems with "surprise, frustration, and antagonism," seeking "either to deflect these claims or to confront them head on with jaws set" (Schacht and Powers, p. 25). Public institutions likewise often respond to nonissues in ways that are inadequate and inappropriate. Individual sufferers of toxic contamination may turn first to public institutions for assistance only to find that these institutions, like private corporations, help to maintain the definition of nonissue and delay the transition to public issue. Difficulties arise from organizational problems specific to chemical contamination as well as from more general bureaucratic pathologies.

Many of the problems of public institutions resemble those affecting private institutions, as recognized by a growing number of theorists and as shown by empirical studies (Gross, 1980; Murray; Stone, 1980; Roberts and Bluhm). The similarities between private and public institutions appear in both organizational processes and substantive responses. To stress the parallel problems, I examine both private and public institutions with the same set of analytic categories. I argue that social institutions have four basic difficulties that contribute to causing chemical disasters and delaying public recognition: inadequate detection systems, inadequate understanding of the problem, inadequate internal communication systems, and inadequate external coordination systems.

These analytic categories reflect an emphasis on information, especially on problems in the collection, interpretation, and flow of information. But the problems are not simply cognitive difficulties in understanding the events. The organizational interests of both private and public institutions affect their responses to early reports of toxic contamination. The corporate responses are affected by financial and organizational interests in protecting profits, growth, and survival. And public bureaucracies have their own interests and constituencies that create problems in detection, understanding, and communication, especially when the interests of public agencies overlap with those of the responsible private organizations. Analysis of these four factors helps explain how social institutions prolonged the phase of nonissue in our cases and how they structured the initial responses of the sufferers of toxic contamination.

Detection Systems

The more inadequate the detection systems, the more likely that the phase of nonissue will be prolonged, as private institutions remain unaware of their effects on society and public institutions depend on private companies for information about problems.

Private institutions are structured to monitor only certain social effects, those effects most likely to influence narrow organizational interests. No organization can monitor all possible external effects; some choice is therefore necessary, based on likelihood of occurrence, probable costs or benefits, or ease of measurement. Companies tend to focus their information gathering "on such matters as industry sales volumes, capital costs, market shifts, and competitor behavior. Their information nets are simply not designed to haul in other, 'softer' data, less immediately relevant to profits, growth, and prestige" (Stone, 1975, p. 202). Corporations monitor social data they know from experience will affect their performance. The measures used for detecting problems are known as "scanning mechanisms" in organizational theory.

Apart from monitoring standard kinds of information with scanning mechanisms, corporations usually wait for problems in production or consumption to come to them. Complaints serve as a feedback mechanism on which producers depend for information. In the language of organization theorists, corporations depend on a "motivated" search, which means a search is stimulated by a problem and depressed by a solution (Cyert and March, p. 121). But a narrowly defined problem can result in an inappropriate solution. The specific complaint may be addressed, while the underlying problem continues to worsen.

Inadequacies in the detection systems of our three cases existed first at the level of production. Failures of information about production helped cause the contaminations, since they reduced corporate abilities to detect a problem and prevent its worsening. Both the Kanemi Company and Farm Bureau Services kept quality-control samples of their products, but did not normally subject them to tests for toxicity, so that the detection systems in place could not identify the problems that occurred. The two American companies, Farm Bureau Services and Michigan Chemical, lacked accurate inventory systems to monitor usage of toxic chemicals, signal potential dangers, and prevent possible mix-ups. ICMESA/Givaudan maintained a quality-control system for dioxin levels in the production of trichlorophenol, but the company did not closely monitor temperature or install an automatic cooling system in the production process. These inadequate scanning mechanisms contributed to the inability to detect potentially catastrophic problems at an early stage, which in turn contributed to the persistence of the contamination as a nonissue and to the creation of the chemical disaster.

Inadequacies in the detection systems existed also at the level of consumption. The corporations depended on others to discover problems and contact the company. That passive mechanism delayed discovery and prolonged the phase of nonissue. The Kanemi Company did not detect problems with its dark oil until it received complaints about chicken deaths, and then did not attempt to determine possible problems with its cooking oil, waiting instead until such information arose independently. Farm Bureau Services, after Halbert's first complaints, did not initially contact other users of the same feed to determine if similar problems existed. ICMESA/Givaudan tested dioxin levels in trichlorophenol but did not test contamination levels outside the factory on a regular basis.

All the companies in our cases, except Farm Bureau Services, possessed some information about the hazardous nature of the chemicals that gave rise to disaster. Kanegafuchi Chemical, Michigan Chemical, and ICMESA/Givaudan knew they were dealing with toxic chemicals, including some substances with potentially serious human health consequences. The corporate detection systems, however, were not designed with adequate precautions to avoid gross contamination incidents. Even the Kanemi Company received some information from Kanegafuchi Chemical about the toxicity of Kanechlor; and managers and workers in the cooking oil production unit knew that the law prohibited any foreign substances in goods intended for human consumption. The question of whether the Kanemi Company received sufficient information from Kanegafuchi Chemical became a major issue in the complex litigation around the poisoning.

Cost considerations generally constrain the design of corporate detection systems to problems considered likely to occur in production or consumption. Information-gathering systems cost money, and expenditures must be justified within for-profit organizations. Even in the imperfect markets of the real world, competition does occur between companies and between managers within a single firm, serving to reduce information collection in order to prevent cost inflation (Wolf, p. 76). Because of the ongoing expenses, companies do not monitor the consumption of all products. As a result, opportunities are missed to detect and prevent low-probability, high-cost events, such as chemical disasters.

In all of our cases, however, companies detected—even without external complaints—that a problem of toxic contamination had occurred. In the ICMESA case, the company did not need a sophisticated detection system to determine that an explosion at the trichlorophenol reactor had sent forth a chemical cloud from the factory. In the Kanemi Company, some managers knew, without a complicated detection system, that a substantial amount of heat-transfer agent had leaked into the cooking oil. In these two cases, even crude detection systems identified the existence of problems with toxic contamination. In Michigan, employees of Farm Bureau Services reported some unfamiliar product names in magnesium oxide shipments, which could have served as a warning signal. As long as these detection systems were contained to the private sphere, as non-issues, the "motivated" searches by corporations were neither highly motivated nor particularly effective. Corporate responses to the available information remained inadequate, as discussed below in the section on understanding.

The government agencies in the three cases also had detection systems that succeeded to a degree. Their systems identified private companies that had repeatedly or flagrantly violated health and safety laws. Those violations, however, did not result in measures to prevent the related problems that led to disasters. These failures to prevent disasters reflect the difficulties that public institutions confront in detecting high-risk industrial processes and in compelling changes in corporate behavior. The failures in detection contributed to maintaining the chemical disasters as nonissues.

In Michigan, the U.S. Food and Drug Administration had cited both Farm Bureau Services and Michigan Chemical several times prior to the PBB disaster for cross-contamination, improper labeling, and sloppy procedures in production (U.S. General Accounting Office, pp. 3, 8). Despite the pattern of problems at Michigan Chemical, the FDA later maintained, "The violations encountered in the late 1960s were not the same as those related to the [PBB] incident. . . . Thus it would not have been possible to predict and prevent the PBB mix up merely on the basis of the earlier unrelated violations" (FDA, 1977). The FDA sought to defend its inability to "predict" the PBB mix-up because the earlier problems were not directly related to PBBs. Yet, greater government intervention in the company's production and labeling activities might have changed overall corporate procedures and thereby helped to "prevent" the chemical disaster. Increased surveillance of problem companies (those with multiple violations) might help identify and correct problems that contribute to causing chemical disasters.

The public authorities in Italy also had identified the ICMESA factory as a problem facility with multiple citations. In the fall of 1976, the province of Milan published 157 pages of documents that showed continual and unsuccessful efforts by provincial officials over twenty years to compel ICMESA to control its wastes. The history was punctuated by repeated violations, delays in treating waste water, and possible air pollution. But the province concluded, defensively, that its trained technical staff knew nothing about ICMESA's production of trichlorophenol (Provincia di Milano, p. 8). The province not so subtly implied that it should not be blamed for ICMESA's dioxin cloud. The Parliamentary Commission recognized the province's efforts to control ICMESA's pollution, but concluded that "the provincial administration did not avail itself of all powers conferred upon it by current regulation and laws on water and air pollution" (*Commissione Parlamentare*, p. 87). The province had detected problems with ICMESA, but those problems remained a nonissue, out of the public and political domains.

Even the detection of gross violations, like Kanemi's dark oil problem and mass killing of chickens in the spring of 1968, did not result in increased governmental monitoring of the private firm. Discovery of the toxic contamination of one product did not lead public officials to detect (or even suspect) problems with related products. More troubling is the fact that detection of a problem may not even produce a solution for that problem. The difficulties of enforcing regulations and obtaining improvements are well known. Detection of a problem, according to one former United States environmental official, is only the first step: "Regulatory law enforcement, from the time a violation is detected onward, is a mess. If an agency is lucky enough to detect a violation, it is often able to do little more. If jawboning fails to produce compliance, regulators must either give up or litigate, and litigation is uncertain, slow, and costly" (Drayton, p. 1).

In our cases, middle-level governments (states in the United States, regions and provinces in Italy, and prefectures in Japan) often possessed structures for the detection and control of chemical contamination. But those structures rarely functioned as effective detection systems. These public administrations lacked systems of sufficient complexity to comprehend, much less to control, the production processes of modern industry. The capacities of such public systems suffered from problems in personnel, equipment, funding, and design. At the national level, public administrations may have had enough technical expertise to detect problems, but they lacked jurisdiction over local areas or they lacked the ability to monitor the activities of private manufacturing firms.

The administrative efforts at pollution control in Italy, for example, have been restricted "within the limits fixed by law and the statutory regulations, by chronic financial difficulties, and by a lack of specialized personnel, suitable equipment, specific research and standardised methods for measuring data" (dell'Anno, p. xiv). In the Seveso case, the Lombardy Regional Committee against Air Pollution, an official body, received a report in March 1975 from ICMESA, which included chemical products manufactured by the company. Among them was trichlorophenol (ICMESA). The committee's technical staff, however, did not recognize that production process as potentially dangerous until after the explosion. The committee in Lombardy, as in other regions, worked under strict limitations of funding and personnel, especially an inability to provide salaries competitive with private industry (Conti, p. 183). Since the central government does not provide operating expenses for these air pollution committees, funds "must be obtained from local authorities, which are already burdened by considerable commitments and are nearly all in deficit" (dell'Anno, p. 21).

Michigan, after an environmental crisis with mercury in 1972, created a Governor's Committee for Health Implications of Pollutants, an interdepartmental clearinghouse on pollutants, with the purpose of detecting and responding to potential hazards quickly. This structure, however, never sounded the alarm on PBBs until after the problem became a public issue and never served to coordinate agencies involved in PBB contamination (Climo). The failure of that clearinghouse demonstrates general problems of interagency committees, discussed below in the section on external coordination. It also shows the failure of a detection system when confronted with an unanticipated contaminant. One is reminded of Gold's principle: "Nothing succeeds as planned" (Heller).

Even if public administrators recognize the need for more resources to detect chemical problems, they cannot easily obtain increased funding from the legislature or central government without a clearly demonstrated need—which often means a crisis. The Michigan Department of Public Health, for example, increased its chemical detection capabilities through a series of environmental crises with DDT, mercury, PCBs, and PBBs (Humphrey interview). Paradoxically, administrators *need* crises in order to expand the resources and capabilities of detection (and other) systems and thereby improve the probability of preventing other crises. This need for crisis is related to a broader "learning curve," as both organizations and individuals increase their ability to recognize and respond to chemical problems. It represents more general patterns of institutional and individual adaptation to changes in the environment and to new problems.

The inadequate detection systems made public authorities dependent on private institutions for information about potential problems. In all three cases, public officials identified the withholding or misrepresentation of information by private companies as a major factor that delayed the responses of public institutions. The asymmetry in information between public and private institutions represents a general problem in the control of toxic chemicals. As Doniger wrote, "Often the parties with the most accurate information on the costs and effects of regulation have an incentive to withhold or misrepresent that information" (p. 18). When public authorities come to depend on private companies for information and guidance, the company's private interest in viewing the problem as limited can be transferred to the public sphere. This point demonstrates an important way that structural conditions can determine the direction of policy.

Finally, the detection of "minor" problems results in only "minor" responses. In public bureaucracies, officials tend to respond to signs of problems by taking the least steps down the path of least resistance (policy by least steps) and by allocating the burden of uncertainty to those who seek change. As shown in another study of environmental policy, an episodic crisis, something called a "disaster," often is necessary to produce a significant response from public agencies (Krier and Ursin, pp. 251–77).

For the victims of our three chemical disasters, the failure of public institutions to detect the toxic problems had devastating consequences. Undetected problems persisted and worsened. Given the incentives of private companies not to notify public officials, the victims themselves often had to prod and pressure public institutions to recognize the problems. But even when a public agency detected a problem, it did not immediately understand the problem or its consequences.

Understanding of the Problem

The more inadequate the understanding of a problem, the greater the tendency to respond with routine and inappropriate procedures and thereby prolong the phase of nonissue.

Once organizations recognize that a problem exists, they begin to search for an explanation and a resolution. That search and response process is affected by a tendency for organizations to apply simple explanations and to respond with routines. Organizations search for "simple-minded" explanations, beginning with a simple model of causality until driven to a more complex one (Cyert and March, p. 121). Organizations then respond with routines, or habitual actions, even if changed conditions make the usual response inappropriate (Simon, p. 89).

Inadequate understanding also results from other deficiencies, especially inadequate detection and internal communication systems. Indeed, not having enough information and not circulating that information reinforce simplistic understanding and simple responses. In such circumstances, private and public institutions tend to view a problem in isolation and not as part of larger systems of production and consumption. As a result, organizations experience what Ansoff called a "lag response" before they understand the full strategic consequences of information that shows the existence of a problem. The typical response is for corporate management first to change its operations, next to reorganize in some way, and finally, when the true cause of a problem is understood, to shift its attention to strategy (Ansoff, p. 25). In our three cases, the Kanemi Company showed the longest lag between first evidence and final diagnosis of the problem; ICMESA/Givaudan showed the shortest.

At the Kanemi Company, management prolonged the phase of non-issue through concerted efforts at cover-up. Company officials actively denied: the repeated indications of contamination, the disappearance of large quantities of Kanechlor in early 1968, the discovery of the chicken deaths in March of that year, evidence of some toxic agent in dark oil in June, and finally, reports of human illness associated with cooking oil in October 1968. Throughout this period, company officials made every effort to conceal their actions and took only minimal steps to determine if a problem existed in its cooking oil. The lag period for the Kanemi Company lasted about nine months.

Farm Bureau Services responded more appropriately to reports of problems during the phase of nonissue. It stopped production of Ration 402 pellets in December 1973, several months after Halbert's first complaints, and recalled the pellets in January 1974, after the evidence of repeated experiments with mice. The company also contracted for feeding tests with cattle and for chemical analysis of the suspected feed. But identification of the contamination was not complete until late April 1974, partly because of the company's strategy of analysis. Farm Bureau Services concentrated its efforts on looking for normal contaminants, such as pesticides and heavy metals. The company apparently did not check Halbert's feed to determine if all the expected components had in fact been added (*Michigan Farm News*, 1 July 1974). Such an analysis would have indicated the absence of magnesium oxide in the required amount, showing that no magnesium oxide had been added and suggesting a search for a nonnormal contaminant that might have been added instead. The lag period for Farm Bureau Services lasted about eight months.

ICMESA/Givaudan management responded most quickly to the chemical disaster, closing off the trichlorophenol production area in the factory and taking samples of possibly contaminated materials outside the factory. Almost immediately, the company's management understood the likelihood of dioxin contamination. Yet company officials did not inform Italian authorities about dioxin until ten days later. It should be noted, moreover, that ICMESA managers did not use information they possessed prior to the explosion to reduce the likelihood of an accident by installing automatic cooling equipment or to reduce the consequences of a possible accident by constructing an overflow tank.

How can we explain the differing lag responses of the three companies? A major factor was the quality of information available to the managements and to outsiders. The sudden explosion at ICMESA provided visible evidence of a problem in the factory and in the surrounding community, something that management could hardly ignore or deny to local residents. The feed problem at Farm Bureau Services, on the other hand, appeared more ambiguously, with no immediately obvious explanation of how the feed had become poisonous. At the Kanemi Company, the contaminant was less visible than at ICMESA but also less ambiguous than at Farm Bureau Services, especially to Kanemi managers who knew that a leak of heat-transfer agent had occurred. This suggests that quality of information alone did not determine corporate responses.

Corporate resources to understand chemical contamination also affected the length of lag response. Although larger corporations in the case studies possessed more appropriate resources, one can imagine a large corporation with inadequate technical expertise, as well as a small company with highly specialized resources for analyzing chemical contamination.

Among the companies considered here, ICMESA/Givaudan, with Givaudan's chemical research laboratories (including highly trained personnel, advanced equipment, and funding), had the greatest potential to understand the consequences of the factory explosion and possible dioxin contamination. They also had the easiest job: Givaudan company managers and scientists knew from previous industrial disasters with trichlorophenol production that dioxin contamination had probably occurred. Farm Bureau Services, with less technical resources, knew they had a problem by January 1974 and made various efforts to understand the contamination puzzle (mainly by contracting outside specialists), but they lacked an effective strategy or prior experience that could have significantly reduced the period of search. Kanemi Company officials, with the least technical resources available, never seemed to comprehend the potential consequences of chemical contamination and made little effort to analyze the contamination puzzle. The Kanemi Company lacked adequate information about the chemical used as a heat-transfer agent and about the possible health consequences of leaks.

The lag response was also affected by the relationship between the producers and consumers of industrial chemicals. The Kanemi Company and Farm Bureau Services had both purchased materials from other firms, which hampered the user's ability to understand. Only in the Italian case did a single company, ICMESA/Givaudan, serve as both supplier and user, and was itself a chemical company. The more prepared a company was to deal with chemical contamination, the sooner the puzzle became solved and the shorter the lag period.

The lag response was critically affected by corporate business interests. Even after a company understood the problem, the organization still sought to keep the solution private, thereby prolonging the phase of nonissue. An emphasis on the capacity to process information can only partly explain the lag response of private corporations. The role of business interests delayed the transition from nonissue to public issue, as corporations sought to contain the chemical problem as a private trouble.

In all three cases, the companies sought to continue business as usual. Production came first and did not stop until outside forces compelled a halt. Even at ICMESA, managers did not cease operations in other parts of the factory until the union shut down production, an action soon supported by official government measures. At the Kanemi Company, the impulse for corporate survival and for continued production motivated the management's persistent denial of any problems, which continued until the problem became a public issue and government temporarily halted operations. Farm Bureau Services continued its normal feed production and sales operations, even when it realized serious feed-related problems existed with one product and some customers. Business interests produced an unwillingness to acknowledge a problem and thereby promoted ineffective responses. Those responses reduced corporate costs in the short run but increased costs in the long run.

Organizational theorists Cyert and March stated that private firms faced with a problem consider alternatives sequentially: "The first satisfactory alternative evoked is accepted. Where an existing policy satisfies the goals, there is little search for alternatives. When failure occurs, search is intensified." In general, "the organization seeks to avoid uncertainty by following regular procedures" and does not predict problems but reacts to feedback (p. 113). The organization therefore has a tendency to "move from one crisis to another" while relying on standard operating procedures to make decisions (p. 102). These routines "permit the organizations to deal more effectively with previously experienced situations than could an individual considering the situation without experience, but they normally retard adjustment to strikingly different situations" (p. 107). Similar patterns also hold for public bureaucracies.

In our three cases, public authorities relied on routine procedures because they lacked alternative programs, especially diagnostic facilities and emergency plans for chemical contamination. The structures and strategies of public institutions contributed to inadequate understanding of the problem and to inappropriate responses.

In Japan, for example, the Ministry of Agriculture and Forestry in 1968 employed no research scientist specializing full-time in toxicology. Dr. Kohanawa, the veterinarian who performed the dark oil experiment, had studied some toxicology, but his specialty was isotopes; he knew little about chickens; and he lacked sufficient research funding, analytical equipment, and personnel. Until 1970, the National Animal Health Research Laboratory, to which he was attached, lacked a separate department for toxicology, and Dr. Kohanawa worked on poisoning problems as a hobby (Kohanawa interview). Given the lack of support systems, it is remarkable he was able to carry out the experiment at all.

In Michigan as well, the laboratory that performed the initial tests on Halbert's feed lacked the capability to identify or isolate a toxic agent. The veterinarian who tested Halbert's feed on mice explained, "We ran every test we could here. . . . This is 99.9% a regulatory laboratory. We usually check to see that the percentage of fat in hamburger is what it should be and that there is no added preservative. Only once in a great while we run a test to see if something is toxic." But he added, "Once the mice died, we were quite sure there was something in the feed" (Gatzmeyer interview).

Halbert encountered similar problems with officials at the U.S. Food and Drug Administration. "Their attitude was that they were not a research agency but a regulatory agency, and unless we knew what was wrong, there was little they could do to help us" (Halbert, p. 40). The farmer-scientist concluded, "In dealing with a new, unidentified contamination, one discovers that there is no existing mechanism for new problems to enter the agency structure—quickly. Instead one encounters the attitude that every disease and possible form of contamination has already been discovered and is adequately dealt with in the agency's 'cook book.' If the problem does not occupy a page in the 'cook book,' it does not exist" (p. 99).

In 1976, the Lombardy region possessed few technical resources in the area of chemical contamination. The province of Milan had more expertise in chemical questions, but it took some time before the provincial expert was notified of the events in Seveso. He quickly diagnosed the problem not by analyzing samples of contaminated soil for some unknown chemical, but by logic; he studied the factory's industrial processes and possible contaminants that could be produced. His strategy for diagnosis significantly shortened the process of organizational search. The example demonstrates the important role that an innovative and motivated individual can play in defining a problem as a public issue.

Most professionals within the public institutions, however, tended to respond with routine procedures. Veterinarians in Japan and Michigan, for example, responded routinely by initially considering infectious diseases. Next, scientists in public agencies searched for residues of known agricultural chemicals and heavy metals, the "normal" contaminants. The public authorities tended to look for what they knew how to detect, rather than analyze the problem in a systematic way.

Public authorities in Seveso also responded initially in routine ways. The local health officer sent word of the problem to Milan by mail, requiring five days to travel the twenty kilometers, rather than using the telephone. Seveso's mayor, who lives in a well-to-do neighborhood slightly separated from the main part of town, did not visit the affected district until nearly one week after the cloud, and then only at the urging of a journalist, Mario Galimberti. The Lombardy Region at first treated the problem at Seveso as serious but not requiring extraordinary measures like evacuation. That perception changed only through the insistence of LaRoche's representative, Dr. Reggiani, and the persuasion of his maps and the letter from Givaudan's managing director.

The resources and responses of public systems were designed for common accidents, not for the low-probability, high-risk chemical disasters that occurred. Indeed, one could not expect public agencies to be fully prepared with material resources for the unexpected; the costs would be prohibitive. But public institutions could design better strategies, within resource constraints, for identifying and understanding potentially catastrophic chemical problems.

The inadequate understanding of the problem by public institutions produced inappropriate measures and prolonged the phase of nonissue. Especially in the Yusho and Michigan cases, the authorities viewed the problem as a private trouble that did not require active intervention by the government. At Seveso, the authorities responded more actively, in part because of the nature of contamination, but they did not comprehend the implications of dioxin contamination until two weeks after the explosion. The following observation about the United States applies also to Italy and Japan: "In general, local level organized responses to chemical disasters are not as good as those to nonchemical disasters. Responses tend to be more delayed, uncertain and erratic; take place with little knowledge and understanding about the situation; and are often fragmented and uncoordinated, especially in the early phases. . . . Overall community preparedness, specifically for chemical disasters, is either nonexistent, poorly developed or merely a paper plan in most communities" (Disaster Research Center, p. 4).

Internal Communication Systems

Within organizations, the greater the gap between potential problem discoverers and problem resolvers, the greater the difficulties in discovering, defining, and responding to toxic contamination.

Private and public institutions have a segmented structure, which distributes information and power unevenly into specialized units but which must be coordinated for a unified organizational effort. Organization theorists Paul R. Lawrence and Jay W. Lorsch called these two components "differentiation" (into specialized units) and "integration" (into the corporate whole). Different problems and conflicts require different combinations of differentiation and integration within the organization (pp. 8–13).

Within private corporations, the structural gaps between workers and managers can contribute to delaying the discovery and response to toxic contamination. The gaps retard the flow of information between potential discoverers of problems, usually workers, and mandated resolvers of problems, usually managers. These gaps between workers and managers reflect the hierarchical organization of the workplace. Management tries to control the workplace partly by controlling information (Coye). Restricting the flow of information about toxic chemicals, however, can undermine the potential for workers to serve as a first line of prevention against disasters and obstruct workers' ability to protect their own health and safety. "The goals of occupational safety and health are not adequately served if employers do not fully share the available information on toxic materials and harmful physical agents with employees. Until now, lack of this information has too often meant that occupational diseases and methods for reducing exposure have been ignored and employees have been unable to protect themselves or obtain adequate information from their employers" (U.S. Occupational Safety and Health Administration).

In our three cases, workers could have acted to prevent or contain the chemical disasters if managers had provided workers with more information about hazardous chemicals. Workers did not know the names of chemicals they made or used, or the possible health consequences of products and by-products. At ICMESA, workers knew that batches of trichlorophenol required testing in Switzerland, but only after the explosion did they learn the name of the contaminant and its high toxicity (Argiuolo interview). Workers at the Kanemi Company were uninformed about the chemical composition or toxicity of Kanechlor and its consequences for animals or humans. Employees at Farm Bureau Services lacked adequate job training in the commercial names and technical components of different feeds. The workers' lack of information contributed to the failure of prevention, the delayed discovery of problems, and the prolonged transition from nonissue to public issue.

In addition, information about potential problems did not effectively move up the corporate ladder. According to court depositions, employees at Farm Bureau Services noticed and reported the appearance of a new trade name in the warehouse, Firemaster instead of Nutrimaster; the supervisor responded that it was just another name for magnesium oxide and to keep adding it as required (Courter and Lehnert, p. 12). It is uncertain whether the Kanemi Company's president knew about the contamination of cooking oil at the time it occurred; information about the Kanechlor spill may have been contained at a lower level in the company.

ICMESA/Givaudan transmitted information up to management more effectively than did the companies in the other cases. Its ability to respond came partly from the nature of the contamination—a clear chemical release into the external environment. But the response also came from the plant managers' awareness of the likelihood and potential consequences of dioxin contamination and from the centralized corporate structure of a chemical company and its subsidiary. Even with the communication that occurred inside ICMESA/Givaudan, however, the corporation still needed external pressure for the problem to become a public issue.

In corporations, there exists "a natural tendency for 'bad news' of any sort not to rise to the top" (Stone, 1975, p. 45). Information must be filtered and selected as it moves up the organizational ladder, and most corporations lack incentives to pass up bad news, especially with an uncertain interpretation. But a crisis, especially when visible and public, requires that someone bear the bad news to top management.

The internal communication systems of public institutions also contribute to the persistence of nonissues and the delay in recognizing social problems. The structure of a public institution critically affects what research is done, how research is interpreted, and who learns the results. When research is carried out within a single bureaucracy, "it is quite likely that in anticipation of adverse reactions from other parts of the agency, research objectives would be structured so that potentially embarrassing lines of inquiry were not pursued at all" (Ackerman et al., p. 149). On the other hand, when the "thinking" and "acting" components are placed in separate agencies, research can become less focused on practical questions of implementable policy and less concerned with follow-up study (pp. 148–49).

Internal communication problems in public agencies occurred in both the Michigan and Yusho cases. In both incidents, an experiment was designed to answer a simple question: Was a specific feed toxic to animals? Positive responses in both cases, however, produced different agency reactions, because of the ways in which the question was asked and because of the agencies' structures for "thinking" and "acting."

In Japan, the experiment occurred within a single bureaucracy, with a direct connection between the "thinking" and "acting" components, but the question was asked in such a manner that "potentially embarrassing lines of inquiry were not pursued at all." The experiment was part of an official inquiry by the Ministry of Agriculture and Forestries on the quality of the feed and its additive, the dark oil by-product from rice-bran processing. The research was integrated into broader policy questions of the ministry, resulting in official action on the feed and on dark oil. But the scientist who carried out the experiment did so outside his normal duties. Both he and upper-level bureaucrats asked only limited questions about the feed and arrived at limited conclusions. They narrowly interpreted the results as a feed problem, without viewing it as part of a complex system in which feed for animals was connected to other industrial processes and products, and ultimately to food for humans. Contamination of the cooking oil remained a nonissue.

In Michigan, gaps between the "thinking" and "acting" components meant that no official conclusions were drawn from the experiments, even for the feed. The feed experiments on mice resulted from an individual request by Halbert and not from an official inquiry by the Department of Agriculture. The experiment was not integrated into the agency's policy. The scientist who demonstrated the feed's toxicity did not consider the problem as part of his normal duties or of a larger system. Detection and limited understanding of the problem did not result in any official action.

As far as can be determined, higher officials in the Department of Agriculture were not formally notified about the feed experiments or their conclusions. Internal communication problems within the department contributed to keeping the feed problem as a nonissue until the mass media made it a public issue.

In the Seveso case, internal communication between different sections of the public health bureaucracy worked slowly but relatively effectively. The local health officer's letter to the provincial authorities followed normal bureaucratic procedures. The director of the Milan Provincial Laboratory for Hygiene and Prophylaxis, who received the letter, proceeded to collect additional information, visit the town and the ICMESA factory, and identify the production process responsible. Back in his office, the director reviewed the scientific literature to identify dioxin as a likely chemical contaminant. He then confirmed possible dioxin contamination with factory officials, who admitted it "with a certain reticence" (Cavallaro, p. 225). In sharp contrast to the other two cases, the government scientist not only confirmed the existence of a problem but also named the chemical contaminant.

Internal communication in public agencies worked better in the ICMESA case, in part because of the nature of the toxic contamination and in part because of the nature of the public agencies. By the time the provincial authorities in Milan became involved, animal deaths, plant damage, and human illness had clearly appeared in Seveso, and the chemical disaster had already surfaced as a front-page story in newspapers. Moreover, the provincial official in Milan was formally responsible for the protection of public health, and he actively pursued the problem and informed other officials of his findings. In contrast, in the Yusho and Michigan cases, the public agencies responsible were agriculture bureaucracies not principally concerned with public health; the feed experiments occurred while the problems were still nonissues; and the scientists who carried out the experiments were not primarily responsible for those tasks as part of their normal duties.

The incentive structure within an organization thus can reinforce the internal gaps between the people who understand a problem and the people who can act on the problem. In both private companies and public agencies, bureaucratic specialization produces fragmentation, which is supported by incentives within an organization. "As part of the division of labor inherent in every bureau, each official's incentives become focused mainly upon how well he does his specialized job. Any contributions he makes to other sets of tasks are considered of secondary importance in assessing his rewards and punishments" (Downs, pp. 104–5).

External Communication and Coordination

The greater the problems of external communication and coordination between private and public institutions and among different public agencies, the longer the problem remains a nonissue and the more inappropriate the responses.

For private corporations, inadequate external communication is probably the most serious flaw with regard to toxic contamination. Limited external communication by private companies deprives public agencies and toxic victims of accurate information about chemical contamination. The withholding of information by companies contains the problem as a nonissue and delays efforts to provide redress to the victims.

One business executive wrote succinctly, "Disclosure, in the context of corporate operations, has historically been a peripheral consideration to a company's basic business activities. First the company acted; then its actions were either disclosed or not disclosed" (Clausen, p. 63). The goals of private corporations to protect profits and growth and to limit financial liability encourage a conservative response to perceived threats. Managers do not want to provide legitimacy for possible threats to the interests of the organization. Corporation lawyers enhance the fear of releasing information that might open the way to financial liability, a policy that supports a reactive approach to problems.

Corporations also seek to avoid government interference in the organization and therefore resist the disclosure of information not already possessed by public authorities. This separation of public and private is supported ideologically by the neoclassical model of the firm. In neoclassical theory, the market provides adequate control of the firm, through perfect competition and information, and the consumer exercises control over firm behavior through choice in the market. Government intervention therefore is not necessary, because the market creates adequate public accountability for private firms. In practice, corporate officials seek to withhold information from public authorities in order to maintain private control over managerial decisions and to avoid possible harm to the firm's competitive position and profits.

In our three cases, the private companies initially withheld information from outsiders, responding evasively, defensively, and deceptively to questions. Company officials waited until government scientists had identified the contaminant with certainty (Michigan and Yusho) or near certainty (Seveso) before publicly admitting the existence of a problem or providing necessary information about the problem. In all three cases, company officials knew more than they told public officials or toxic victims. Corporate denials and evasions delayed effective attempts at cleaning up or containing the toxic contamination and contributed to keeping the problem defined as a nonissue.

Kanemi Company officials most persistently denied problems and concealed information in the face of mounting evidence of toxic contamination. Despite multiple contacts with government officials from the spring until the fall of 1968, company managers never mentioned (as far as we know) possible contamination of their rice-bran cooking oil. They continued to reject accusations that their dark oil had been harmful. Yet, during this time, someone in the company altered the cooking oil production records and destroyed daily records for the deodorization equipment. Someone knew that a leakage of heat-transfer agent had occurred, with potentially serious implications, and that the problem required cover-ups and falsifications (Katō Y., 1989, p. 140).

Executives at Farm Bureau Services stubbornly denied serious problems with their feeds, especially to Halbert, while making private efforts to reduce the damage and identify the toxic agent. These executives apparently did not officially notify the Michigan Department of Agriculture about the decision to recall all orders of Ration 402, although a state field inspector reportedly was notified about the halt in production of Ration 402 (Armstrong 1977, p. 1628). Some contacts occurred between Farm Bureau Services and the Department of Agriculture regarding Halbert's feed problems, but did not result in any public actions. In addition, the company's recall of Ration 402 was of limited effectiveness (Egginton, p. 38). The cooperative did not officially notify state authorities that it suspected some toxic contamination of its feeds, nor did it inform its members or its customers. Information about the problem was kept private and confidential, prolonging the phase of nonissue until Halbert identified the contaminant as PBBs.

ICMESA/Givaudan managers did not deny the problem so adamantly, in part because of the public visibility of the damage. But even ICMESA managers initially made only tentative efforts to inform local authorities in vague terms about a potential danger. The letter from ICMESA/Givaudan director Waldvogel, written about two weeks after the explosion, included several evasive and ambiguous statements. One suspects that the letter intentionally did not mention the word "dioxin," even though the company had confirmed the chemical's presence outside the factory in the first samples taken after the explosion, and even though the company had already trained two Italian experts in measuring dioxin.

The resistance of these companies to providing accurate and full information to outsiders can be explained partly by inadequate corporate understanding of the contamination problem. But understanding was a necessary not a sufficient condition. Even when corporate managers understood the details of a problem, they still decided to withhold their information.

Each corporation sought to protect its own interests by maintaining the problem as a nonissue, even though the end result was counterproductive. Givaudan managers were most forthcoming in providing information. They perhaps recognized that notifying the public authorities was not only required by law but was also necessary to reduce human exposure to dioxin contamination. Lowered exposure could reduce damages and thereby contain corporate liability and serve long-term corporate interests. Givaudan managers may also have understood that the ICMESA problem was already in the process of becoming a public issue. The Kanemi Company and Farm Bureau Services carefully guarded information to prevent the problem from becoming public. It is unlikely, in these two instances, that the decision to keep the problem a nonissue in fact served long-term corporate interests. In both cases, earlier action by public authorities could have reduced not only the level of damages but the ultimate corporate liability.

The failures of the corporations to communicate with outsiders became a major focus of subsequent criticism by the victims. They criticized inadequate corporate actions to reduce or prevent contamination and inadequate corporate information to victims and public authorities about the problem. Efforts by companies to maintain the problem as a nonissue, when executives knew the social nature of the contamination, subsequently undermined the public credibility of the private institutions. During this period, victims received inappropriate or inadequate medical care, if any; when paid, compensation applied to only a few cases of extreme harm; and cleanup by the private corporation tended to be ad hoc and sometimes counterproductive, making the toxic contamination problems worse. Contradictions between a corporation's private knowledge and its public actions created in victims a sense of moral outrage.

Public bureaucracies also experience serious problems of external communication and coordination, which contribute to official inaction and the persistence of nonissues. With public authority divided among different specialized organizations, each agency has its own "policy space" or "territorial zone." Agencies are highly protective against threats to this territory, especially threats affecting the central policy space (Downs, pp. 215–16). At the same time, the boundaries of policy zones can leave gaps on the periphery claimed by no organization.

As part of their protective strategies, public agencies guard "their" information from outsiders and use it as a source of power. For some public institutions, collecting and controlling information seems to become an end in itself, almost as important as the agency's formal mandate and a key incentive affecting the behavior of personnel (Wolf, pp. 74–75). Instinctive hoarding and timely release of sensitive information are recognized techniques in the bureaucratic politics of public agencies. For a problem like toxic contamination, which crosses several agency turfs and also falls into peripheral gaps, the problems of bureaucratic coordination are inevitable. These problems have serious consequences in delaying the transition from nonissue to public issue, while allowing toxic effects to worsen.

In Seveso, a narrow definition of agency tasks and a rigid adherence to agency boundaries combined to achieve specific bureaucratic acts, while leaving broader preventive goals unfulfilled. Officials from the National Association for the Control of Combustion inspected the inside of the ICMESA factory in 1971, 1972 (twice), 1973, 1974, and 1975, performing pressure tests of the chemical reactor that exploded in 1976. But the inspectors did not know that the reactor produced trichlorophenol, nor did they notice the "safety valve" attached to the reactor and opening directly to the sky. After the disaster, one agency inspector lamely explained that industrial secrets did not permit his agency to determine the process occurring in the reactor (*Commissione Parlamentare*, pp. 89–91). The Parliamentary Commission criticized that institutional fragmentation: "The practical consequence of this situation is that each agency tends to view an area circumscribed to its own limited sector of competence, with no global consideration of the problem of safety" (p. 91).

In the Yusho case, the lack of official communication between the Ministry of Agriculture and Forestry (responsible for feed) and the Ministry of Health and Welfare (responsible for food) prevented recognition that the dark oil incident contained serious public health implications. Officials of the Ministry of Agriculture and Forestry stuck adamantly to their own turf, neglecting to contact any public health officials and fending off tentative efforts to learn about the chicken poisoning. According to one report, unofficial contact did occur. A researcher at the National Institute of Preventive Hygiene learned of the dark oil incident and in August 1968 tried to obtain information from the Ministry of Health and Welfare, which knew nothing, and from the Ministry of Agriculture and Forestry, which refused to provide him with materials. He abandoned the problem at the closed doors and has regretted it ever since (Kaga, pp. 160–61).

An identical gap occurred in Michigan between the agriculture and public health bureaucracies. Officials of the state Department of Public Health learned of the PBB contamination not from the Department of Agriculture but from an article in the *Wall Street Journal*, despite the prior existence of an interdepartmental committee created to avoid such bureaucratic gaps. Public bureaucracies often attempt to deal with gaps between organizations by forming interagency committees (Gross, 1964). But these committees frequently are ineffective because of goals that exceed capabilities, an overgrown membership, overly institutionalized staff and procedures, and advice produced for someone with no authority to act or no desire for advice (Seidman, p. 180). Such committees "are the crab grass in the garden of Government institutions. Nobody wants them, but everyone has them. Committees seem to thrive on scorn and ridicule, and multiply so rapidly that attempts to weed them out appear futile" (p. 171). In our cases, the existing interagency committees had no impact on facilitating communication among public agencies to shorten the period of nonissue.

Some kinds of external communication, especially between public agencies and private companies, can prolong the phase of nonissue. These contacts can influence a public agency's detection and understanding of a problem as well as its actions and inactions. Ties of friendship, political power, and mutual interest between public and private organizations can transfer private perceptions to the public sphere. In short, private institutions can affect what gets on and what stays off the public agenda. The relationship between regulator and regulated can reach the point of regulatory capture, in which the public agency ends up protecting private interests more than public interests (Bernstein; Plumlee and Meier).

Such ties of power and constituency existed in each of our cases. In the Yusho case, the long-standing relationship between the Kanemi Company and the Ministry of Agriculture and Forestry, and the substantial political influence of the company president, contributed to the cautious, kid-gloved attitude of government officials in investigating the poisonous dark oil. In the Michigan case, the Michigan Farm Bureau's close ties at many levels to the Michigan Department of Agriculture perhaps influenced state veterinarians who visited farmers, like Halbert, to assume that the problems resulted more from management mistakes than from contaminated feed, thereby prolonging the phase of nonissue. And in Seveso, connections between the managers of ICMESA and the municipal government of Seveso complicated efforts to regulate the chemical factory prior to the disaster. In all three instances, ties of interest and power prolonged the phase of nonissue.

These examples of external "coordination" between public and private organizations illustrate the general problem of structural conflicts of interest. When a public agency is supposed both to regulate and to promote an industry, failures are bound to occur, usually on the regulatory side, particularly if the industry is a supporting constituency of the agency.

These structural conflicts of interest also persist into the phase of the public issue, when the problem becomes the victims' struggle.

How long a chemical disaster remains a nonissue thus depends on both the nature of the contamination and the nature of the organization. A highly visible problem can compel responses from both public and private organizations. In all cases, the responses of public and private institutions were shaped by organizational problems as well as by interests. Our cases suggest different weights for these factors: the responses of public agencies are explained more by organizational theories of bureaucratic pathologies, while the responses of private companies are driven more by business interests and intentional actions.

Victims and Nonissues

Toxic victims initially have little access to any of the three dimensions of power presented in chapter 1. In addition, they are obstructed by social institutions that use the second dimension of power (the capacity to set the agenda and maintain nonissues) to keep the issue off the agenda and the third dimension (the ability to shape the social construction of reality) to contain the victims through symbolic means. As long as the problem remains a nonissue, victims are relatively powerless, as isolated individuals, psychologically conflicted, and practically disadvantaged.

Sufferers of an unexplained deterioration in health or environment experience a searing conflict between a desire to keep their affliction private and secret and an urge to find a social explanation and fellow sufferers. These conflicting emotions represent classic problems of persons affected by social stigma (Goffman, p. 5). Some individuals' responses prolong the phase of private trouble; others' promote the transition to public issue. But in all sufferers, the dilemma of going public remains an enduring personal conflict and can contribute to keeping an individual's problem private, sometimes for years.

In our cases of toxic contamination, the sufferers' impulse to contain the problem as a private trouble arose from social stigma attached to a strange illness and from economic threats associated with that stigma. Those perceptions inhibited collective response and public action. In motivation and form, their behavior represented a form of "passing," a pattern of hiding information perceived as discrediting, while presenting oneself as normal (Goffman, pp. 73–91). A similar pattern has been observed for social minorities in Japan and other countries (De Vos and Wagatsuma). A classic example in Italy is the southerner who seeks to pass for a northerner.

Some victims, however, seek to identify the cause of their suffering.

They need an explanation, not only to find a cure for their physical ills but also to give meaning to their suffering. In general, the survivor of a traumatic event needs to find meaning or significance in the event and thereby in life (Lifton, 1979, p. 176). Victor Frankl designed his own form of psychotherapy, called "logotherapy," based on the fundamental principle of a "will to meaning." He proposed that the "striving to find meaning in one's life is the primary motivational force in man" (p. 154). Similarly, in social psychology, attribution theory tells us that a human being is a "perpetual problem solver" (Shaver, p. 24). People seek to identify the causes of behavior they observe in order to increase their understanding and their ability to predict (p. 4). Attributions of causality and responsibility, however, can also be distorted for purposes of self-protection (p. 110).

In each of our cases, someone among the afflicted took the role of problem solver, of explainer. The problem solver needed to overcome the psychological tensions around social stigma and to address the threat of economic losses. A focus on internal causes (self-blame) prolonged the phase of private trouble, while an emphasis on external causes (other-blame) hastened the transition to public issue. These individual struggles with the psychological and practical conflicts of a nonissue marked the victims' first steps toward creating sources of power to use in obtaining redress.

Social Stigma

Toxic contamination transforms the individual as a physical and social being, marking the individual with stigma and thereby encouraging the individual to keep the problem private.

The appearance of strange physical symptoms changes one's image, in one's own eyes and in society's eyes. From a symbol of what is normal, healthy, and basically good, one becomes a symbol of something abnormal, diseased, and even evil. The growing personal and social uneasiness around such unexplained changes marks the beginning of private troubles for the victims of toxic contamination.

In our three cases, individuals first experienced physical deterioration in themselves or their surroundings. The most scarring changes appeared in the Yusho incident, when a severe skin disease overcame entire families. In Michigan, a physical transformation initially affected the cattle, destroying the animals' healthy appearance and economic viability but also raising deep-rooted fears that similar symptoms would arise in humans. In Seveso, children's skin erupted in red rashes, while green trees suddenly yellowed and domestic animals sickened and died.

Transformations of the body are known to produce symbolic transformations in the community, with important social consequences. As anthropologist Mary Douglas wrote, "The human body is always treated as an image of society . . . and there can be no natural way of considering the body that does not involve at the same time a social dimension" (p. 98). A pattern of labeling and stigmatization has been recorded for various "strange" diseases: for mental illness (Szasz), tuberculosis and cancer (Sontag), venereal diseases (Brandt), and handicapped persons in general (Goffman). Recent fears about AIDS have similarly created patterns of stigmatization and discrimination, raising deep cultural fears about "contagion, contamination, and sexuality" (Brandt, p. 234).

In such situations, the sick person (or person in a "sick" environment) comes to represent incurable disease and physical deterioration, a symbol of decay and death for the individual and collectivity. The sick person feels torn between the society's dominant image of a "good" individual and the person's own image of a "spoiled" self. According to Talcott Parsons, designating illness as "illegitimate" and "deviant behavior" is functional for society by reinforcing the motivation not to fall ill (p. 133). But the sick person, in being labeled as illegitimate, experiences enormous conflicts about illness.

The physical and social transformations of toxic contamination call into question self-worth and self-image. In such situations of social stigma, sufferers commonly experience "strong feelings of powerlessness and depersonalization" (Schur, p. 15). Persons subjected to stigma "suffer a severe loss of self-esteem as well as a restriction of their social options. Quite simply, it is very difficult to maintain a favorable view of yourself if other people see you in a negative light and treat you accordingly" (Schur, p. 15). Sufferers seek to keep their problem private out of a desire to protect self and family from becoming a symbol of disease, a symbol of "undesired differentness" or stigma (Goffman, p. 5).

In our three cases, some sufferers of toxic contamination were treated as less than equal and tried to avoid social contacts. For instance, in the Yusho case, families experienced and feared discrimination in employment, marriage possibilities, and daily human interactions. As a result of their unexplained illness, the Kamino family felt that others treated them as "something less than human." They tried to cope with their stigmatization by turning away from society. Other Yusho sufferers felt similarly.

In Michigan, Halbert carefully managed information about his private troubles, telling some people and not others about his herd's health problems, trying to avoid social stigma and protect his farming and family reputations. Other dairy farmers avoided potential stigma even more effectively (before the problem became public), by not complaining as persistently as Halbert, by selling unproductive cows for hamburger, and by changing feed and trying to overcome the problem. They accepted their difficulties as private troubles and did not rock the boat. Halbert's persistent complaints to Farm Bureau Services were resented by some company officials, who wanted him to leave the problem up to company experts and who treated him as something of a stubborn troublemaker.

In Seveso, social stigma did not play a major role in keeping the problem private. The nature of toxic contamination at Seveso hastened the discovery of the problem and made it a public issue before the victims could even consider keeping it private. Nevertheless, Seveso victims, like those in the other two cases, experienced social stigma. Once the problem became public, town residents resented the attention given to the children with severe chloracne, arguing that their scarred faces should not symbolize the healthy condition of most people in Seveso. Even Seveso residents who showed no physical symptoms suffered discrimination at hotels, at border crossings, and in social transactions throughout Italy. The social perception of invisible contamination produced real personal consequences.

The "tainting" of individual and family represents a combination of actual invisible contamination by toxic chemical with a symbolic invisible contamination by social stigma. The invisible toxic tainting produces social discrimination, creating a secondary social victimization that followed the primary toxic victimization (Lifton, 1976, p. 17). Sufferers of toxic contamination at Love Canal, a chemical waste dump in Niagara Falls, New York, similarly complained about being marked by social stigma, with painful psychological consequences (*Boston Globe*, 7 October 1980).

In our three cases, the symbolic tainting encouraged sufferers to contain their problem as a private trouble. It isolated the afflicted from each other, delayed public recognition of the sufferers as a group, denied them medical care and monetary compensation, and allowed the contamination to worsen. Social stigma associated with unexplained illness helped to maintain the chemical disaster as a nonissue and the victims' status as powerless.

Economic Threat

Invisible contamination, as a form of social tainting, affects not only people but property. In the phase of a nonissue, many toxic sufferers seek to keep their problems private in order to avoid possible economic losses associated with fears of contamination, especially from job discrimination or from reduced sales of goods.

Victims who perceive public recognition of their problem as a threat to

personal economic interests are likely to act in ways that keep the problem from going public. Some Yusho sufferers, for example, kept their
symptoms as private as possible to protect their jobs. In Michigan,
Halbert believed that a government agency would close his dairy farm if
his trouble became public knowledge. He therefore withheld information
and delayed contacting the Food and Drug Administration for months. In
Seveso, in the early period, small-scale farmers with dying animals and
crops reacted as the Michigan farmers did, seeking to avoid public disclosure that might harm sales.

The urge to maintain privacy and avoid possible economic losses persisted long after toxic problems became public issues. In the Yusho case,
some persons with light physical symptoms continued in the late 1970s
to conceal their illness, sometimes even from spouses and children (Kōno
interview). In Michigan, some dairy farmers allegedly affected by PBB
contamination evaded public recognition to avoid economic and social
costs. And in Seveso, local businessmen sought to reduce public visibility
of the contamination problem when sales of locally produced furniture
and agricultural goods dropped sharply.

But toxic victims faced an economic dilemma: despite the benefits of
avoiding public recognition, in order to obtain compensation for their
injuries, they often needed public recognition to create awareness of the
problem and to put pressure on social institutions. As long as the problem
was a nonissue, only a few victims might get minimal compensation. As
the personal costs of toxic contamination increased, victims became more
willing to risk paying the costs of public recognition in order to increase
the probability and level of redress. The threshold of personal costs before
a sufferer was willing to go public varied with individuals. But coping
with these economic threats required that victims explain the causes of
their injuries.

Self-blame

Faced with a troubling enigma, individuals often search for an explanation of it in their own behavior. Self-blame stresses internal over external causes and thereby isolates individual sufferers and contains their
potential power to influence social institutions.

What have I done to bring this damage on myself? The phenomenon of
self-blame has been observed for other illnesses and social problems.
Victims of AIDS, for example, confront powerful social attitudes that
attribute the cause of their affliction to homosexuality rather than to the
retrovirus and that view the disease as "self-inflicted" and deserved
(Brandt). Similarly, victims of venereal infections in the early twentieth

century were divided into the innocent ones, especially familial cases of women and children, and the guilty ones (Brandt). Even the survivors of the Hiroshima atomic bomb have suffered from feelings of self-blame (Lifton, 1967, p. 274). The attribution of self-blame, as an explanation for illness, can obstruct the ability of sufferers to recognize external causes and can delay the transition from nonissue to public issue.

A correlate of self-blame is society's effort to blame the victim. William Ryan defined blaming-the-victim as "justifying inequality by finding defects in the victims of inequality" (p. xiii). For our cases, the justification is accomplished by attributing the causes of unexplained illness to the sufferers themselves. Another interpretation of the phenomenon of blaming-the-victim is that people want to preserve the belief in a "just world" even when confronted by injustice, and do so by asserting that the victim "deserved what he got" (Lerner).

Individual adoption of self-blame, especially when supported by social processes of blaming-the-victim, presents a formidable barrier to understanding by sufferers that their private troubles have social roots, or in our cases, that their illness resulted from external toxic contamination. Moreover, viewing a problem as an internally caused private trouble produces a simple conclusion and solution: change the victim and not the society (Ryan, p. 8). The definition of the problem thus constrains the realm of solutions. The phrasing of social problems in terms of personal attribution or situational attribution determines to a large degree the proposed policy solutions (Shaver, p. 133).

In two of our cases, initial explanations concentrated on internal, private causes and denied or downplayed external, social causes. In the Yusho case, sufferers worried about internal causes that attributed skin symptoms to genetic factors or to diseases such as syphilis or leprosy. In Michigan, when Halbert's cows fell sick, he meticulously examined his entire dairy operation, considering external agents such as contaminated feed and possible sabotage by a former employee, but he was most tormented by the thought that his own actions might have caused the problem. In both cases, the process of self-blame prolonged the phase of nonissue.

Other-blame

Although victims of toxic disasters often try to keep their trouble private as long as possible, practical, physical, and economic problems motivate them to search for others to blame.

The consequences of a chemical disaster create incentives for toxic victims to make their troubles public and to seek an external cause for their problems. Sick victims need a correct diagnosis of their illness in order to receive appropriate care. If they lose their job or income from the disaster, victims then have difficulty paying for medical care and even daily living expenses, prompting them to seek compensation. Victims also commonly seek some way to clean up the harmful effects of toxic contamination. The personal costs incurred by toxic victims push some individuals to relinquish self-blame and to seek an external cause and other actor that can be held responsible. The sufferers' efforts to find this social explanation, rather than an internal or personal explanation, help transform the private trouble into a public issue.

In each case, some victims began the search for external causes. In the Yusho case, Kunitake performed the role of problem solver, piecing together the mystery to find a common food, pushing doctors at Kyūshū University to analyze the cooking oil, and finally taking an oil sample to a government health center for analysis. In Michigan, Halbert became similarly obsessed with the task of identifying the cause of his problems as toxic contamination from an external source. In the Seveso case, the ICMESA union initially took the most active role in seeking a social explanation for the unusual cloud and strange happenings in the factory and its surroundings.

But in cases of toxic contamination, the searchers for external explanation work against enormous odds, with little power and few resources, lacking information, experience, and contacts. The obstacles to constructing an adequate social explanation come partly from the victims' own psychological reactions; but they also come from powerlessness. As discussed in the first section of this chapter, obstacles arise from the responses of social institutions, which seek to avoid the blame for toxic contamination and the anticipated economic and political costs of redress. Social organizations respond in ways that prolong the phase of nonissue and delay the transition to public issue. One article on toxic contamination concluded with a warning to victims and others: "The best advice for hacking through the bureaucratic jungle was provided by the witch who called upon Macbeth to 'Be bloody, bold and resolute . . .'" (*Lancet*, p. 21). In addition, one might add, the victims of chemical disasters need allies. Here, the role of the mass media is critical.

Mass Media and the Transition to Public Issue

The mass media play a central role in setting the public agenda, and thereby exercise the second dimension of power. The media can determine which private troubles become public issues, and when and how. Media attention can transform a nonissue into a public issue and can affect the distribution of resources in society. This important function of the media has been recognized and analyzed for many social problems. Barbara J. Nelson, for example, examined the role of the mass media in creating the issue of child abuse. The mass media, she wrote, "exist at the boundary between the private and the public. Their task is to discover, unveil, and create what is 'public'" (Nelson, p. 51).

The power of the mass media comes also from their use of the third dimension of power, the realm of symbols. As Todd Gitlin put it, "The media specialize in orchestrating everyday consciousness—by virtue of their pervasiveness, their accessibility, their centralized symbolic capacity. They name the world's parts, they certify reality *as* reality" (Gitlin, p. 2). The media define an issue with a set of symbols, to which others respond.

In our three cases, coverage by the mass media transformed nonissues into issues. The media created and defined the public issues of toxic contamination. Their coverage provided communication among the elite and notified the public of a significant event. How the media reported the event differentially affected governments, companies, and victims. For the victims, coverage by the media compelled both private and public institutions to move from inaction to action. But the mass media reported the problems only after significant delay in our three cases. Indeed, the media did not actually "discover" the problems but received tips of a story or requests to report an incident. The cases confirm that most reporting (even most investigative reporting) combines chance and a good network of information sources. In most instances, information seeks the reporter and not vice versa (Gans, pp. 116–18). For this reason, the media also contributed to prolonging the phase of nonissue.

As many studies have shown, the mass media are not an autonomous force in society. They do not simply report "the facts" but create "the news" through social processes of selection, exclusion, and emphasis. The news results from an interaction between bargaining within the media company and bargaining outside the media company (Sigal). The processes of making the news are embedded in media organizations, in the people of those organizations, and in their relationships with outside sources and organizations, especially in government (Reich, 1984a, p. 148). The sources of information, and how a reporter chooses sources, often determine the form and content of a story. Government leaks, routine sources, press releases, and collaborative stories all affect what gets into the news. These processes can make public issues but can also perpetuate nonissues.

Communication among the Elite

The more media coverage, the greater the likelihood that a problem will be defined as a public issue among the elite.

Major coverage by mass media creates awareness among public officials of a social problem. The generally secretive approaches of public bureaucracies and private corporations create problems of communication among organizations and prolong the phase of nonissue. The mass media help overcome that institutional fragmentation by serving as a means of communication among the elite about important social problems. The mass media define issues for the public agenda that require elite consideration.

Media attention in the Michigan case was initially delayed by official efforts to contain information. According to an official internal memorandum, the state Department of Agriculture decided to be "very guarded in the release of any information to the outside," delaying early coverage by local and state media organizations (Chen, p. 55). Indeed, the first public information about PBB contamination on 8 May came from the *Wall Street Journal*, not from a Michigan newspaper. That paper learned of the story from a contamination victim and regular reader, Rick Halbert (p. 55).

The Yusho case suggests that low-profile coverage cannot overcome the bureaucratic obstacles to recognition of a public issue. In April and May 1968, several newspaper articles described the chicken deaths in western Japan. The *Asahi* article of 11 April 1968 reported the probable cause of the chicken deaths as dark oil added to the feed but then minimized the potential public health implications. Another story in May protected the corporation responsible for the poisoning by referring only to the "K-Company" in a section of Kitakyūshū. The news media did not investigate the chicken deaths and thereby discover human illness. Newspapers published little about dark oil until 11 October, the day after the *Asahi* scoop on the human illness of Yusho. Public health officials, if they read the articles about sick chickens, did not formally consider possible human consequences.

The reporting of the chicken deaths illustrates the more general principle that media coverage is shaped both by the source of information and by the journalist's interpretation of that information for the expected audience (Gans, pp. 80–81). An *Asahi* reporter stationed in Kyūshū in 1968 explained, "It is not news if chickens die. The chicken story appeared in local papers, but in small articles. It was not a big incident and occurred entirely in the countryside. To attract attention, the event must reach a certain threshold and be dramatic. These are the structural limits of newspapers and reporters. Events do not become a big problem until they affect people" (Nishimura interview). The reporters accepted the official story from the affected companies and local governments and relied on the normal interpretation of what constitutes news. The media treated the problem as a routine matter, of concern to chicken farmers but not to general consumers. Chicken farmers and feed manufacturers perceived their interests in maintaining the chicken deaths as a private problem in order to avoid difficulties in selling their chickens or their feed; and the media collaborated. The media thus helped maintain the contamination of cooking oil as a nonissue and delayed the discovery of human victims.

Media scoops often contain both uncertainties and mistakes. Edward Jay Epstein noted that the "inherent pressures of daily journalism severely reduce the possibility of verifying a leak or disclosure in advance of publication" (pp. 11–12). Uncertainty is a basic element of news reporting, given the time constraints, the competition, and the pressures to get the story out. The *Asahi* article on 10 October 1968, for example, did not mention the name of the company that had produced the poisonous rice-bran oil. A reporter explained, "We had no basis for saying that it was actually the Kanemi Company, and we did not know the cause of the disease. But if you wait for certainty, you can't write articles" (Nishimura interview). The breaking story for Seveso's toxic cloud similarly did not have all the facts.

But a media scoop can create a public issue. In our cases, those stories, with all their uncertainties and inaccuracies, served to notify public officials that the mass media had found a nonissue that should be a public issue, that a problem requiring authoritative action existed, and that a public inquiry by the media had begun.

Transition to Public Issue

The more media coverage, the more the problem is redefined as a public issue and the sufferers as victims.

In our three cases, mass media reports of toxic chemical contamination suddenly brought private and public institutions into the public spotlight. Press reports redefined the nature of the problem and the roles of participants. Media decisions to report an event as an important public issue became a self-fulfilling prophesy. Reports of toxic contamination as a front-page article or top television news story effectively transformed the event into an important public issue. Coverage by one media organization brought coverage by others. A snowballing of public recognition compelled action by the responsible corporations and governments. Public scrutiny by the mass media raised the stakes and placed a new urgency on the responses of social institutions. Media reporting catalyzed the transformation of nonissues into public issues and introduced a period of public action.

Chapter 6

Public Issue: Victims' Struggle

In responding to the transformation of toxic contamination from non-issue to public issue, institutions in our three cases generally sought to contain the public issue's scope, while the victims strove to expand the scope. In their search for redress, the victims needed group action to increase their power over social institutions, but group action in turn imposed additional conflicts and costs on them. In this chapter, I examine how individuals and institutions responded to the transformation to public issue and explore the internal and external problems confronted by toxic victims in their efforts to obtain redress.

I first examine the responses of social institutions to the public issue, because these responses provided the context within which the victims struggled for redress. In their responses, institutions proceeded through three stages: definition of the issue, limited administrative action, and legitimation of policy. In our cases, private and public institutions showed a range of responses in attempts to protect their interests and contain the issue. I then analyze the victims' two forms of group action—organization and protest—used as means of empowerment to assert their private grievances as a public issue. Through group action, the victims competed with social institutions over control of the public agenda (the second dimension of power) and over the symbols of the public issue (the third dimension of power). When the problem became a public issue, some victims created new organizations, with names, members, and objectives, to pressure private and public institutions on the substantive problems of redress. Some victims adopted methods of protest, the intentional and public disruption of social institutions, to increase their power in society. The two forms of group action yielded both empowerment and costs for the victims who participated, thereby structuring the dilemma of group action.

The dilemma of group action represents a subproblem within the field of collective action. I use the phrase "group action" to refer to both organization and protest and to avoid confusion with the term "collective action." The study of collective action seeks to explain when and why individuals act together to achieve some common goal or interest that could not be satisfied through individual action. A major problem is the free rider, the individual who takes no action but still gains access to the collective good achieved, which is theoretically available to all group members. The theory of collective action depends on the concept of economic rationality, the narrow self-interest of individual participants, to explain the conditions under which people work together. According to Mancur Olson, "Unless there is coercion or some other special device to make individuals act in their common interest, rational, self-interested individuals will not act to achieve their common or group interests" (Olson, pp. 1–2). In this chapter, I examine how victims' groups in our cases dealt with the problems of free riders and economic rationality in seeking to obtain redress.

In the phase of public issue, a major goal of both social institutions and toxic victims is to control the issue's definition, because of its importance as a source of power. But this competition for control does not occur on a level playing field. Institutions begin with significant advantages in establishing and maintaining the definition of a public issue. Most victims start with few resources and without an existing organization yet must confront well-established social institutions. Those victims who belong to existing organizations begin with a head start in asserting their grievances to public and private officials. In this phase, the victims struggle mostly on their own in efforts to control the emerging public issue and to reassert control over their own lives and environment.

Conflict over the issue's definition becomes a conflict of symbols and of power, as different groups seek to make their view of reality "stick" in society (Berger and Luckmann, p. 109). As with other public issues, specific problems can be defined in various ways and the accepted view of reality affects who gets what. Different meanings and values can be attached to the same images of toxic contamination and thereby produce different symbols, with diverging consequences for the victims' redress. A group's ability to get its symbols accepted in society provides an important source of power by shaping public consciousness of the issue.

In our three cases, victims attempted to expand their symbols of toxic contamination into sources of power, while private and public institutions sought to contain the symbols and restrict the power of victims. These processes unfolded at different paces in the three countries, but underneath the differences lay fundamentally similar patterns among both institutions and victims in the three societies.

Social Institutions Respond

The dominant strategy of private corporations is to protect their interests by containing the public issue but also by dealing with some substantive problems of redress. While corporations seek to minimize the immediate financial costs of a chemical disaster, they also strive to reduce the uncertain long-term liabilities. Reducing long-term costs, however, requires up-front expenditures, which conflict with the goal of minimizing immediate costs. In addition, companies seek to protect their public reputations, maintain their current clients and customers, avoid lawsuits, and defend the potential for future growth and profits. In general, these interests motivate corporations to contain the public issue. A company may seek to expand an issue's scope if the issue can be shaped in ways that shift the focus of attention away from the corporation. The techniques for coping with these problems have developed into the field known as issues management, "the newest management profession created for the benefit of senior executives" (Heath and Nelson, p. 7).

Public institutions also have their own interests in relations with external constituencies, in maneuvering for public funds, and in the individual career goals of agency leaders. These interests create well-known patterns of bureaucratic politics. Far from acting wholly in the interests of victims, public institutions seek also to manage public issues in order to meet their own organizational interests. Some public agencies seek to contain an issue because of its negative consequences for important constituencies. Other public institutions, with different interests, seek to play up aspects of an issue to exploit advantages.

Public institutions, like the victims of a toxic disaster, have conflicting impulses about whether to expand or contain an issue. Their dilemma is that expanding an issue may help generate resources for the organization, but the process also carries risks and costs. Political leaders are known to create and maintain power by encouraging public anxiety and then placating it through policies, rhetoric, and gestures. Cycles of anxiety and reassurance can provide both symbolic and material rewards for constituents (Edelman, 1971, p. 147). But expanding a public issue also makes public institutions more likely to be blamed for mishandling or causing the problem, with the potential for damaging the institution and its interests.

The three stages of institutional responses to a public issue—definition, action, and legitimation—represent a progression with important symbolic and material consequences for the victims. The institutional responses create obstacles that the victims must struggle to overcome through their efforts at empowerment in group action to obtain redress.

Definition of an Issue

After a problem becomes a public issue, the initial strategy of withdrawal and denial, which characterized the phase of nonissue, no longer serves to protect the institutions' interests. When institutions can no longer keep an issue off the public agenda, in the second dimension of power, they seek to control the issue's definition. Public revelation, initiated by others, compel corporations to change their initial strategy and to respond to the substance and symbol of the public issue. Institutions' efforts in this direction, especially in situations of uncertainty, reflect their awareness of this process as a source of power.

The ability to control others can be achieved by "controlling, influencing, and sustaining *your* definition of the situation" (Hall, p. 51). The corporate definition of the public issue often portrays the problem as limited in scope, as worsened by the responses of public administration, and as exaggerated by the emotional reactions of victims. A primary concern is damage control to the company, spanning the financial, political, and public-image dimensions. The success or failure of this strategy in our cases depended partly on the relationship of the company to the chemical disaster.

In Japan and Michigan, the companies that supplied the chemical, but were not the primary locus of disaster, succeeded initially in keeping some distance from the issue. Kanegafuchi Chemical maintained its position of public silence and Michigan Chemical made only minimal public statements, allowing the focus of the issue to remain elsewhere. Michigan Chemical defined the contamination issue as limited in scope and not the company's fault, portraying the company in the best light possible. By implication, these two companies defined the public issue as a nonissue for them, and by silence they hoped to keep the definition that way.

The other companies in our three cases, as the proximate locus of disaster, could not escape public attention or the public issue so easily. They defined the issue to protect corporate interests in ways that combined information, misinformation, and disinformation. None of the companies followed a policy of full disclosure. ICMESA/Givaudan at first defined the issue as a serious problem but one that could be handled through appropriate public action to contain damages and that could be fully redressed through corporate resources and insurance. Farm Bureau Services initially responded to the public issue by defining the issue as limited in scope and as already solved, representing wishful thinking that soon proved false. In the Yusho case, the Kanemi Company's president, while not admitting legal responsibility, promised to assist the victims to the best of the firm's ability; he then attempted to shift the blame to the chemical company.

In each case, corporate definitions of the issue stressed reassurance—to the victims and to the public authorities—that the company would do everything possible to meet its social responsibility. Corporate definitions sought to portray the company's intentions as socially beneficial, in attending to the problems of redress, while seeking to limit the company's potential liability and not admitting legal fault. The definitions portrayed the disaster as an unexplained accident and the company's efforts at redress as voluntary acts of charity and good faith. Those statements of good intentions, however, served the broader corporate goals of protecting both public images and private interests. Moreover, the promises required a long time to fulfill even partially and did not apply to all victims.

Public officials who confront a public issue similarly seek to define the issue in ways that reduce threats to organizational interests. For many government officials, two major concerns are to avoid public "panic" and to allay public fears. As with private companies, public agencies seek to provide reassurance. Public officials commonly predict panic behavior and then use that fear to justify a problem's definition—even though public panic rarely occurs (Jackson, p. 218). To avoid the predicted panic, they seek definitions that focus on solvable aspects, downplay controversial elements, and meet the approval of key constituencies. They commonly define the issue to fit their available organizational resources and their constituencies' private interests. Public bureaucracies, with the cooperation of the responsible private companies, portray the issue as under control and as limited in scope in order to create symbolic boundaries around the problem. The involvement of the private companies, because they are seen as knowing the most about the problem, results in cooperation that approaches collusion in protecting the mutual interests of private and public institutions.

In our cases, public officials initially defined the issue to fit the available resources and routines, as more or less normal public issues. In Michigan, Department of Agriculture officials repeatedly defined the contamination as an agricultural problem and not a public health problem. In Japan, the Ministry of Health and Welfare defined the contamination as food poisoning and not a pollution disease. And in Italy, regional health officials defined the contamination initially as a serious problem but not one requiring evacuation. These definitions contained the boundaries of the issue and restricted the scope of administrative measures in the provision of redress. Public officials defined the issue in ways that a priori limited the administrative actions and confined the questions asked.

The administrative definition of the issue created symbolic boundaries around the contamination issue, boundaries that connected or isolated the issue in society. In our three cases, but especially in the Michigan and Yusho cases, the definition isolated the issue, stressing the need for private rather than public resolution of the problems of redress. In Michigan, the first meeting of public and corporate officials to discuss the contamination problem and edit a press release (on 17 May 1974) showed the extraordinary sensitivity of public administrators to the symbolic impact of particular words to define the public issue. Moreover, the discussion of which words to choose—how to define the problem— illustrated the vulnerability of public officials to the influence of the private companies.

This sensitivity to symbolism derives from the material consequences of bureaucratic definitions for the problems of redress. In cases of toxic contamination, the definition provides the basis for specifying the eligibility criteria for redress, which are then implemented in administrative actions taken by social institutions. This process for toxic contamination belongs to the broader class of problems of how societies redistribute resources to disabled people (D. Stone). In general, public authorities define official categories of disability so as to serve two conflicting purposes: to provide and to restrict access to resources. Containing inherent conceptual ambiguities and social tensions, the definitions become the focus of political conflict.

In our three cases, administrative definitions used specific numbers, representing levels of toxic contamination, as symbolic boundaries around the issue. The numbers and symbolic boundaries then became rigid, and their basis in uncertain scientific data became forgotten. As Robert Socolow commented about "golden numbers" in another environmental controversy: "A number that may once have been an effusion of a tentative model evolves into an immutable constraint. . . . Apparently, the need to have precision in the rules of the game is so desperate that the administrators seize on numbers (in fact, get legislators to write them into laws) and then carefully forget where they came from" (Socolow, p. 15). An organization becomes committed to a number not only for specific institutional interests but for broader social and political reasons. Protecting the number comes to represent protecting organizational integrity as well as constituents. Organizational commitment to a golden number leads public officials to deny scientific uncertainty.

In Michigan, for example, the Department of Agriculture adamantly supported the 0.3 ppm level of contamination as safe and persistently defined cattle contaminated at a lower level as healthy. The department's defense of the 0.3 level arose not only from its assessment of scientific evidence but also from a structural conflict of interest. The department was mandated to protect both the common consumer and the agricultural industry, but the department leaned more toward the industry through its close ties to the Michigan Farm Bureau.

In Italy, the health authorities similarly defended their golden numbers on the dioxin-contamination levels used to justify evacuation or non-evacuation. Public officials used measurements of dioxin in soil to draw the boundaries for zone A. Even though the levels varied greatly within a short distance, the public authorities still drew lines and defended their decisions with tenacity. The scientific uncertainties of the sampling, the measuring, and the interpretation were all lost in public statements about the contamination levels and their implications. Indeed, the golden numbers used to draw the lines even became forgotten; what remained were golden lines that set the stage for political conflict.

These examples illustrate how the definition of a public issue tends to become frozen as the official position of a bureaucratic agency, and thereby to resist change. Organizations need clear guidelines as part of standard operating procedures to make decisions. In disaster circumstances, organizations hold onto official definitions with even greater obstinacy to provide a sense of security amidst the chaos. Changing the institutional definition often requires severe shocks to the system.

In all three of our cases, once the governmental authorities established an issue's definition, it became difficult to change. Public agencies acquired a bureaucratic commitment to their definition of the issue, which required substantial power to alter. In Michigan, for example, health officials became committed to a definition of PBB contamination as an agricultural problem and not a health problem, as affecting cattle and not people. Officials recognized some farmers with ill health but refused to admit a connection to PBB. The Michigan Department of Public Health continued to defend its position that no "PBB syndrome" existed, unwilling to admit publicly the uncertainties and problems of the department's initial epidemiological survey. That position constrained the administrative actions taken and resulted in a policy that lasted three years, a policy of no medical care or assistance to farm families who had consumed relatively large quantities of PBB-contaminated products.

Similar instances of the persistence of bureaucratic definitions occurred in the other two cases. In Japan, the health authorities adopted an official definition of a Yusho patient, based on symptoms, and then resisted any changes in the criteria. In Italy, as noted above, the boundaries of contamination around ICMESA became locked in bureaucratic definitions, which the authorities refused to alter. While some flexibility existed in both cases in early periods, once the bureaucracies adopted an official definition, they stuck to it and stubbornly resisted change.

Administrative Response

An institution takes administrative action to provide limited concessions to some victims in ways that are intended to deal with some substantive problems of redress but also to create symbolic reassurances and restrict the issue's scope.

Based on the issue's definition, private companies use administrative action to avoid or delay litigation, reduce negative publicity, and minimize the company's overall liability. When confronted with a public issue for which a company believes it is partially responsible, and is likely to be held legally responsible, management often offers compensation for some damages through an out-of-court settlement. Various incentives encourage companies to settle. In most cases, settlement costs much less than a court-ordered damage award. In the United States, an additional incentive to settle out of court is that federal tax law allows a corporation to deduct the cost of a settlement but not the amount of a court decision (Soble, p. 714). The prospect of negative publicity from court cases encourages corporations to settle consumer disputes, even when they might win the cases in trial (Upham, 1976, p. 518n). Finally, a corporation's willingness to provide these payments can be encouraged by insurance coverage for pollution damages, which reduces direct costs to the firm. The corporate decision to offer limited concessions to some victims, as an administrative action, thus reflects calculated efforts to reduce the overall financial and social costs to the company.

In our cases, the three proximate companies all offered concessions to some victims. Givaudan announced limited concessions in early August 1976, when LaRoche director-general Jann proclaimed that the company carried insurance for such problems and would compensate all damages that could be demonstrated. In Michigan, Farm Bureau Services began negotiating out-of-court settlements in the summer of 1974 with the most highly contaminated farmers, large dairy farmers with major accounts, including some of the company's most important customers. In the Yusho case, the first concession from the Kanemi Company's president was his agreeing to meet with victims. At the meeting, he yielded to requests to provide some medical expenses for Yusho victims, with possible compensation in the future, but only if the factory resumed business as usual.

These administrative responses by private firms had contradictory consequences. On the one hand, they provided some measure of redress to a group of victims, who tended to be those most severely and obviously contaminated. This distribution of resources met some real needs and also served to defuse some demands by victims for redress, thereby serving to contain the issue. On the other hand, for persons not included in the official definition of victims to receive resources, the availability of limited redress created incentives to intensify their demands. The exclusion from redress provided the basis for future group action, thereby serving to expand the issue.

Public officials, similarly aware of both symbolic and substantive consequences, take administrative actions designed to contain the issue's scope to the official definition as well as to respond to real problems. The process of proposing administrative measures provides symbolic reassurance that government is doing something, thereby delaying or inhibiting the mobilization of victims (Edelman, 1971; Piven and Cloward). At the same time, administrative action results in a distribution of material resources, which helps quell potential mobilization. Public officials resist some policy measures (such as expanding the boundaries of contamination) because they are seen as symbolically promoting public alarm, while they adopt others (such as providing health examinations) because they believe the actions will calm public unease.

In our three cases, public officials adopted measures that served the symbolic goal of reassurance by setting "acceptable" levels of contamination and identifying "official" victims of contamination. Action to monitor the toxic levels in the environment (Seveso) or in animals (Michigan) and to examine people in health centers (Yusho) set official boundaries around the public issue. In all three cases, the administration of medical examinations served the symbolic purpose of containing the issue more often than the analytic objective of understanding the problem or the public health objective of providing care for the victims. Public authorities collected medical information that was inappropriate or was never fully analyzed. But the process of collecting information served to reassure people that government was doing all it could.

The administrative measures adopted by public officials, who were suddenly confronted with the complexities of toxic contamination, also served real needs. Given the paucity of data available when the toxic problems became public issues, officials desperately sought information about the scope of the contamination. They were required by law to protect the public food and water supplies and the public health. But once public officials established the issue's definition, they tended to resist administrative actions or interpretations that might question or discredit their definition. The approach of "Seek not and ye shall find not" helped to contain the issue's scope. Inaction affirmed the official definition.

Both private and public agencies used the official definition to decide on administrative actions and then used those actions to confirm and reinforce the definition. The definition implied an expenditure of resources by private and public institutions; and both institutions sought to limit and contain those costs. To fortify the symbolic boundaries around the issue, organizations relied on legitimation.

Legitimation

An institution seeks to provide legitimacy for its definition of a public issue and its administrative response by using such means as specialized legitimators (such as technical experts), specific institutional symbols, and broader cultural symbols (Berger and Luckmann, pp. 94–95).

Legitimation is a general process of explaining and justifying a particular distribution of resources, especially of power and wealth, in society. When policies favor one group of interests over others, legitimation becomes essential (Mueller, p. 179). Both private and public institutions need legitimation and can adopt various strategies to achieve it. Under normal circumstances, a company must legitimate the process of private accumulation. Under disaster circumstances, the company seeks to legitimate the social wrongs resulting from private accumulation, a more challenging task. The company must respond to public criticism, justify its role in causing the problem, and defend its efforts to resolve the social problem.

In our cases, a company chose its form of legitimation depending mainly on the social relationship between victims and victimizers. Givaudan, as a foreign company in Italy dealing with victims outside the organization, relied on technical experts who were Italian nationals to legitimate the company's definition of the issue and its actions. Legitimation through technical experts was also a common technique used by public authorities, as discussed below. In Michigan, Farm Bureau Services confronted victims who belonged to the farmer-owned cooperative, and the company sought to legitimate its activities by appealing to the loyalty of its "members" and to the organizational myth of serving its farmers. Through this appeal, the cooperative sought to avoid litigation from its member-victims and to reduce its liabilities. The Kanemi Company, as a firm dealing with victims outside the organization but still within Japanese culture, relied on broad cultural symbols to legitimate its actions, especially through the company president's public apology to victims and his offers of token compensation and care.

These symbols of legitimacy served, for a time, to protect each company's interests. The symbols provided a source of power (in the third dimension) to influence other actors involved in the public issue. As we shall see below, however, the symbols eventually became targets for the victims to attack through their organizations and through protest; the symbols became objects for the victims to redefine and capture as their own sources of power. The companies then needed new strategies and defenses for the phase of political issue.

Public administrations confronted by a public issue commonly rely on expert advisory committees to provide legitimation, especially when the agency "wishes the outside world to believe that a policy was adopted for technical rather than political reasons" (Primack and von Hippel, p. 34). Expert committees bring with them the symbols of science, embodied in individual scientists who are portrayed as rational problem solvers unswayed by emotions, values, or interests. The legitimation provided by expert committees works "even when the decisions being defended fly in the face of the information and analyses the advisors have provided—as long as the advice itself is kept confidential" (p. 34). The symbol of the rational scientist, if skillfully used, can provide a solid source of legitimacy for an issue's definition and for administrative actions, and can thereby serve as a source of power for the institution.

In our three cases, public administrations organized expert committees soon after the problem became a public issue and used the committees as a central means of legitimation. The experts served to define and defend the issue's boundaries and to legitimate the administration's control of the situation. They also served to deny legitimacy to alternative definitions, thereby establishing a single "correct" definition. In Italy and Japan, the public administrations early on created expert committees to define the public issue and contain its implications. In Michigan, the governor initially relied on experts in bureaucratic agencies and moved to establish an external committee only after the problem reached the phase of political issue.

Public officials also imported individual experts to enhance the appearance of independence and legitimacy. But an outside expert who was indeed independent posed a threat of uncertainty, since such an expert could undermine rather than support administrative legitimacy. Because of that uncertainty, governments sought to use outside experts who were controllable and predictable. The role of Dr. Poland in Seveso exemplifies how a public administration can use an external expert to bolster its legitimacy. The role of Dr. Selikoff in Michigan illustrates how a public administration can perceive an independent scientist as threatening to its legitimacy.

In the process of legitimation, public officials can use expert committees to create the illusion of action. One former American bureaucrat explained, "In my experience, nothing was simpler than to set up an advisory group. It started wheels turning, it bought time, it was a surrogate action, and it produced a kind of structural grandeur. It implied someone was taking charge of the problem, and perhaps that things would work out. This is the way of government" (cited in Primack and von Hippel, p. 31). The proliferation of advisory committees in countries around the world (Organization for Economic Cooperation and Development; Pempel) reflects their central symbolic role in the bureaucratic processes of policy-making, in providing public reassurance about governmental action on an issue.

Public officials also seek legitimacy through the generation of hard data. In addition to the golden numbers described above, public agencies generate statistics in order to support the official definition and administrative action. Even when the numbers themselves have little meaning, hard data convey the impression of scientific legitimacy and certainty, which reinforce the symbols of govermental control over an uneasy situation. As Murray Edelman wrote, "Statistics are so effective in shaping political support and opposition that governments sometimes publicize statistics that have little or no bearing on an issue creating anxiety, either because none that do have a bearing are available or because the pertinent ones point in the wrong direction" (1977, p. 31).

In our cases, public agencies used hard data to set boundaries around the issue, to legitimate the government's activities, and to argue that all was under official control. In Michigan and Italy, public officials publicized the number of contaminated animals dead or slaughtered, providing a public body count of the government's activities. In Italy and the Yusho case, public officials provided a public body count of the number of official victims. These hard data initially provided some official legitimacy, but they eventually became the objects of victims' criticisms and protests.

Organizations of Victims

For those affected by toxic contamination, forming an organization requires a transformation in identity: sufferers of an individual illness, caused by unknown internal agents, must come to perceive themselves as victims of a shared disease, caused by external human acts. Along with the new identity emerges the awareness of a community of victims and the expansion of the issue's scope. Redefinition of one's illness as a public issue assists the change in consciousness. But reports of a public issue in the mass media do not always or simply result in a victims' group.

A victims' group provides the potential for empowerment for the purpose of achieving redress. It can operate in all three dimensions of power, through direct control, agenda setting, and symbol manipulation. Organization provides a supportive self-help mechanism for fellow sufferers; by promoting solidarity with other sufferers, a victims' organization generates a positive identity and consciousness along with a sense of efficacy. For the victims, organization provides one step toward countering the ability of social institutions to define, act on, and legitimate the public issue.

Victims of a toxic disaster turn to organization when their individual attempts fail to compel institutional action (the first dimension of power). A victims' group can create sources of power by affecting the public agenda and by manipulating public symbols (the second and third dimensions of power). The group provides a social reinforcement to the victims' new identity and new "symbolic universe," and thereby constitutes a countervailing force to the institutional definitions and legitimation (Berger and Luckmann, pp. 92–128). Social reinforcement strengthens the victims' belief in an external cause and helps victims resist society's definition of themselves as socially contaminated. For individuals with stigma, organization provides the means to create a "counter-community" to refute the objective and subjective identities proposed by social institutions (p. 166).

The establishment of a formal group, then, creates "incipient counter-definitions of reality and identity" (p. 166). The process of redefinition, carried out through the organization, gives the victims a common sense of mission and accomplishment. It moves them toward achieving personal redress and restoring social justice; it provides the basis for reducing their powerlessness. The victims' group helps counter the institutional definitions of the public issue as a limited problem, as easily solved, or as a nonissue. It provides the potential for victims to change the public agenda in society.

In the symbolic domain of their struggle, victims seek to transfer the burden of social stigma from themselves to those labeled as the victimizers, the private or public institutions. The victims work to change the images of toxic contamination from symbols of internal blame, which maintain the status quo, to symbols of institutional blame, which can transform the status quo. The victims learn to use the redefined public symbols as sources of power and as instruments in bargaining with more powerful social institutions.

But the processes of organizing can reduce as well as enhance the potential power of victims. Victims can be split into separate groups; the energy of victims can be channeled into activities that reduce social influence; and a hierarchy of victims can arise, with leaders who seek their own objectives. How an organization operates critically affects its power on established social institutions. Cloward and Piven argued persuasively that "merely drawing people into organizations will not yield much power," and therefore recommended "organizing for crisis" (p. 137), a strategy discussed below in the section on protest.

Limitations on the power of organization derive also from the fact that not all victims join. Some victims perceive the costs of joining to be greater than the likely benefits, including psychological, social, and economic aspects. In some instances, victims obtain what they consider acceptable redress through individual action, without joining other victims in organization. Finally, some groups impose entrance barriers, which prevent individuals who do not meet specific criteria from joining. These barriers can prevent individuals from obtaining any redress, even if they have sustained injury.

In our three cases, forming an organization to serve the victims' interests and to increase their power was not an easy process. Creating an organization imposed both material and psychological costs on the victims. Someone needed to assume the initiative in gathering together the victims, in deciding on strategies and structures, in collecting information, funds and materials from members, in managing personality and other conflicts, and in taking actions directed at both private and public institutions. All those activities required time, energy, and resources. The scale and the distribution of these organization-building costs varied among the cases, and also within each case, depending on the nature of the chemical disaster and the circumstances of the toxic victims. Below I examine the factors that facilitated and hindered the victims' struggle for organization in the phase of public issue.

Public Recognition

Public recognition of a human-caused disaster creates an explanation and identifies a victimizer, and thereby provides a focus for organization. With public discovery of a specific victimizer, the victims can perceive a common enemy, an important symbol for mobilizing groups into action (Fagen). Public explanation of the disaster as caused by human activity also affects the victims' feelings about the injury and what to do. It creates the potential for moral indignation. As Schopenhauer wrote, "Suffering which falls to our lot in the course of nature, or by chance, or fate, does not seem so painful as suffering which is inflicted on us by the arbitrary will of another" (cited in Allport, p. 138). For toxic contamination, public recognition is a necessary condition for organization, but it is not sufficient.

In our three cases, public recognition reduced major barriers to forming a victims' group by establishing who and what caused the problem. It reduced the sense of isolation associated with a nonissue and provided sufferers with a social identity as victims of a shared affliction. It enhanced a common victims' consciousness and created a potential for organization.

Public disclosure of toxic contamination established the problem as a human-caused disaster and not a nature-caused disaster. Human-caused disasters differ from nature-caused disasters mainly on questions of responsibility and intentionality. While preindustrial society regarded nature-caused disasters as purposeful divine acts, today we view such events as neutral, unintentional phenomena. Indeed, when we recognize human actions as triggering and exacerbating the destructive forces of nature, such as the lethal flood at Buffalo Creek (Erikson), we come to view many "natural" disasters as human-caused. Human-caused disasters provide a specific human agent responsible for the damage, with various possible degrees of intentionality from stated purpose, to unintended consequence, to accident (Barkun, pp. 225–26). But these "subtle gradations visible to outsiders are unlikely to be perceived by survivor-victims. They adopt a simpler causal view: if human actions created the crisis, those actions must have been intentional" (p. 226).

The cases in this book show that attributing intentionality and blame is not that simple. The assignment of responsibility is a political and social process. Public recognition, by making possible the organization of victims, gives victims the potential to enhance their power and to become active participants in the process of attributing blame.

Public recognition, however, brought conflicting emotions to the victims, which affected their ability to organize. On the one hand, it created a sense of relief and euphoria among victims when they believed that they had solved the problem or that the problem had ended. Some Yusho victims recalled such feelings on discovering the cause of their illness. They learned the cause of their suffering, found a community of fellow sufferers, and expected a quick cure for their illness. Similarly, in Michigan, Halbert felt "something approaching euphoria" at finally identifying the cause of his problems, at solving the vexing puzzle (Halbert and Halbert, p. 133). These feelings contributed to positive views about the future and to efforts at organization.

On the other hand, for people who had expected no problem, who found themselves suddenly caught up in a chronic disaster that had just begun and seemed likely to persist, the dominant response was not euphoria but despair. The euphoria of Yusho victims rapidly dissolved into hopelessness, as the victims learned that no quick fix existed and that toxic contamination involved complex medical and social problems with no simple solutions. Public recognition of the problem also heightened social stigma and economic threats; in so doing it intensified the conflicts victims experienced about going public. When Halbert realized that public disclosure was the beginning and not the end of his problems, he felt an enormous sense of depression, connected to the fear of losing his farm, his business, his cattle—life as he knew it. At Seveso, when residents learned that the toxic cloud was not the "normal" pollution from ICMESA, that it was a disaster involving one of the most toxic chemicals known to man, they felt no sense of euphoria but anger, fear, and despair. They sensed that much worse would follow, probably illness and perhaps death. For these victims, public recognition brought feelings of hopelessness and depression, which retarded efforts at group action and contributed to feelings of powerlessness.

To join together victims needed some belief in their ability to influence the course of events. Public recognition produced contradictory effects, creating the potential for organization but also psychological tensions that delayed and obstructed group formation.

Geographical Distribution of Victims

The greater the density and proximity of toxic sufferers, the greater the ease in organizing a formal group. When a new community of victims overlaps with an existing community, the process of organization is significantly facilitated.

Physical proximity and high density of victims assist the process of organizing a formal group, just as they facilitate the process of discovering a common problem. Both processes are eased by practical reasons of closeness and social reasons of predisaster relationships. Moreover, when group members live within a bounded geographical area or have a high rate of social interaction with one another, the cohesiveness of the group is enhanced (Gamson, p. 36). Where social ties have to be created, the process of organization requires more time and effort, despite physical proximity and a shared identity as victims. It requires major individual initiatives and enormous investments of energy to contact strangers and organize a group. The occurrence of a toxic disaster within an existing community therefore increases the likelihood that an organization of victims will arise quickly. The process of empowerment is simplified for victims.

In Michigan, the wide geographical distribution of PBB-contaminated farmers across the entire state impeded the formation of a victims' group, as it similarly retarded recognition among individuals of their shared problems. The victims lacked a common forum. The public hearings held in March 1976 in farm areas created that forum for farmers with low-level contamination who had not received any redress. For this group, the hearings created an opportunity to recognize common problems as victims and served as a catalyst to organize group action.

Following those hearings, when Michigan farmers finally began to organize, geographical factors, especially the clustering of victims, shaped the process of organization. Groups formed around key organizers who lived near one another, which in Michigan meant within forty or fifty miles. Some group members lived closer, almost as neighbors, although they had not previously known each other. Often, organizers and group members lived within the same telephone-service district and thus avoided long-distance telephone charges. Physical proximity in Michigan thus served as a contributory factor in promoting the organization of victims, but not as a primary causative or sufficient factor.

Previously existing communities played a more immediate role in the Seveso and Yusho cases. In Seveso, after the evacuation, the victims moved to two separate hotels, each of which formed its own organization based on existing leadership in the community. In the Yusho case, the outbreak of victims in existing communities also facilitated group formation. The first group formed on 14 October 1968, only four days after the *Asahi* printed its scoop report of the poison-oil disease. The core of the group worked for the same company and lived in the same company-owned apartment complex. In other areas where victims lacked previous social ties, especially in Tagawa and Kitakyūshū, victims' groups took somewhat longer to form. Even in these areas, however, geographic factors shaped organizational processes. Territorial boundaries and identities contributed to the formation of victims' groups; but they also, as we see below, contributed to the splintering of victims' groups.

Organizational Expertise

Victims with more organizational expertise have fewer difficulties in forming a group. Geographically based disasters increase the possibility of an overlap between the new community of victims and an existing organization. But geographically based disasters that occur in areas with a low distribution of organizational skill put the victims at great disadvantage in creating a formal group to seek redress.

Like most political and economic resources, organizational skill is not equally distributed in society (Cloward and Piven, p. 74). Rich people tend to have more than poor people, and urban residents often have more organizational experience than rural residents. The distribution of other resources usually considered necessary for creating and running an organization, such as money, professional expertise, and personal ties to officials, also tend to be skewed away from the poor and rural people in society (Cloward and Piven, p. 80). The uneven distribution of these skills, experiences, and resources helps account for different degrees of delay in forming a victims' group and for different abilities in using the three dimensions of power.

A clear example of organizationally skilled victims who formed a group on their own initiative occurred in the Yusho case. The Fukuoka city victims, according to another victim, were "organization people," who worked for a large company, included some persons active in the company union, and "knew how to make groups" (Kamino interview). Once the Fukuoka city group formed, it served as a model for other victims to follow, especially in Tagawa and Kitakyūshū, allowing them to create organizations faster than might otherwise have happened.

Striking differences in organizational skill also existed between urban and rural settings in Japan. The urban workers of Kitakyūshū and Fukuoka cities lived worlds apart from the rural fishermen in the Gotō Archipelago in 1968. In the archipelago town of Naru people had little experience with the large organizations of modern urban society. One victim explained that prior to the Yusho incident, Naru residents had little understanding of political parties, since candidates for town assembly ran as "unaffiliated," although all ran within the sphere of the conservative Liberal Democratic party. The island of Naru did not even have its own lawyer or hospital (Kuroiwa interview). This gap between urban and rural experience belies the oft-mentioned cultural homogeneity of Japan. Naru fishermen required an education in modern organization to overcome their disadvantages and to understand the complex conflicts and organizations that invaded their lives. Although the victims gained that education through their struggle, their initial lack of organizational expertise contributed to the delay in recognizing the problem and to the delay in acting on the problem. Even forming an organization, for them, depended on the assistance of outsiders.

In Seveso, the victims possessed not only organizational expertise but their own organization, formed before the factory explosion. The second major evacuation (2 August 1976) matched almost precisely a district of Seveso that had organized itself in 1975. The area was newly settled, still semi-rural, ignored by the municipality, with few roads paved at the time of the explosion. In January 1975, residents there decided to form a neighborhood committee (*comitato di quartiere*) and in the municipal elections of June 1975, they elected a representative to the town council (on the Social Democratic list but declared Independent Left after the election). The representative, Giacomo Corna, explained that his election was possible only because Socialist and Communist voters in the district agreed to vote for him and not for their respective parties. After the dioxin problem became a public issue, Corna became the evacuees' direct representative in the town council (Corna interview). In that sense, the victims' self-help organization existed prior to Seveso's dioxin problem.

Social Structure of Victimization

The fewer predisaster social ties between victims and victimizers, the easier it is for victims to organize. In toxic contamination, the social structure of victimization can overlap the existing social structure in various ways, depending on the pathway of exposure. If the social structure of victimization overlaps existing ties of obligation and dependency between victims and victimizers, sufferers feel restricted in their public actions and group efforts against the private company responsible. Those restraints are enhanced if both sides expect the social relations to continue after resolution of the contamination problem. Existing social ties inhibit the sufferers' ability to assert public symbols of power essential to a formal victims' group, especially the symbols of self-as-victim and other-as-victimizer.

In Michigan, farmers poisoned by PBBs found themselves in a painful dilemma: their representative body, farm supplier, and general advisor, the Michigan Farm Bureau and its affiliated Farm Bureau Services, had become their adversary, poisoner, and victimizer. Normally, for political and bureaucratic action, Michigan farmers took their complaints to the Farm Bureau, Michigan's "most important farm organization in the twentieth century" (Dunbar, p. 619). But in the PBB case, farmers found a conflict of interest at Farm Bureau between its organizational interests and its members' interests, which reduced organizational responsiveness. The ongoing social obligation and financial dependencies between individual farmers and the organization initially restrained the victims' ability to voice their complaints, to form a separate organization. The company's definition of the problem and its legitimation of policy—especially its use of the corporate myth of serving the farmer—became significant but not insurmountable obstacles to PBB victims in Michigan. The structure of victimization and the institutional context thus delayed the mobilization of Michigan farmers as PBB victims.

In the Yusho case, by contrast, victims lacked such restraints. Few victims had any connection to the Kanemi Company, which was marginal in the local, regional, and national economies. The victims could more easily confront the Kanemi Company directly and publicly and mold the company into a symbol of evil for mobilization and for confrontation. The Yusho case suggests that although the structure of victimization in consumer-product situations complicates and delays the discovery process, the social distance between producer and consumer can facilitate the mobilization and organization of victims once the case becomes a public issue.

Seveso residents also did not feel constrained by the social structure of victimization, at least in relationship to ICMESA. They were not dependent clients (like Michigan farmers) or independent consumers (like Yusho victims) but were neighbors in the bordering town. Though ICMESA was located in Meda, the wind blew the toxic cloud over Seveso. The relative lack of strong social obligations between victim and victimizer in Seveso facilitated an early mobilization of victims.

Cultural Context

Culture provides a context within which victims approach the task of organization and from which they select appropriate symbols. As a system of interacting symbols that give shared meanings to a social group, culture plays an essential role in our cases of toxic contamination, providing people with a universe of meaning, within which they define, implement, and legitimate reality. At the same time, culture does not determine the actions of individuals or institutions. Nor is culture immutable; the symbols can be changed and redefined, as different groups seek to control the meanings attached to key public images. Our cases demonstrate the inadequacy of broad cultural themes as explanations for patterns of organizational development.

Japanese culture is commonly described as promoting a strong group consciousness, in which the individual totally serves the organization as an end in itself. But the Yusho case shows conflicting evidence about the role of group consciousness in Japan. Victims in Fukuoka city quickly formed an organization, but victims in some other areas did not become formally organized for one year or more after public disclosure. The cultural promotion of a strong group consciousness did not alone result in the formation of Yusho victims' groups, nor can it explain why victims' groups formed quickly in some parts of Japan and not in others. The case demonstrates that group formation is not an automatic process in Japan and that group consciousness by itself does not produce a group. Other factors, especially existing social ties and the distribution of organizational expertise, are more important than general cultural propensities in explaining the complex process of group formation for Yusho victims. Another cultural theme found in writings on Japan is the reluctance to litigate (Kawashima). That aversion to litigation is often tied to the desire to mediate a dispute rather than engage in head-on conflict, a pattern connected to Japan's alleged consensus-oriented attitude (Haley, 1978, pp. 359–60, n. 1). Yet in the Yusho case, Fukuoka city victims went to court almost immediately after public disclosure and on several occasions. In December 1968, the group filed criminal charges against the Kanemi Company president. In February 1969, it filed a complaint with the Fukuoka office of the Ministry of Justice against the doctors in the Yusho Research Group, charging the doctors with irresponsible treatment of victims and violation of the victims' rights. Also in February 1969, the group filed its suit for civil damages against both companies. For this group, the alleged cultural propensity against litigation had little effect.

In other areas of Japan, victims filed civil damage suits much later, beginning in December 1970. Japan's alleged cultural discouragement of litigation, then, did not alone impede the move to legal action, nor can it explain why groups resorted to litigation at different times. Other factors, especially differential access to legal and political resources, played more important roles than general cultural orientation. As legal scholar John O. Haley argued, Japanese do not feel a special cultural aversion to litigation, but they do confront significant institutional obstacles to litigation, especially a lack of lawyers and judges, overcrowded courts, long delays in completing a trial, as well as structural obstacles to "prompt and efficient justice" (Haley, 1978, 1982). In the Yusho case, the victims' group in Fukuoka city showed a strong group consciousness (supposedly consistent with traditional culture) along with a great propensity to litigate (supposedly inconsistent with traditional culture). This contradiction illustrates the problems of using broad cultural traits to explain the behavior of toxic victims.

But the subculture of Japan's antipollution movement did provide Yusho victims with a context to understand and act on their problems. By the late 1960s in Japan, controversy over pollution had become a major political issue, and the antipollution movement had created symbols of pollution victims well known to many Japanese (Huddle and Reich). Conflict over Minamata disease, the infamous case of mercury poisoning in southwestern Kyūshū, set the symbolic stage on which the Yusho disaster played. Many Yusho victims, living on the same island of Kyūshū, learned models of victims' groups from the precedent of Minamata disease. That subculture of the antipollution movement facilitated the process of organization for Yusho victims. The subculture provided them with symbols and strategies that served as sources of power to contest the definitions of Yusho provided by social institutions, to get their problem on the public agenda, and to compel redress from social institutions. One specific strategy, the demand for a personal and public apology by the president of the responsible corporation, connected with Japanese cultural traditions and shaped the targets of protest and corporate responses, as discussed below.

In the Michigan case, culture provided contradictory impulses for victims, encouraging both group action and individual action. Americans are commonly considered a group-forming people. In cultural stereotypes, Japanese are highly committed to an existing group, while Americans are more dedicated to forming new groups. According to Tocqueville, "Americans of all ages, all stations in life, and all types of dispositions are forever forming associations" (p. 513). But in Michigan, the first formal group of farmers affected by PBBs did not appear until the summer of 1976, two years after public disclosure. That delay argues against the group-forming cultural generalization.

A contrary cultural pattern, the farmers' strong values of individualism, provides a context to explain that delay in organization. One Michigan historian wrote, "The farmer of a century ago was a rugged individualist, and not infrequently an eccentric. As agriculture became less of an art and more a science, he lost some of his intense individualism, his suspicion of new-fangled ideas, and his self reliance" (Dunbar, p. 618). The PBB poisoning demonstrated all too convincingly the farmer's search for, and acceptance of, "new-fangled ideas" and technologies (magnesium oxide for dairy feed to increase production) and the resulting decrease in his "self reliance." But the farmer's belief in rugged individualism lived on, usually tied to a Republican rural conservatism. Halbert's personal struggle to solve his problem exemplified the lost but romanticized intense individualism of the Michigan farmer, as well as its relative impotence when confronted with complex modern organization. Those individualistic values discouraged organization and group action.

Within the cultural context of conflicting group and individualist values, structural factors in Michigan determined the victims' approach to organization. The social relationship of victims and victimizers acted to restrain the potential power of Michigan's rural tradition of group action. In the late nineteenth century, farmers banded together in their own organizations, such as the Grange, to overcome powerlessness against the railroads and the farm-machinery manufacturers (Dunbar, p. 618). But in the PBB incident, farmers had to confront their own organization, the Michigan Farm Bureau, which had grown to assume many powers of the railroads and farm suppliers and to control more than serve the individual farmer (Berger).

The cultural context in Seveso provided a generally positive orientation toward organization. The Seveso area belongs to a largely "white" district, where the influence of the Catholic church prevails and secular leftist ideas are weak (in contrast to a "red" district, where the Communist party dominates). Seveso residents lacked clear models of pollution victims' groups as in Japan, but they also lacked the values of stubborn individualism found in Michigan. Residents placed a positive value on organization, while their basically conservative political beliefs placed a positive value on existing social institutions. The victims' direct political representation in the Seveso town council (through Giacomo Corna) also supported the cultural context for organization. In their initial response, ICMESA victims organized themselves into a group but continued to use and depend on existing community leaders and structures.

Conflicts among Victims' Groups

Rarely does a single organization arise to include all victims. Multiple victims' groups form, and conflicts emerge among the groups over questions of strategy, leadership, policies, and affiliation, thereby limiting the effectiveness of organization.

Within the formal victims' organizations of our cases, debate over the strategy for redress became a major focus of controversy, as competing leaders advocated different positions. The question of strategy often involved settlement versus litigation versus protest. In the Yusho case, in the small town of Tama no Ura, bitter hostility persisted for years between two factions of a group that split apart, over a disagreement about whether to settle out of court or to confront the company in litigation. Conflicts over strategy also occurred among farmers in Michigan as some settled privately, others quietly sold their cattle and started over, and still others fought the PBB issue in court and in public. In Italy, different strategies also formed lines of conflict among victims, as some evacuees from the southern area of zone A fought to have their homes cleaned for reentry, while others decided never to return to their homes.

As the victims' groups became increasingly formal organizations, they acquired their own internal rigidity, reflecting a transition from "movement" to "institution" (Alberoni). The leaders in control came to represent a particular strategy and to suppress dissent. In Seveso, several families that did not want to return to their homes and challenged the leadership's strategy found themselves effectively ostracized from the victims' organization. In the Yusho case, Kamino Ryūzō resigned as director of the confederation of victims' groups because he considered the organization overly bureaucratic and therefore ineffective. Similarly, the PBB Action Committee in Michigan experienced increasing rigidity that excluded some victims from active participation and constrained the group action mostly to litigation.

Victims also were divided by public administrative policies. Official decisions about "safe" levels of contamination and official definitions of disease due to contamination divided those victims who had already received compensation from those who had not. Official victims often sought to prevent others from becoming defined as victims. Michigan farmers were divided by the level of PBBs in their cattle. Residents near Seveso were divided by the lines of dioxin levels and the boundaries of zone A. Yusho victims were divided by official criteria for qualifying as a certified patient of the oil poisoning. In each case, dissatisfied minorities of victims arose, people who believed they had problems caused by toxic contamination but not recognized by the authorities.

These dissatisfied minorities challenged the official definition of the situation, seeking to change the lines of cutoff and the shape of policy. They became a source of group action that helped transform public issues into political issues. When peripheral victims were excluded from the benefits of redress and from membership in existing victims' organizations, they looked elsewhere for support in their efforts to redefine the public issue and alter administrative boundaries on the toxic problems, and to change the institutional definitions of the public issue. Often, as discussed in the next chapter, peripheral victims aligned themselves with individuals and groups outside the main political currents, with those who sought potential political issues among dissatisfied minorities.

Opposing interests and conflicting groups arose among victims as a consequence of these institutional decisions. In Michigan, an organization of already-compensated victims, Concerned Michigan Farmers, argued in 1977 that farmers with low-level PBB contamination continued to complain for personal economic reasons and not for reasons related to PBB-damage problems. In Seveso, a group of property owners and local businessmen, worried about long-term economic impacts on the town, opposed efforts to enlarge zone A and supported efforts to return the district to "normality." In the Yusho case, some victims complained that other victims (who received official recognition later) had not suffered as much, had not worked sufficiently for the victims' organization, and through their claims might make the Kanemi Company go bankrupt.

The legitimacy of the peripheral victims' claims of injury and demands for redress became public controversies in each case. From the viewpoint of social institutions, both public and private, persons outside the official boundaries of the contamination were not victims. They were perceived and represented as seeking to become official victims for psychological or economic reasons: they wanted to join in sharing the benefits of redress.

From the perspective of existing official victims, the peripheral claimants appeared as free riders seeking to enjoy the benefits of redress without paying the full costs. They were portrayed as threatening to break the bank of redress (especially in the Michigan and Yusho cases) or increase the number of pieces from a limited pie, in either case, forcing a cutback in benefits for existing victims. Tragically, official victims came to depend on the continued economic success of the victimizer, and they resisted efforts that might undermine the victimizer's financial well-being. The peripheral victims, on the other hand, came to see themselves as unjustly discriminated against by public and private institutions—and by official victims.

Conflicts over strategy, leadership, and policy became embodied in organizational affiliations. These conflicts among victims' groups weakened the potential power of the victims by dividing their efforts. Whether intentional or not, administrative policies of both public and private organizations enhanced conflicts among victims' groups, working as efforts to divide and conquer. Sometimes the victims themselves created the divisions, sometimes the social institutions did. As one group of victims turned against another, the authorities could better maintain social control. That task was assisted by the increasing organizational rigidity of some victims' groups, which became involved in administrative actions and controlled through negotiated concessions.

Organization served important symbolic and material purposes for victims, placing them on the political map for dealing with the chemical disaster and giving their demands for redress enhanced legitimacy and power. Organization gave victims access to the power to mobilize bias and manipulate symbols. But organizations also had limitations in their trend toward internal rigidity, exclusion of potential members, and creation of conflict among victims' groups. Protest provided another form of group action and empowerment, which helped victims overcome some of those limitations in obtaining redress.

Protest by Victims

Protest is a central strategy for dealing with the "problem of the powerless," the lack of political resources to bargain with the elite (J. Q. Wilson). Protest uses institutional disruption to mobilize the powerless and force concessions from the powerful (Piven and Cloward). By employing group defiance of social norms, protest operates in the first dimension of power to exert influence over others. Power is exercised through "the application of a negative sanction, the withdrawal of a crucial contribution on which others depend," when the protesters cease to conform to expected social roles and create institutional disruptions (p. 24).

Protest also operates in the third dimension of power, by conveying important symbolic messages to mobilize third parties. The powerless commonly rely on that broader process of mobilization (Piven and Cloward, p. 24n). The symbolic dimension of protest is directed at the mass media, to create reportable events and thereby shape social reality, and at other potentially sympathetic third parties. Protest seeks to mobilize sympathetic "reference publics" and enlist their power to change social institutions and policy (Lipsky). But both protest as institutional disruption and protest as third-party mobilization create new conflicts for victims.

Origins of Protest

Victims move to protest when they believe that other avenues of appeal and other paths of communication have failed. As Cloward and Piven persuasively argued, people resort to "breaking the rules" when they believe they have no other alternatives and no other resources for influencing the decisions of social institutions (p. 86).

An economic rationalist would argue that people choose protest when they believe the costs of protest are less than the costs of other means (such as litigation or legislative appeal) to achieve the anticipated benefits, and, moreover, that the expected benefits would exceed the likely costs of protest. In contrast, Cloward and Piven stress the calculation of comparative effectiveness more than the estimation of probable costs and benefits in the choice of protest. Conflicts over strategy derive in part from differences among victims and groups in the weights they give to effectiveness and to costs in considering protest as a means to achieve the objectives of redress.

In our three cases, the transition from discovery to protest was not a simple process. It required many months; and even then, not all victims joined in public disruptions. Going public in protest as a toxic victim required courage and a willingness to take risks, a disillusionment with existing institutions, but also an optimism and hope that protest would accomplish something. In Michigan, protest did not begin until at least one year after the contamination's discovery. In Japan, protest began several months after public disclosure. In Italy, the first planned protest by victims occurred three months after the explosion.

Why did Seveso residents respond so quickly with a public disruption at the time of the second evacuation on 2 August 1976, just three weeks after the toxic cloud? On one level, the victims exploded in anger at the trauma of evacuation, at being separated from the homes they had built by hand, from the gardens that fed them, from all their material possessions and their community. The evacuation stripped them of their existing social identity and imposed a new, degrading identity as contamination victims. On a more immediate level, they exploded in anger at the ineptness of public administrative actions: the confusion caused by an uncoordinated policy of evacuation, which kept them waiting for three hours in the hot sun. Under that anger festered a deeper resentment over the three weeks of delay before the public authorities decided to evacuate the area.

Lombardy's regional health director argued that the victims' early outburst of resistance resulted from the nature of toxic contamination and from the nature of required administrative actions, not from any administrative ineptness. He insisted that similar demonstrations would occur "if someone in a white suit came to American citizens to evacuate them with absolutely no belongings." Such resistance would happen, he claimed, in "any population, no matter how much faith it had in institutions" (Rivolta interview). This interpretation reflected the general denial by public officials of any inadequacy of administrative actions, except perhaps in public relations. As the Lombardy health director put it, "The region was not deficient in taking measures, but did not sufficiently continue to inform people every day." He admitted only that a lack of official information contributed to confusion (Rivolta interview). But his statements cannot explain the lack of public resistance in the first evacuation and its presence in the second evacuation. That difference is more persuasively explained by administrative ineptness during the second evacuation. Similarly, at Love Canal the evacuations in 1978 and 1980 did not by themselves cause resistance, but administrative confusion and ineptness over policies of evacuation did result in public outbursts of rage by victims (Holden).

The Seveso case illustrates that protest and organization have a complex relationship. The initial public disruption at Seveso resembled ad hoc resistance without an organization more than an institutional disruption planned by an organization. Indeed, the public authorities planned the gathering, which turned into the first protest. This event shows that protest occurs along scales of premeditation and organization. In some cases, protest can be unanticipated, even by its participants; in others, organizations carefully plan and control the protest and regularly rely on this strategy.

Planned protest, distinct from ad hoc resistance, arose as some victims lost faith in the legitimacy of normal institutional means to resolve the contamination problems. Some victims concluded that administrative actions had failed to resolve the pressing problems of redress and that the institutions would continue to fail. In Michigan, low-level PBB farmers confronted testing delays and confusion as well as official denials of PBB contamination and PBB-related illnesses. In the Yusho case, victims confronted official denials of responsibility, lack of financial assistance or compensation, and failure of medical treatment. And in Seveso, victims decided on protest after "many promises never kept," as victims' representative Corna explained in November 1976. "They kept us continually with the sword of Damocles over our heads, first promising mountains and seas, deluding us that we would be returned to our homes in a few months, then telling us that the time would be prolonged into years. In this condition, it is not possible to continue living; our nervous system is being shattered" (*Giorno*, 15 November 1976).

Victims expected the public authorities to intervene, provide redress, and restore the previous social conditions. Victims made these demands based on their vision of social justice. But gaps emerged between the victims' expectations and their perception of events. Those gaps provided motivation for protest, pushing victims to use any source of power they could mobilize in order to affect the decisions that controlled their lives.

Symbols and Protest

To expand the scope of a public issue, victims develop images of their suffering into public symbols that can be used in protest. People who are powerless reach into their own social circumstances for images that can be molded into symbols of protest (Piven and Cloward, p. 20). In protest, the powerless transform symbols of suffering and oppression into sources of power and opposition. They change symbols that maintain their stigma and isolation into symbols that restore their health and wholeness, creating a symbolic reversal.

In our cases, the victims redefined the images of toxic contamination into symbols of protest. They used their scarred bodies or afflictions as public symbols to demand the impossible: a restoration of their original condition, a total and complete redress. They demanded that private and public institutions return their bodies and their communities to good health, that they be decontaminated socially and physically. The victims of other chemical disasters, such as those at Love Canal, made similar demands for social and physical decontamination (Holden). Those feelings can be understood by other people, by nonvictims. In all three cases, the victims demonstrated an intuitive sense about which images would generate the most powerful symbols in society to emphasize their persistent affliction and suffering.

The initial protests by victims evolved around central images of toxic contamination. Michigan farmers, in an early protest, brought sick cattle contaminated by low levels of PBBs to the hearing held by the state Department of Agriculture in May 1975. The first protest by Yusho victims involved appeals on city streets, in which victims displayed their scarred faces. Seveso residents, in their first planned protest, blocked the highway, stopping the flow of automobiles and normal life, and then reentered their still contaminated homes in zone A. In all three instances, the protests asserted images of toxic contamination that compelled other citizens to recognize and feel the victims' suffering. The symbols of protest expressed the victims' sense of injustice at being arbitrarily selected to suffer and then denied redress. The symbols warned nonvictims that they too could suffer a similar injury.

The protesters in our three cases focused on symbols of life and death: dying cows in Michigan; severely scarred faced in the Yusho case; the evacuated, dead zone A in Seveso. Victims chose symbols of life and death not simply because the symbols expressed the pain of their suffering but because they represented ultimate forms of power: to restore the dead to life and to threaten the living with death. In choosing symbols that represented unjust death, the victims challenged a basic legitimacy of the social order: the legitimation of death and its integration into social existence (Berger and Luckmann, p. 101). The symbols around death also served to unite and mobilize the victims, to help them overcome feelings of despair and hopelessness. The choice of these images reflected a broader pattern of using death symbolism in movements for political and social change (Lifton, Katō, and Reich, pp. 277–89).

The victims used these symbols in dramatic social disruptions so as to create reportable events that rose on the public agenda and to challenge the authority of private and public institutions. The use of death symbolism enhanced the intensity of protest: a farmer's shooting of his own cows in Michigan, a mass rally to commemorate the deaths of two Yusho victims, a public occupation of Seveso's zone A and intentional exposure of self and others to deadly poison. These disruptions challenged the legitimacy of institutional policies and actions and sought to redefine the public issue in favor of the victims' interests.

The symbols of protest expanded the scope of the public issue by connecting the particular issue to broader ideas in society. In Japan, victims connected Yusho to the national concept of "pollution disease," while the public authorities defined Yusho as an isolated case of food poisoning among private parties. In Michigan, victims expanded the administrative definition of a private agricultural problem to the victims' definition of a public health consumer issue. In Italy, victims linked the public issue of dioxin contamination to the broader community and its economic interests, beyond the administrative focus on just the contamination. By redefining the issue and by connecting it symbolically to other issues, victims expanded the number of participants in the public issue and helped move the problem to the political realm.

In expanding the issue's scope, victims rejected the language of social institutions and asserted their own symbols, using the third dimension of power. While symbols of death figured prominently in protest, victims needed to redefine other images as well, to create their own symbolic universe. They needed their own counter-definition of social reality, one under their own control. A Yusho sufferer explained why he and other victims did not use the Japanese word for contamination (*osen*) to describe their disease: "*Osen* means dirty, defiled, contaminated. It is used by the authorities to minimize the damage. . . . But in Yusho, families have been destroyed, the integrity of the human body destroyed. This is *hakai*—destruction, ruin. People have been killed by something violent and furious. It is tragedy" (Kamino interview). Similarly, Seveso residents aggressively rejected the toxic "mark" placed on their community by signs on the highway that warned of dangerous contamination.

But no toxic sufferer lightly assumed in public the symbol of self-as-victim. Some victims feared that public action and protest would bring reprisals at their job (Ujino interview). Others feared social discrimination, losing friends, and being marked as different. The fear of permanent contamination, however, also moved victims to planned protest and to attempts at symbolic reversal in order to advance their demands for redress, social justice, and restoration of a healthy life.

Organization and Protest

Absence of stable organization and leadership can increase the likelihood of protest, while formal organization tends to use protest to gain recognition but then tends to avoid or ritualize protest.

Victims do not always first organize and then protest. Indeed, most formal organizations, especially larger ones, tend to reduce and contain protest. Though seeming counterintuitive at first, this organizational pattern has been explained by Murray Edelman: "Formal organization institutionalizes action, formalizes signals of support and opposition, [and] reduces uncertainty about others' behavior" (1971, p. 177). Consequently, organization inhibits more unpredictable and militant action by individuals and promotes patterns of ritualization and cooptation. Formal organization uses planned protest to gain official recognition and put issues on the public agenda but then usually decreases its reliance on protest and employs other channels of communication and pressure.

The absence of organization, on the other hand, "means absence of a chance for effective communication and for establishing an accepted basis for coexistence" (Edelman, 1971, p. 26). A group without social recognition and without stable organization and leadership will express its anxiety and its interests in some manner. Those expressions of anxiety often take the form of protest as ad hoc resistance, as occurred initially at Seveso. Indeed, when conflict occurs between diffuse, ill-defined groups, social control becomes "virtually impossible" (Dahrendorf, p. 226).

Our three cases provide examples of protest preceding organization and also of organization containing protest. In the Yusho case, Kamino Ryūzō resigned from the main victims' organization because he felt it focused too narrowly on litigation and neglected direct action against the company and the government. In Michigan, protests by farmers preceded the victims' organization, the PBB Action Committee. Farmers who displayed dead cows at the March 1976 protest at the state capital did not belong to a formal group. Once the PBB Action Committee became organized, however, public acts of protest declined, as attention focused on legal action and public hearings. At Seveso, several large protests of victims occurred in the first year, but such activity subsequently subsided, as most victims became involved in settlement negotiations and as cleanup operations progressed in the southern part of zone A.

Both public and private institutions prefer to deal with victims in organizations rather than in protests. This preference underlies both the leverage and the limitation of protest by toxic victims. In their struggle for empowerment, victims need protest, but they also need organization.

Targets of Protest

"Institutional roles determine the strategic opportunities for defiance, for it is typically by rebelling against the rules and authorities associated with their everyday activities that people protest" (Piven and Cloward, p. 21). How social institutions respond to toxic contamination affects their prominence as targets of protest. But the political beliefs of victims also influence the direction of protest, especially the choice between focusing on private or public institutions.

Victims in our three cases chose different targets of protest among private and public institutions. In the Yusho case, although some victims protested against prefectural and central authorities, most focused on the companies, mainly the Kanemi Company but also Kanegafuchi Chemical. In the Italian case, the reverse happened. Some protests were directed against ICMESA and Givaudan, but most concentrated on regional or municipal administrations. In Michigan, no protests were made against

Michigan Chemical, some against Farm Bureau Services, and most against the state government.

Among public institutions, victims chose protest targets based on the institution's degree of authoritative involvement: the more involvement, the greater the likelihood of becoming a target of protest. ICMESA victims never demonstrated against the central authorities in Rome but did protest against the Seveso municipal government on several occasions and against the regional Lombardy government most often. Once the region located its special office in Seveso, that institution, both authoritative and convenient, became a magnet for protest, leaving the municipal and regional governments almost ignored by demonstrators. In Michigan, farmers protested most often against state agencies in Lansing, and they traveled to Washington only on rare occasions and in small groups, usually to lobby or testify and not to protest. Only in the Yusho case did victims travel relatively often to Tokyo to demonstrate, reflecting the greater role of the central authorities in managing the Yusho incident and the greater centralization of power in the Japanese political system.

For private institutions, the victims' choice of protest targets was affected most by the administrative policies of institutions, especially policies for compensation. Companies that involved victims in settlement proceedings reduced the probability of becoming a target for protest. Once ICMESA/Givaudan began private settlements with local residents, protest against the company by those victims decreased. Similarly, Michigan farmers in court or in negotiations avoided public protest, since they did not want to jeopardize potential settlements. The Kanemi Company also used out-of-court settlements, as in Tama no Ura, to contain demonstrations by victims. But companies also adopted different strategies to cope with increasing political conflict, and the choice of a strategy affected the company's probability of becoming a target of protest, as discussed in the next chapter.

The political beliefs of victims affected whether they focused on private or public institutions as the target of protest. Conservative political tendencies fostered protests against public institutions and against left or centrist governments. Progressive political tendencies fostered protests against private institutions and against right or centrist governments. When conservative and progressive politics focused protest on a single government institution, private institutions could more easily recede from public scrutiny.

In Michigan, most PBB farmers held conservative Republican political beliefs, with a deep distrust of government and bureaucrats. Those beliefs justified demonstrations against state and federal agencies but delayed protests against a private company. The farmers' worldview reinforced the structural factors that discouraged protest against Michigan Chemical and supported the dependency relationship that reduced direct protest against Farm Bureau Services. Indeed, the victims' conservative orientation made them reluctant to protest against anyone. That reluctance decreased, however, as they experienced persistent difficulties with private and public institutions.

In Seveso, toxic victims held conservative beliefs that focused protest more on regional authorities than on local town officials. Victims in Seveso opposed regional policies for both economic and ideological reasons. Conservative business groups in Seveso protested against the region because of its policies that threatened important economic interests in the town. Groups associated with the Catholic church protested against the region because of its policies that liberalized abortion. These conservative beliefs were reinforced by local perceptions of these two levels of government. While Seveso residents perceived the mayor as a member of the community, they viewed the region almost as a foreign force (Conti, pp. 26–27), thereby facilitating and justifying protest against the region.

More generally, the Brianza area of Lombardy, in which Seveso is located, is known for its strong clerical tradition and its weak Socialist and Communist organizations, patterns that affected the targets of protest. The clerical tradition gained additional support from the presence in Seveso of an important seminary located near ICMESA and near zone A, and from important church figures with ties to that seminary. Thus, although ICMESA/Givaudan presented a good symbol of protest for the left, as a multinational company from Switzerland, the lack of a well-organized left in Seveso meant that most protesters against the company came from outside, especially from nearby Milan. Moreover, those protesters represented the new left more than the old left, since the Socialist and Communist parties were involved in public administration at both provincial and regional levels, thereby helping contain protest, especially against the public authorities. As a result, the Lombardy region became the favorite target of protests from conservative Seveso groups and from new left political groups, and protests against ICMESA/Givaudan declined in prominence over time.

In the Yusho case, the victims held a range of political beliefs, both progressive and conservative. The labor union experiences of Yusho victims in Fukuoka and Kitakyūshū cities promoted an early focus on the private corporations. Some victims, with a progressive orientation, helped organize protests against the Kanemi Company. Those protests gained additional impetus from the position of the Kanemi Company's president as an important right-wing figure in Kitakyūshū, making him a better target for the left. The public administration's definition of the

Yusho problem as private, to be negotiated between victims and victimizer, and previous pollution cases such as Minamata, in which victims
demanded that company officials personally acknowledge responsibility,
both supported a direct approach of protest against the Kanemi Company.

Conservative rural sufferers of Yusho also demanded that Kanemi
Company officials take responsibility for their acts. They followed a cultural orientation in Japan that stresses symbolic and actual forms of
taking responsibility for social wrongs. In Japan, then, a conservative
orientation promoted a focus on the private institutions. While victims in
Italy and the United States also believed that corporate officials should
take responsibility for their acts, Japanese victims placed a special stress
on personal responsibility and public contrition, reflecting a cultural tradition. This cultural expectation affected corporate strategies to deal with
social conflict, as discussed in the next chapter. The focus of protest on
the Kanemi Company was also promoted by the moral position taken by
Kamino Ryūzō, based on his faith in Christianity, which became expressed in a personal protest against the company.

Effectiveness of Protest

Protest enhances the power of toxic victims but also encourages institutions to take measures to control protest.

What does protest accomplish? Protest empowers toxic victims and
enhances their bargaining power with social institutions. By creating a
"publicly visible disruption in some institutional sphere," victims can
redefine an issue, challenge institutional legitimacy, and compel administrative action (Cloward and Piven, p. 99, italics in original). This strategy of "organizing for crisis" presses private and public institutions to
respond (p. 137). "Public trouble is a political liability; it calls for action
by political leaders to stabilize the situation" (p. 99). For private institutions as well, protest is often necessary to get the attention of corporate
managers, to put an issue on the corporate agenda, and to compel administrative action (D. Vogel, 1978, p. 207).

But protest also increases institutional incentives to contain the protest
and victims' power. The containment efforts come from private and public organizations but also from the victims' own organizations. Only
through stubborn commitment to moral principles and dogged resistance
to institutional policies can victims persist in protest, as demonstrated by
the personal crusade of Kamino Ryūzō outside the gates of the Kanemi
Company factory. In most cases, protests by victims succumb to personal
exhaustion or institutional containment, or both.

In all three cases of this study, protest proved effective in placing the victim's redress on the public agenda and in expanding public acceptance of the victims' symbolization of reality. In both these dimensions of power, protest increased the pressure on social institutions to expand the amounts and kinds of redress and to extend the boundaries of the official victims. Protest was used more frequently by peripheral victims, the more powerless among the powerless, those outside the main victims' organizations, and those excluded from the early disbursements of redress.

The victims' use of institutional disruption as a means to increase their power raised the incentives for the targets of protest to design ways to defuse or deflect protest. Social institutions responded with concessions intended to make protest less effective and more difficult for victims to pursue. But the concessions also included redress that met some of the victims' demands.

Through the mass media, protest attracted the attention of other groups in society. It served as a call for help to achieve redress. It assisted in the victims' efforts to expand the issue's scope by widening the participants in the conflict. As institutional policies of containment made protest more difficult, victims increasingly needed outside allies to continue their struggle for redress. These circumstances structured the victims' dilemma of political alliance and advanced the transition to a political issue.

Chapter 7

Political Issue: Society's Conflict

As the victims organize and protest and the scope of the public issue expands, various organizations—governments, companies, political parties, social movements, and the mass media—incorporate the problem into their repertoire of issues. Their differing perspectives produce a crescendo of controversy that transforms the victims' struggle into society's conflict and turns the public issue into a political issue.

What distinguishes a public issue from a political issue? Although most analysts do not make this distinction, separating a public from a political issue emphasizes Schattschneider's notion of scope. A public issue, that is, a problem perceived to involve a group of victims rather than isolated individuals, enters the political sphere when its scope expands beyond the victims to involve other organizations, which posture on the issue, jockey for positions of power and advantage, and compete in the political arena to control the issue's causes and solutions. A political issue reflects a social conflict broader than the victims' struggle; it includes protest by interested nonvictims.

Outside organizations, approaching a political issue with their own organizational objectives, use the issue for political gain, to change the balance of power in society. As the late Terry Dolan, a prominent new right organizer in the United States, put it, "I think the social issues [such as abortion and school busing] are really political tools more than questions for political debate"; they provide a means for conservative Republicans to "destabilize" the Democrats and delegitimate the liberals (*Boston Globe*, 4 March 1981). In Italy, that perspective is known as *strumentalizzazione*, a word used to accuse someone of ignoring an issue's substance while exploiting the issue as a political instrument for personal or organizational gain.

This chapter examines how victims and their supporters approach toxic contamination as a political issue and how social institutions respond. I examine the victims and their supporters first because in the phase of political issue, they gain the initiative and compel social institutions to respond. I am especially interested in how victims and supporters form alliances, expand the issue's scope into the political domain, and use conflict to increase the alliance's power. Next I analyze private and public institutions' strategies to protect themselves from the costs of expanding social conflict by reintegrating conflict into normal social processes. Institutional responses to the political issue move through stages of dissociation, confrontation, and diversion.

For the victims, expanding the issue into politics and expanding the conflict in society offer the promise of increased power. But that strategy also creates the risk of reduced power, for the process produces dependencies on supporter organizations, which carry their own baggage of controversies, ideologies, and strategies. Moreover, some alliances create conflicts among victims as well as protests against institutions. Victims thus confront the dilemma of political alliance, the dilemma that they gain power only through a sacrifice of autonomy.

Victims, Supporters, and Conflicts

Victims initially do not want politics or political resolutions for their toxic problems. They want redress and justice, because they have been harmed. But victims gradually come to recognize that to obtain redress, they need to expand the issue into the political domain. Because of the limits of their own organization and protest, as discussed in the previous chapter, the victims need outside help in expanding the issue. Allies can help empower the victims to put pressure on social institutions, influence decisions about policy, and obtain adequate redress.

The victims' allies resemble Michael Lipsky's "reference publics" and sympathetic "third parties." Victims do not, however, make alliances simply through Lipsky's notion of protest as "unconventional public displays." More important are bargains, sometimes explicit and sometimes not, between victims and supporters about what each party expects to gain from an alliance. Lipsky distinguished between "activating reference publics" and "alliance formation" but assumed that the goals of groups in an alliance are similar (p. 1146). That assumption greatly underestimates the tensions that arise among allies.

The bargaining between victims and supporters structures the victims' dilemma of alliance. Victims sense that reducing their powerlessness requires an alliance but also that creating an alliance can lessen their control. A symbiosis develops in which victims and supporters each use the other for different purposes, not always in balanced ways. Similar problems with alliances occur in other areas of political action, such as the alliance politics of international relations (Neustadt, pp. 115–51). In our cases, competing alignments of victims and supporters reflected the existing patterns of political competition and caused dissension among the victims themselves. Along with helpful symbolic and material support, outsiders brought new conflicts.

As victims recognize their own powerlessness and realize they need allies, they scan the horizon to see where they can build coalitions, with which groups, and at what costs. They begin the search with other victims and then extend it outward, often toward the mass media, social movements, and political parties. Each potential ally asks, What can I gain? In bargains made with allies, victims may have to give up part of the redress they believe they deserve. The decision to join a coalition is not, however, the purely economic calculation described by Mancur Olson. The decision to join can result from outrage at the victimizers and outrage at political bargaining. For some, the drive to get justice and revenge overcomes some costs of alliance that other people would not tolerate.

Victims' Search for Allies

The effectiveness of the victims' search for allies depends on the victims' personal, organizational, and political connections. Some victims have an initial advantage in building alliances with nonvictims. They may have personal ties to experts or people who can help. They may belong to organizations that can provide support. Or they may know persons who can obtain political assistance. Socially marginal people who lack such connections encounter greater difficulty in coping with the stresses of victimization, as Kai T. Erikson found in his study of the survivors of the Buffalo Creek flood.

In each of our cases, personal ties provided an important pathway to allies for some victims. When Kamino Ryūzō needed a lawyer, he contacted an acquaintance who had previously worked at the coal mine where Kamino was employed. That lawyer, with connections to the Japanese Communist party, initiated legal steps that evolved into the mammoth trial in Kitakyūshū, associated primarily with Communist party lawyers. This litigation was accompanied by broader political protests of Yusho victims and support groups. In addition, the legal strategy contained political objectives in that it named the city government of Kitakyūshū and the central government as plaintiffs in the trial. Another

Yusho victim in Fukuoka city contacted an old friend who worked as a lawyer. That lawyer, with ties to the Democratic Socialist party, organized the smaller trial in Fukuoka city, with an emphasis on obtaining compensation through conventional tort liability. This trial did not include the city or central government as plaintiffs and was not coordinated with external political action or a social movement.

Using organizational connections to recruit allies accelerated the process of locating nonvictim supporters. The victims employed at Kyūshū Electric Power Company, for example, received support from their union organization. In Seveso, one worker at ICMESA contacted a physician who had worked with the local union on occupational health problems. They both belonged to the same new left political organization, the Workers' Movement for Socialism, a link that provided the basis for rapidly creating an outside support group.

Appealing to political representatives was a third path for recruiting allies. In Michigan, farmers with PBB-contaminated herds initially appealed to their traditional supporters in the Republican party. But as the problem became increasingly politicized, farmers affected by low levels of PBBs (often due to secondary contamination) found Republican politicians less receptive and Democratic politicians more receptive, reflecting the growing partisan nature of the dispute. At Seveso, victims also appealed to their political representatives, especially the local mayor and their city councilman Corna. In that case, the proximity of representative and constituent within the same community, as well as the status of Corna as both toxic victim and direct representative, made politicians much more accessible in Seveso than in Michigan or in the Yusho case.

Social class affects the victims' personal, organizational, and political connections and thereby shapes the search for allies in the same way it affects the victims' ability to organize and protest. The social resources necessary for these activities are distributed away from the lower and more marginal classes. Piven and Cloward argued that the degree of personal harm does not necessarily correlate directly with the intensity of social protest. The ability to respond collectively in protest tends to be associated with people "whose lives are rooted in some institutional context, who are in regular relationships with others in similar straits, who are best able to redefine their travails as the fault of their rulers and not of themselves" (p. 19). The same point applies to the victims' search for allies.

In our cases, the more marginal victims lacked the required connections to facilitate their search for allies. Such victims included dairy farmers in Michigan affected by low levels of contamination, residents of Case Fanfani and Baruccana in Seveso, and fishermen in the Gotō Archipelago in the Yusho case. Their lack of an adequate social network to cope with collective disaster delayed the transition to political issue, as it previously had delayed the transformation of nonissue to public issue and the creation of a victims' group and protest. It also helped exclude them from being defined as victims and from being eligible for redress. They often waited until prospective allies approached them with an offer of assistance. People at the margins of society—those with relatively less power—suffered an added burden when struck by disaster. The less able they were to cope with the normal pressures of modern society, the more they suffered from the additional traumas of a chemical disaster.

Mass Media

The mass media, an important potential aid to the victims in their search for allies, can act directly as an ally as well as provide a forum for debating the political issue. But support from the mass media depends on organizational and geographic factors.

Through their reporting of an issue, the media can alert social movements, political parties, and other organizations to the opportunities that can be gained through alliances with toxic victims. But the media do not act as autonomous forces in society. A reporter's choice of sources affects both the content and the form of the news (Sigal, p. 189). Gans observed: "As a result, news organizations are surrounded by individuals and groups wanting to get their messages into the arena with a maximum of helpful and a minimum of harmful publicity. If necessary, they will use their power to pressure journalists for this purpose" (p. 249). In chemical disasters, the victims, their organizations, and their supporters seek to get their story into the mass media; and they compete against similar efforts by social institutions. To the extent that the victims can persuade the media to present their version of events, the victims can gain power and the responsible social institutions lose some.

The relationship between toxic victims and mass media involves a complex form of exchange, of give-and-take, fraught with both symbiosis and tension and complicated by different interests and objectives. Victims initially resent the intrusion of journalists into their private lives. But the victims learn that the media can be used to increase their power vis-à-vis dominant groups in society through the media's influence on the political agenda and the media's manipulation of public symbols, using the second and third dimensions of power. Journalists meanwhile seek to use their contacts with victims both to advance their careers through the reporting of top stories and to fulfill their personal values and beliefs about society. Finally, media companies strive to use reporters and victims to make the product more newsworthy and to increase sales.

Powerless groups, such as toxic victims, usually begin without the necessary resources to influence the mass media. Most toxic victims lack personal ties to reporters and lack experience in shaping the news. In contrast, government officials and company representatives have well-established contacts, routines, and resources to get their story in the media. They know how to present material, in press releases and information packets, to facilitate the reporters' tasks and thereby shape the end product.

Victims need to develop media skills and exchange relationships with reporters; they learn to trade information and pictures for access and influence. Victims depend both on meeting sympathetic reporters who search for the victims' story and on creating events that attract reporters. Sympathy can result from a reporter's career goals, professional ideals, or political beliefs. One reporter observed, "Many reporters subscribe to a credo that goes like this: 'The role of a newspaper is to print the news and raise hell'" (M. C. Jensen, p. 56). Raising hell can help victims get political support for their position. In addition, some journalists and some media companies view their role as ally or supporter for less powerful groups in society, to provide voice for their grievances and to compel the elite to respond. Victims need to identify, assist, and cultivate these journalists and media companies, as part of the search for allies and empowerment.

The opportunity for victims to meet sympathetic reporters is affected by geographic location. A problem that occurs near a major urban area has a better chance of receiving consistent and prominent attention from the media, as well as from other groups, than a problem located far in the periphery. Among our cases, Seveso received positive media attention because of its location on the outskirts of Milan, which put the disaster near the hometown of one of Italy's major national newspapers, *Corriere della Sera*, and near a national center of media activity. Reporters from national and international media stationed in Milan could easily visit the contaminated area around ICMESA. Moreover, the proximity to Milan made the dioxin contamination appear as a threat to the city, making the disaster more difficult for the media to ignore. Proximity encouraged media organizations to give high priority to the chemical disaster and therefore to assign reporters and media resources to cover the issue.

In the other two cases, the problems occurred far from national centers, and therefore received less national media attention. The location of Yusho victims in western Japan became an obstacle to making the problem into a major political issue. Stories reported in the Kyūshū edition of the *Asahi* newspaper, for example, sometimes did not appear in the Tokyo edition. The PBB contamination in Michigan confronted even more difficult obstacles. Reporting of the problem by the *New York Times* or *Washington Post* occurred mainly after the visit of Dr. Selikoff in November 1976, and then only briefly. The Michigan case appeared only fleetingly as a political issue on the national scene.

In all three cases, however, media attention persisted at the local and regional levels. Individual journalists became specialists in the chemical disasters and reported regularly on the problems. Their articles, which criticized public officials and private companies for not assisting the victims, contributed to making the problems into intense political issues.

Whether the mass media by themselves create political issues and provoke social conflict is a controversial question. The mass media have been criticized for both expanding and containing social conflict (Gans, p. 295). Our cases demonstrate these contradictory effects. Institutional disruption, as an event out of the ordinary, easily makes the news. By reporting the social disorder of protests, the media provide an important form of support to toxic victims. Conversely, once social order has been "restored," the media resist reporting the same old problems, making it difficult for victims to get public voice. Victims find that although their initial protests received attention from reporters, later complaints about persistent problems may go disregarded by the same reporters because they are no longer considered newsworthy. That difficulty, in turn, creates an incentive for more intense forms of disorder to attract attention to the issue and to create news. In this process, victims seek to expand their protests to include external supporters.

The mass media thus are fickle but necessary allies for toxic victims. Although reporters generally view themselves as objective reflectors of reality, they act more as socially constrained creators of reality. Journalists who become committed to the victims' perspective can be transferred to other beats by editors seeking to restore what they regard as balance to the coverage. Media companies may refuse to report on problems of toxic contamination because editors consider the problems not to be news or to be contrary to the company's interests. As part of their political battle with more powerful social institutions, victims need the collaboration of some journalists and some media companies to place the victims' perspective and interests in public debate. Developing the skills to bargain with the media requires time and experience. Victims need other allies, as brokers and supporters, to help in dealing with the mass media and in appealing for redress.

Social Movements

Social movements, because of their expertise in creating and managing political issues, can provide victims with important forms of material and symbolic support in the phase of a political issue.

A social movement is defined here as an aggregation of persons or groups that seeks to transform society and individuals through joint actions and that is less formally organized than established organizations such as political parties, trade unions, and interest groups (Heberle; Alberoni). One scholar identified fourteen forms of "new aggregations" or social movements (Melucci, p. 150), and the fourteen probably do not include all forms (Marconi, p. 110). Movements possess resources, skills, and experience in political conflict. Adept at collecting and presenting information and at creating and using symbols, social movements can assist victims in presenting demands, pressuring institutions, and obtaining results.

Toxic victims and supportive social movements exist in a symbiotic relationship of mutual dependency and mutual exploitation, but not always on equal terms. Social movements bring to an issue their own goals and rhetoric. Supportive movements seek to help victims but also to use them as a source of legitimacy for the movement's larger social vision. Victims, on the other hand, seek to use social movements as a source of legitimacy and power for their specific demands.

The mobilization of a social movement to assist toxic victims depends primarily on the degree of overlap between the issue's definition and the movement's objectives. Movements cannot mobilize resources on all issues and therefore must select which issues to push. Consequently, different problems of toxic contamination receive different degrees of political attention from social movements at different times. The interaction of a social movement with an issue reminds us that official definitions build confining walls around an issue. Even a social movement, which is less bureaucratic than a government agency or a private company, observes its own organizational boundaries in deciding which issues to pursue and confronts its own rigidities against revising the dominant definition of an issue. A movement has its own policy sphere; and if that sphere is not directly touched by the victims' issue, the movement may refrain from acting. On the other hand, when the sphere is disturbed, a movement can push an issue into the political domain and produce effective support for victims.

Given the nature of our three cases of chemical contamination, one might have expected the environmental movement in each country to become directly involved in creating political issues out of the contamination problems. But environmental movements either did not become involved or did so only after great delay. These examples demonstrate that a social movement can experience its own organizational lag in recognizing and acting on the political potential of a public issue. The consequences of this lag are often a delay in attention to redress for victims.

In Italy, a social movement mobilized almost immediately around the issue. The women's movement and the antiabortion movement rapidly transformed the abortion problem at Seveso into a major political issue locally and nationally (Tognoni and Torri). The abortion problem was propelled into national political focus by strong national movements on both sides of the issue and by the importance of abortion as a major political issue in recent events (contributing in early 1976 to the downfall of the Italian central government). The direct overlap between the definition of the issue and the objectives of various movements and organizations brought Seveso's dioxin problems quickly into the ongoing political controversy over abortion. Protests by the feminist movement pushed the regional government to adopt a policy to provide abortions to women possibly exposed to dioxin and to set up a consulting clinic in the town of Seveso. Mobilization of the feminist movement resulted in important changes in policies to provide care for women in Seveso, especially with respect to who should receive care and what kind of care they should receive.

But Italy's environmental movement did not give the same immediate political attention to Seveso's dioxin contamination. Italy's environmental movement in 1976 was weak on issues of toxic chemicals (Nebbia interview). Established environmental organizations, such as Italia Nostra, focused more on the preservation of natural and historical treasures and gave only perfunctory attention to toxic chemicals (Reich, 1984b, p. 388). Newer environmental groups, however, with an orientation toward antinuclear activism, gave Seveso more attention. A group of experts in Milan formed around *Sapere*, a journal of science and politics, and published a special issue on Seveso (*Sapere*). Another volume of essays on Seveso appeared about the same time and included some of the same authors plus several journalists (Martorelli). The contributors spanned the old and new left, closer to the new ecologists than to the old conservationists. But few of those experts became or remained active in organizing the victims of Seveso.

The group that became most consistently involved with Seveso victims was the Workers' Movement for Socialism, through its People's Scientific and Technical Committee. This group, on the new left, actively sought out political issues in controversies over the boundaries and maps of contamination, over the care and compensation of peripheral victims, and over the competence of political leaders and medical experts. Activists in the group organized protests and sit-ins at the special office in Seveso on numerous occasions and heightened the level of social conflict around dioxin. The group contributed to expanding the boundaries of victims, to include more peripheral victims previously denied official measures of redress.

In Michigan, no social movement immediately picked up PBB contamination as a political issue. The contamination was initially defined and perceived as an agricultural and not an environmental or consumer issue. The environmental movement, notably strong in Michigan and in the United States in 1974, became involved only in minor ways. For example, Environmental Action of Michigan hardly mentioned the PBB contamination problem in its monthly publication, *Michigan Earth Beat.* The organization's director confirmed the general inactivity of Michigan environmental groups on the PBB issue, saying it was due to the traditional conservationist orientation of many groups and to the perception of adequate grass-roots involvement by the PBB Action Committee (Sagady interview). But the Committee was formed only in the late summer of 1976, two years after the PBB problem had become a public issue. Moreover, no scientific or legal experts formed a support group for PBB victims, although several scientists did initiate independent action as individuals.

The mobilization of social movements in Michigan occurred after the issue's scope had expanded from a farmers' issue to a general consumers' and public health issue. In 1977, the lobbying activities of the Michigan Citizens' Lobby and a United Auto Workers group helped push a bill through the state legislature that lowered Michigan's PBB tolerance level and provided additional medical research. Mobilization here resulted in legislation that redefined what was considered safe for consumers and thereby assisted in the process of PBB cleanup; it also promised better care for farmers affected by PBB contamination through epidemiological research and rural clinics.

In the Yusho case as well the issue's initial definition delayed potential support from social movements. The Yusho incident was perceived as a case of food poisoning and not as a case of environmental contamination or a general consumer problem. That definition shifted over time, from the efforts of victims and supporters, and especially after the discovery of widespread PCB contamination of the environment in Japan in 1971. That discovery linked Yusho victims to the growing environmental movement in Japan, providing a new source of organizational support.

The mobilization of the environmental movement in the early 1970s on PCBs and the Yusho case moved the central government to expand its definition of the disease and to increase funds for research on the poisoning. Mobilization of social movements heightened political pressure on the Kanemi Company through sit-ins at the factory gates and on other actors through protests against Kanegafuchi Chemical and against doctors supposed to certify Yusho victims as official patients. These activities of social movements pressed institutions to revise policies for the care and compensation of Yusho victims.

When mobilization of a social movement did occur in our cases, it had important consequences for victims and their demands for redress. In all three cases, mobilization of a social movement in support of toxic victims expanded the scope of conflict, placed the victims' problem within a more general political issue, redefined key elements of the demands for redress, and compelled social authorities to take measures to reintegrate the conflict. The measures taken by authorities represented limited concessions to the victims and the supportive movements. In all three cases, a reallocation of resources occurred to meet rising demands for redress. The institutional responses also demobilized the social movements on these specific issues, as discussed later in this chapter.

Political Parties

Political parties can provide important organizational support to toxic victims because of their central role in managing political issues and controlling social conflict. Political parties maintain reserves of power through which they can mobilize political support around an issue. The availability of support from political parties depends on the structure and dynamics of the political system. The degree of interparty competition determines the probability that victims will receive opposition party support. The party's internal structure affects the development and implementation of party policies and circumscribes the independence of individual party members. Whether the victims fit into an existing party constituency critically influences the party's responsiveness to the victims' problems.

In Japan, vigorous interparty competition (between ruling and opposition parties, as well as among opposition parties) contributed to a relatively rapid movement of opposition-party members to organize support groups for Yusho victims and to arrange public protests. The role of opposition political parties in assisting pollution victims resembles their role in assisting other disadvantaged or marginal groups in Japanese society, such as the outcastes or Burakumin (Wagatsuma, p. 85; Upham, 1987). The support of disadvantaged groups becomes a mechanism for opposition parties not only to attack the ruling Liberal Democratic party and maintain the loyalty of party members but also to compete with other opposition parties in attracting new voters and members.

In the Yusho case, the parties of the left, the Socialist and Communist parties, lacked any administrative or major legislative responsibilities in the local governments and could therefore assume positions of full opposition to the party in power, the Liberal Democratic party. From that position, both parties worked to assist Yusho victims and to turn the Yusho case into a political issue. But, as discussed below, competition between the two leftist parties also created significant conflict among the Yusho victims.

In Lombardy and Michigan, on the other hand, muted interparty competition retarded the main opposition party's transformation of the contamination problem into a political issue. In Italy, the Communist party adopted a position of limited opposition, which closely resembled the stance of the Democratic party in Michigan. Both parties shared responsibility with the ruling party in the regional or state legislature; this pattern reduced strong opposition activities. Both parties became critical of the executive branch and the ruling party only after the contamination problem became a major political issue. But important differences also existed between the Communist party in Lombardy and the Democratic party in Michigan, especially in the origins and the consequences of their limited opposition.

The limited opposition of the Democratic party in Michigan came from local circumstances and local party strategy—from the party's control of the state legislature and from its perception of the issue as agricultural and outside its main constituencies' concerns with labor and urban issues. Party politics in Michigan since 1948 have been characterized as highly competitive compared to other states (Lockard, p. 176), with the Democratic party based in a liberal-labor coalition and the Republican party based in automobile-industry executives, with support from rural areas, small towns, and suburban Detroit (Stieber, pp. 6–8). But until the 1930s, Michigan was a one-party system with the Republicans dominant. In the first three decades of the twentieth century, only one Democrat was elected to statewide office (LaPalombara, 1960, p. 22).

In the PBB case, Democratic party leadership viewed the contamination as a rural farm problem affecting a traditional Republican constituency, and therefore as a Republican issue. The decision of the Democratic speaker of the Senate to appoint the Republican Senator Welborn as head of the special PBB-investigating committee in 1975, along with three other rural Republicans and only one Democrat, indicated that the Speaker understood the PBB issue as a minor problem that affected only a few Republican farmers (Koons interview). Until the issue became connected to Democratic strongholds in the cities, as a consumer and public health issue in 1976 and 1977, the Democratic party did not act on PBBs as a significant political issue. The transformation into a political issue resulted from the protests of farmers affected by low levels of contamination, the actions of an individual Democratic House member (Albosta), and the initiatives of Democratic party staff aides. The aides in particular perceived both the political opportunities of the PBB issue and the human implications for affected farmers and general consumers.

The limited opposition of the Communist party in Lombardy, by contrast, depended more on relations between political parties at the national level. The conciliatory relations between Christian Democrats and Communists that existed at the national level in 1976—the famous "historic compromise"—contributed to restraining Communist activists from organizing dissatisfied victims, as might have occurred ten or fifteen years earlier. As Peter Lange observed, "The outcome of the 1976 elections created a stalemated Parliament requiring some form of [Christian Democrat and Communist] co-operation if a government was to be formed, much less function effectively" (Lange, 1980a, p. 124). Thus, in Italy's Seventh Parliament, the Communist party was involved, for the first time since the Constituent Assembly in 1947, in official governmental collaboration, with party members serving as the chairman of the Chamber of Deputies and filling seven of twenty-five committee-chair positions in both houses of Parliament.

The Seveso case exemplifies the pattern reported by Lange: "Communist behaviour at the sub-national level displayed even more clearly the dogged pursuits of the tenets of the historic compromise." The party chose local government coalitions that embraced cooperation with Christian Democrats rather than control of the government with Socialists. In that way, "the party persisted in its policy, linking sub-national policy to national strategic goals" (p. 124). In the Seveso case, one Communist member of the regional assembly, Laura Conti, spoke out often against, and sometimes for, the regional executive's handling of the dioxin problem, but she was occasionally limited in her criticism by party leaders who wanted to maintain harmonious relations with Christian Democrats. When the historic compromise became too much of a compromise for the Communists in early 1979 and they left the national government (Graziano), the regional party assumed a more aggressive stance on Seveso in opposition to the Christian Democrats (*Giorno*, 22 February 1979).

Differences in the internal structure of parties in the United States and Italy also affected the abilities of politicians to take up the contamination problem as a political issue. Stricter party discipline in Italy resulted in greater control over regional Communist politicians, who might otherwise have been more active in mobilizing the victims and the issue. The Italian Communist party was known to exercise greater control over its politicians than other Italian parties through its "democratic centralism," which one writer characterized as "bureaucratic centralism" in practice (Allum, p. 87).

Political parties in the United States, on the other hand, generally lack effective control over individual members (Burnham, p. 279). The lack of control over individual party members is considered part of a general decline of the American party system (Burnham). The lack of party discipline is especially evident in Congress, as Theodore Lowi argued, where "parties are built manifestly to perform constituent functions and not to perform—indeed, perhaps to avoid performing—policy functions" (Lowi, p. 268). In the PBB case, one of the first politicians in the state legislature to begin actively working on contamination problems was a Republican, who became increasingly critical of the Republican governor on the issue. That legislator began working on the PBB issue as an individual politician to serve his farm constituents, not as a party politician to build a party policy, illustrating Lowi's point about the dominant nature of American parties.

Differences in center-periphery relations within the parties also meant that when Michigan's Democratic party recognized the PBB problem as an important political issue, the state party was not constrained by national party policy. In all instances, when Michigan Democrats adopted the PBB problem as an opposition issue, no constraints existed from the national party (then in power under President Carter). In Lombardy, on the other hand, precarious national relations between Communist and Christian Democratic parties required a more oblique and tentative form of Communist opposition on Seveso, until early 1979 when the breakdown of the historic compromise allowed a resumption of the traditional conflict.

An important consequence of limited opposition is that when major opposition parties did not adopt an issue, other political groups more peripheral to centers of political power acted. In Seveso, groups around the new left, such as the Workers' Movement for Socialism, actively organized and protested in the local area when the larger leftist parties held back. And in the regional assembly in early 1979, elected representatives from the new left political party Proletarian Democrats used disruptive tactics to criticize the region's policies on Seveso (*Corriere della Sera*, 23 March 1979). Similarly, in Japan, new left groups and independent citizens' groups focused on aspects of the Yusho problem disregarded by Socialist and Communist organizations.

In these two cases, peripheral political groups saw opportunities for activism, growth, and power in issues that more established organizations ignored. Peripheral political activists, as issue-entrepreneurs, served the important function of redefining the issues, raising the political stakes, and compelling social institutions (and opposition parties) to give greater attention to redress for toxic victims. The mobilization of peripheral political groups increased the intensity of competition to control the political issue. These activities affected the political agenda and increased the relative power of the victims.

But such peripheral political groups did not always take up the contamination issue. In Michigan, the state's Human Rights party, a party with origins in the late 1960s on the left of the Democrats, did nothing significant on the PBB issue. The party's inactivity can be explained by the issue's definition as a rural problem not of central concern to progressive-liberal politicians who focused on urban issues. Its disregard of the issue also represented the party leadership's lack of understanding of the issue, and the party's declining influence in the mid-1970s (*Detroit News*, 22 November 1976).

The political vacuum around the PBB issue in Michigan was filled by individual politicians. Initially, when Democratic party leaders did not actively pursue the PBB issue, a Republican (Welborn) in the state Senate took leadership and sought to assist his farm constituents. As the issue acquired increasingly partisan overtones, Welborn became less active, and a newly elected Democrat (Albosta) in the state House began working on it. The Democratic Speaker of the House at first sought to curtail Albosta's politicization of the PBB issue by denying funds for his hearings (Egginton, p. 237). But Albosta persisted in holding the hearings— thereby helping to mobilize the affected farmers. Albosta went on to use the issue in his successful bid for a seat in the U.S. Congress in 1978.

A political party's relationship to the victims also affects party policies toward a contamination issue. Victims who represent an existing party constituency are likely to receive political attention. Conversely, a party delays active assistance to victims outside its constituencies. In Japan, the two prefectures of Fukuoka and Nagasaki offer contrasting examples of the ruling party's relations with victims. The Liberal Democratic party in Fukuoka did not adopt an active policy of assisting victims and did not have any prior special relations with Yusho victims. Indeed, some party leaders in Kitakyūshū had close ties with the Kanemi Company's president; these ties tended to retard party support of Yusho victims. Nagasaki prefecture, on the other hand, adopted policies that provided more assistance to Yusho victims than in Fukuoka prefecture. One reason, suggested by local activists, was a constituency connection between the prefectural governor and many Yusho victims in Nagasaki prefecture. The governor's home was in the Gotō Archipelago, on the same island as the village of Tama no Ura. Also, he was a Christian, as were many Yusho victims in the archipelago. These ties as constituents facilitated the adoption of some policies requested by Nagasaki victims. Indeed, the prefectural commission on Yusho problems in Nagasaki included representatives from the victims' organization, something hardly imaginable in the antagonistic political environment of Fukuoka prefecture.

Party constituency can also help explain the actions of parties and politicians in Michigan. The state governor, William Milliken, although a

Republican, represented a moderate, even liberal constituency, with strong urban support. In a Democratic stronghold like Michigan, Milliken had depended on ticket splitting to get elected. Indeed, both Republican and Democratic governors in postwar Michigan politics have depended on voters who break ranks and cross party lines, casting their ballots for the candidate rather than the party (Stieber, pp. 92–93). Republican George Romney's 1962 run for governor of Michigan was the state's first campaign based explicitly on a strategy aimed at ticket splitters (DeVries and Farrance, p. 94). Romney even placed newspaper advertisements with instructions to voters on how to split tickets (Stieber, p. 71). The strategy of downplaying party loyalty and emphasizing personal appeal has become conventional wisdom in American electoral politics.

In the 1978 elections in Michigan, Democrats made an almost clean sweep of the state but lost the gubernatorial race to Milliken (Barone, Ujifusa, and Matthews, p. 417). Milliken could antagonize the Michigan Farm Bureau, as his PBB policies did in 1977, and still win the election in 1978, with 57 percent of the vote, up from 52 percent in 1974. Milliken lost some rural votes because of the PBB controversy, but he more than made up the loss by inroads into traditional Democratic areas near Detroit, the southeastern corner of Michigan with about two-thirds of the state's total vote (*Detroit News*, 10 November 1978). Kalkaska county, where the state first buried PBB cattle, voted two to one against Milliken, after giving him 50 percent of the vote in 1974. "But the angry Kalkaskans cost the governor only about 1,200 votes" (*Detroit News*, 10 November 1978). An indication of Milliken's urban support across party lines is that Detroit's mayor, Coleman A. Young, then vice-chairman of the Democratic National Committee, hailed Milliken's reelection (*Detroit News*, 9 November 1978).

In Italy, the Lombardy Region's political leaders adopted a relatively active stance toward the Seveso problem partly because it involved their constituents, with Seveso and the surrounding Brianza largely in the pocket of the Christian Democratic party. But the government with the largest interest in the victims as constituents was the local municipal government of Seveso. For those politicians, the issue of dioxin was not something that could be ignored. By contrast, the Communist party's lack of a strong constituency in the Seveso area contributed to the party's reluctance and difficulty in mobilizing the issue locally (Galli interview).

How did these varying positions of political parties affect the expansion of conflict in society and the provision of redress to victims? In all three cases, political support to victims tended to come more from opposition parties than from ruling parties. Established opposition parties took a position of "opposition within the system" as a "pragmatic, bargaining opposition over governmental policies and leadership," while more peripheral opposition parties took a position of "opposition outside the system" as a "principled, uncompromising opposition to basic aspects of the existing structure of power and authority" (Leiserson, 1973, pp. 394–95). The position of within-the-system opposition resulted in tendencies to contain conflict within social institutions, to use mobilization in carefully controlled situations, and to avoid challenges to the party's legitimacy as a political actor in the system (Katznelson and Kesselman, p. 179). The position of outside-the-system opposition resulted in tendencies to organize protests and expand the conflict, especially in substantive areas neglected by the more established parties.

In Japan, the Communist party stressed opposition in the courtroom, through litigation, and outside the courtroom, through demonstrations. In this strategy, litigation became a mechanism to raise basic moral questions about the structure of society, and mobilization provided the means to reach broader audiences with the message, as has occurred in other cases of political litigation in Japan (Upham, 1987, p. 216). The combination of litigation and mobilization is a common strategy in other countries as well (Scheingold, pp. 203–19) but holds "little or no promise of 'fundamental change'" and is more likely "to reinforce the existing order" (p. 218). The other major party involved in the Yusho case, the Japanese Socialist party, worked mainly through labor organizations and combined mediated negotiation with controlled demonstration. That strategy constituted even less of a challenge to the dominant order of society and similarly served to contain conflict and to place the victims' demands within a predictable institutional context.

In Italy, the Communist party from mid-1976 to early 1979, the period of the historic compromise, was not an opposition party, both in Lombardy and at the national level (Graziano), and the party's strategy on Seveso stressed administrative collaboration and negotiated settlement. Even in early 1979, when the party returned to the opposition nationally, it still did not stress significant mobilization in Seveso. Similarly, the labor confederation in Milan became involved in the Seveso case not so much to mobilize local residents as to mediate their settlement demands, using union lawyers to assist victims in negotiating with Givaudan. When the Communist party was out of the opposition, other political parties and groups filled the void: the Radical party in Parliament, and the new left parties and groups in Lombardy. These organizations spanned opposition within and outside the system and worked to expand the conflict and mobilize the victims rather than to contain either.

In Michigan, individual Democratic politicians and their aides provided assistance to the victims' group, the PBB Action Committee, mainly through information, legislative hearings, and legislative proposals. Some political aides informally advised victims on organizational techniques and strategies. But the party generally tended to contain rather than to expand conflict; its position resembled that of the Italian Communist party more than that of either Japanese opposition party. Although individual legislators provided some assistance in negotiations, neither the Democratic nor the Republican party provided PBB victims with significant organizational support or resources in negotiations, in demonstrations, in scientific expertise, or in litigation. Perhaps the major contribution of the Democratic party, through the office of the Speaker of the House, was to bring Dr. Selikoff to Michigan, who helped to redefine the PBB issue as a public health problem for all consumers and put pressure on the state to provide better care for PBB victims. His initial visit and continued studies marked a turning point in expanding the PBB issue into the political domain.

The approach of elections also affected the positions taken by political parties. In 1978, as elections approached in Michigan, the Democratic party took a more aggressive and openly critical position. Similarly, in Italy, as elections approached in 1979, the Italian Communists became more publicly active on Seveso as a political issue. As Piven and Cloward wrote about protest, "Even serious disruptions, such as industrial strikes, will force concessions only when the calculus of electoral instability favors the protesters" (pp. 31–32). Similarly, toxic victims are more likely to receive support from political parties, including from the opposition, in the campaign season and when the electoral calculus of political competition favors the victims.

Conflicts among Victims' Allies

Conflicts among victims' allies result from competition among support organizations, reflecting the structure of political competition in the society.

Differences in strategy—protest versus litigation versus settlement—become embodied in competing organizations and in competing alliances of victims and supporters. The lines of competition commonly follow the divisions among political parties, with choice of strategy related to structure of organization. But competition among supporters also produces conflicts among victims and a fragmentation of victims' groups. That situation forms the dilemma of political alliances, which provide support to victims but also divide and fragment them, thereby weakening their potential power.

The embodiment of different strategies in competing political alliances and organizations appeared most sharply in the Yusho case: support groups affiliated with the Communist party focused on litigation and the courtroom struggle; support groups affiliated with the Socialist party stressed negotiation with government and company officials for settlement; and support groups affiliated with an independent citizens' movement or the new left stressed direct confrontation and direct negotiation with public and private officials.

The split in strategies of the Communist and Socialist parties is explained by the organizational structure and resources of the two parties. The Japanese Communist party had diligently and successfully recruited specialists, especially lawyers and doctors, into its ranks. Beginning in the mid-1960s, the Japanese Communist party followed a program of expanding its organizational strength and its electoral support, in a process of deradicalization (Kim). That program involved organizing experts to meet the short-term material demands of floating groups in the Japanese electorate, including farmers, small-business owners, and city residents (Totten). Another group that received organizational attention from the Japanese Communist party was pollution victims.

The Japanese Socialist party, on the other hand, was much less successful in enrolling specialists and remained bound to its labor union base. J. A. A. Stockwin wrote that the Socialist party's continued dependence "upon trade unions both for its supply of electoral candidates and for much of the logistics of local organization remains the Achilles heel of the JSP, and has inhibited the growth of a broader and more independent organizational base" (p. 153). When Yusho victims began to consider litigation, the Communists could offer support while the Socialists waffled. Similarly, in other pollution tragedies in Japan, lawyers affiliated with the Communist party provided assistance in litigation to the victims (Huddle and Reich, pp. 125–26, 281).

The competition between Socialists and Communists in the Yusho case reflected a standard pattern of competition between the two parties in Japanese social movements. The same pattern can be found in the antipollution movement (Huddle and Reich, pp. 275–82), antidiscrimination movement (Rohlen; Hah and Lapp; Upham, 1987), peace movement (Totten and Kawakami), and labor movement (Thurston, pp. 138–43). The competition between these two parties in the Yusho case was especially acute in Kitakyūshū, an area with a long history of bitter Socialist-Communist antagonisms, and much less prominent in Nagasaki prefecture, an area with a significantly weaker Communist party presence.

The Japanese support groups that emphasized direct action also reflected a basic pattern of political competition: the emergence of an independent citizens' movement in the early 1960s (Takabatake) and of a new left student movement in the late 1960s (Stockwin, pp. 202–4). The group that formed around Kamino Ryūzō stressed a protest strategy and became linked to several environmental and consumer groups, mostly outside the Socialist and Communist party spheres. In the Yusho case, groups with a direct protest strategy also were more common in Fukuoka than in Nagasaki prefecture, because of the stronger student movement in Fukuoka prefecture, especially at Kyūshū University, which had a tradition of student activism since the late 1960s and became a source of support for various antipollution protest groups. This type of support group also appeared in the Seveso case, representing the continued strength and activity of Italy's new left, particularly in the Milan area.

In all three cases, competition among supporters produced conflict among victims. The issue, in short, became an arena for normal political competition. In the Yusho case, victims' groups became embroiled in debates over the appropriate role of political parties in the victims' struggle, especially regarding litigation, and over the choice of which political support to accept. Some support organizations were torn apart by hostility between old left and new left activists or by competition between Socialist and Communist activists. Victims were forced to choose which alignment to follow, with real consequences in the strategies pursued and the redress received.

In Lombardy, the organizational and strategic conflicts were not between the Communist and Socialist parties as much as between the old left and the new left. Prior to 1976, the Italian Communist party had served as the main social arbiter for the opposition (Lange, 1980a), much as the Japanese Communist and Socialist parties competed to do. The Italian Socialist party, however, lacked the electoral strength or public legitimacy to play a major role as mediator for the opposition. The party's role in the Italian left was weak compared to the Socialist role in Japan and in most European countries (Graziano, p. 195). Moreover, in Lombardy from 1976 to 1979, both Communist and Socialist parties belonged to the majority that supported the regional government.

But the Italian pattern did resemble the Japanese pattern in that old left organizations (especially the labor confederation) provided victims with legal assistance for negotiating with company representatives and for some litigation, while new left organizations provided support for political mobilization, public tactics of confrontation, and the creation of political issues. A major difference from Japan, however, was the existence of new left political parties (Proletarian Democratic party) with elected representatives in the Lombardy regional assembly and thus direct access to a legitimate political forum for raising issues.

In Michigan, organizational and strategic conflicts did not follow the political party patterns found in Italy or Japan—certainly not along lines of competition between the Communist and Socialist parties or between the old left and the new left. Parties of the old left have no significant political presence in Michigan; in this respect they reflect the general problems of third parties in the United States (James, pp. 48–50). As various political analysts have argued, the United States not only lacks parties of the left, it lacks any major party based on particular programs or on mass mobilization (Burnham, pp. 287–306). And Michigan's Human Rights party, which might be considered "new" left, played no meaningful role in the PBB issue.

But even in Michigan, choices of strategy became embodied in organizations. Some farmers who settled their claims formed the Concerned Michigan Farmers, which had ties with the Michigan Farm Bureau and represented a traditional rural Republicanism. Some farmers who filed civil suits for compensation formed the nucleus of the PBB Action Committee, which became informally linked to Democratic state politicians and aides, and represented a political activism of conventional lobbying, litigation, and limited protest. In this way, conflicts among PBB victims reflected the most significant form of political competition in Michigan: between the Republican and Democratic parties.

How did the lack of a significant left in Michigan affect the politics of PBB? It is difficult to speculate on this question, since the "left" can become involved in various ways or remain uninvolved, as demonstrated by the Japanese and Italian cases. But the existence of a viable left opposition might have provided the victims with direct organizational resources for legal and scientific support rather than leaving them to depend on the individual initiatives of self-motivated professionals (mostly lawyers and doctors). And the existence of a vital new left opposition might have provided more political support to victims for protest and mobilization, giving them greater access to the dimensions of power associated with institutional disruption. The constrained structure of political competition in Michigan limited the victims' access to power and contributed to keeping their activities within accepted boundaries.

But a common pattern did emerge in the kinds of alliances that developed. In all three cases, peripheral victims were willing to accept assistance from more marginal sources of support, from peripheral parties or politicians. In Michigan, farmers who believed that low-level PBB contamination had damaged the health of their herds and of their own families received help from Albosta, not from the traditional farmer organizations, the Republican party, the governor's office, or even initially from the Democratic leadership in the state legislature. In Seveso, dissatisfied residents, those who were not evacuated and believed they had health problems and those who were evacuated but did not want to return to their homes, aligned themselves with the People's Scientific and Technical Committee on the new left. In the Yusho case, persons not initially recognized as official sufferers received support from independent citizens' groups and organizations oriented toward the new left. In all three cases, issue-entrepreneurs outside the main political currents sought out dissatisfied minorities to create political issues and assisted peripheral victims who had difficulty in gaining access to the mainstream sources of power.

In all three cases, the issue became an arena for political conflict and competition. Nonvictim support organizations, especially social movements and political parties, became active on specific problems of toxic contamination, seeking to help the victims but also to help themselves. Victims caught in the cross fire of these organizations had to choose sides, although not all victims actively joined all disputes.

Alliances of victims and supporters produced different degrees of social conflict. Some alliances, as we have seen, expanded conflict through disruption and extracted concessions out of social institutions. Other alliances worked within the accepted social norms and collaborated in institutional efforts to contain conflict in the hope of obtaining a compromise. From the victims' perspective, the main goal in forming political alliances and using social conflict was empowerment—to put social institutions on the defensive. Effective alliances compelled private companies and public agencies to respond with changes in strategy and with offers of redress.

Social Institutions and Conflict Strategies

As the conflicts around contamination enter the political domain, private and public institutions respond to the more complex and volatile social circumstances with a new set of strategies. The issue's wider scope, the entrance of new participants in alliances with victims, and the increasing intensity of social conflict all give greater power to victims, which institutions must address.

Both private and public institutions use strategies of dissociation, confrontation, and diversion to reduce the probability of harm to the institution and its interests. In dissociation, the organization retreats and avoids public involvement in the conflict. Dissociation is a turning inward when faced with conflict—a strategy of stonewalling. In confrontation, the organization challenges conflict directly, with coercion and with symbols, especially with discrediting tactics. Confrontation is a turning outward toward conflict—a strategy of fighting fire with fire. In diversion, the organization uses other institutions to enclose the conflict or to shift the focus of conflict away from the organization—a strategy of passing the buck. Diversion into another institution represents a turning around of conflict, so that conflict that appeared as a threat becomes a form of protection.

Private companies proceed through the three responses, in stages. Dissociation is the normal first response of corporations to political issues. Many corporate executives prefer to avoid and retreat from social conflict and fear that providing information would be more harmful than remaining silent. Confrontation is undertaken next, when the company's managers feel that damage from the political issue exceeds the perceived risks of communication. As corporations become increasingly sensitive to bad publicity and the limits of confrontation, they resort to diversion to contain conflict in other institutions. Using that approach, they seek to shape the political issue as well as public perceptions by indirectly providing information that reshapes social conflict. These three strategies belong to the emerging field of "issues management" for private corporations (Heath and Nelson), which seeks to prevent issues from arising, to combat issues in timely and effective ways, and to redirect the course of issues.

Public institutions also adopt the three strategies but within a context of incentives pervaded by political processes. Douglas Yates has provided a political model of public bureaucracies (1982, pp. 62–100) to identify the multiple ways in which these organizations act politically. Public institutions have incentives to use social conflict as opportunities to increase budgets and personnel; such incentives promote strategies of active involvement. But at the same time, public institutions are reluctant to place too much public pressure on private companies, because of the broader political repercussions with the business sector. As a result, while private institutions seek dissociation and diversion, public institutions often end up with confrontation.

The analysis below shows the continued tensions between protecting interests of institutions and providing redress to victims. In this phase of intense politicization, social institutions are more concerned about strategies for protecting their organizational survival than about mechanisms for providing victims with redress. Strategies for organizational survival can include some provision of redress, reflecting partial overlap between the institutional interests and the victims' interests. This overlap can lead to a resolution of the conflict.

Dissociation

The active effort to retreat from social conflict and seek a public disengagement from the political issue—to construct a moat between the institution and the issue—involves the mobilization of bias and the manipulation of symbols.

Among our private companies, Michigan Chemical followed a consistent dissociation strategy. This assessment is not based on direct evidence, for I lack access to internal company documents, but it can be inferred from the corporation's actions. The company avoided public appearances by its institutional representatives (unless they were specifically requested to appear, as at congressional hearings) and made few public statements after its initial responses to press questions. In the Michigan press, the company did not become publicly identified with an individual and thereby remained a faceless, depersonalized organization. The company presented its position through company lawyers or public relations staff, or through outside consulting firms. The identities of the company's owners, Velsicol Chemical and Northwest Industries, remained out of the public eye as much as possible. When a bill came up in the Michigan state legislature to lower the PBB standard, the company did not challenge the bill directly, as Michigan Farm Bureau did, but employed a politically well-connected local public relations firm to lobby the key politicians (*Grand Rapids Press*, 11 July 1976). Michigan Chemical's strategy of low-visibility contributed to its ability to avoid becoming the focus of social conflict by PBB victims. Taking the strategy of dissociation to an extreme, the company's owners ultimately decided to close its Michigan factory and leave the state entirely.

ICMESA/Givaudan, in the Seveso case, similarly pursued a dissociation strategy when it commissioned an outside consultant to design a social and political strategy to remove the firm from public debate and to dissociate the company from dioxin controversies. Givaudan's consultant combined that dissociation strategy with diversion, effectively insulating the firm from most social conflict.

In Japan, Kanegafuchi Chemical initially reduced its institutional presence and dissociated itself from the conflict. That low-profile dissociation strategy successfully kept the company out of the center of conflict until the first civil trial decision in October 1977 ruled against both the Kanemi Company and Kanegafuchi Chemical. The court decision undermined the company's claim of no responsibility and legitimated the company as a target of protest, making dissociation more difficult. The protest strategy of the victims and their allies, especially their increase in direct protest against the company, combined with the courtroom victory to compel Kanegafuchi Chemical to change its strategy.

The cultural context in Japan also made it difficult for private corporations to follow a dissociation strategy. Some victims persistently demanded to negotiate directly with a specific person in the Kanemi Company, the president. The demand resembled those of victims in other

Japanese pollution cases, exemplified by the demands of Minamata disease victims (Huddle and Reich; Upham, 1987). In those cases, victims
rejected the idea of the private and faceless corporation and called for
public and face-to-face negotiations with company executives. Through
direct protest, victims and their allies transformed the private company
into a political symbol and a public entity. They personalized the institution and expanded the conflict, and thereby increased the political pressure on the company to provide redress.

The Yusho victims' demand that a corporate official take personal
responsibility for the company's crime follows a long-standing Japanese
cultural pattern. Personal responsibility can be expressed in extreme
ways—even in suicide to "apologize by dying" (Lifton, Katō, and Reich,
p. 282). Demands for an apology are often part of demands for an ac-
knowledgment of responsibility. As Frank Gibney noted about Japan,
"The rite of apology is almost as important in the late twentieth century
as it was in the early twelfth" (p. 92). Yusho victims used the demand for
a public apology by a corporate official as a form of public coercion and a
symbol of power. Corporate officials hoped that the public apology
would become a symbolic end to the conflict and allow a return to
business as usual. The stress by some victims on the personal responsibility of corporate officials reduced the ability of Japanese executives to hide
behind the corporate wall and outside the political spotlight and to dissociate the company from political controversy. Kamino's four-year sit-in
at the gates of the Kanemi Company articulated the cultural expectation
that the firm's leaders take personal responsibility, in symbolic ways, for
social wrongs committed by the organization.

Kamino's personal protest arose also from his profound faith in
Christianity and its social legacy in Japan. Christians in Japan constitute a
small minority of the population (less than 1 percent). The universalistic
value system of Christianity contrasts with the dominant Japanese group-
oriented values, and historically, Japanese Christians have been associated
with movements that stressed social activism and moral challenges to the
dominant ideology (Lifton, Katō, and Reich, pp. 17–22). Kamino's protest represented that tradition of resisting Japan's often overwhelming
social pressure to conform; he demanded moral retribution according to
universal standards of social justice (p. 228). With missionary commitment, Kamino preached the moral values of Christianity to the Kanemi
Company and to Japanese society more broadly.

The Yusho case stands in sharp contrast to the pressures on corporate
officials in the Italian and American companies, who did not have to
answer directly and publicly to the victims. The sufferers of contamination in Michigan and Lombardy did not demand direct and public nego-

tiations with the corporate officials, nor did they demand public apologies and public admissions of moral responsibility. In these two cases, private officials were held to public accountability through legal proceedings. In contrast, Yusho victims and supporters used the cultural demand for public apology as a source of power. Pressing corporate executives to apologize served as a potent symbol of the victims' power, as an act of public humiliation and contrition, and as a mechanism for reaching broader publics. Society's expectation of a public apology gave victims support in their struggle for redress. But the demand for apology in Japan did not substitute for formal legal proceedings; it was pursued in parallel, as part of a broader political struggle, sometimes by the same groups, sometimes by different ones.

As with private companies, political risks, limited budgets, and the goal of self-protection encourage public institutions to recede from the direct management of conflicts. But public responsibility and political opportunity make it more difficult for public administrations to pursue a dissociation strategy. Public bureaucrats, when confronted by controversy, seek to advance their interests but also to minimize potential conflicts (Yates, 1982, pp. 103–4). The probable political costs and benefits of addressing a political issue are assessed by public bureaucrats, who then decide whether dissociation is politically feasible or desirable. Among our cases, we find both approaches.

Public administrations in the Yusho case demonstrated the most stubborn insistence on the private nature of the conflict and therefore on a dissociation strategy. This stance arose from the perception that the likely costs of involvement exceeded the incentives and potential benefits, especially in light of limited budgets and potential liability. Central government officials argued that no statutory mechanism existed to provide care to food-poisoning victims and that therefore the private parties must reach a negotiated settlement or a court decision. Central government officials explained that the government did not provide special assistance to Yusho victims because to do so might be interpreted as an admission of liability, which could be used against the central government in ongoing civil litigation (Itō interview). The great attention paid by Yusho victims and supporters to protest against the private companies, rather than the central government, made it easier for officials to follow a dissociation strategy and insist on private not public action. But the government's strategy also delayed for about ten years official assistance to the victims in their problems of care and compensation.

In Michigan, the state government initially pursued a similar strategy of dissociation from social conflict and a similar policy of urging a private settlement of the PBB issue. As in the Yusho case, state agencies in Michigan perceived the costs of involvement as substantial; they similarly sought to limit public expenditures, avoid possible state liability, and reduce political pressure on public institutions. In both cases, governments placed greater priority on conserving public resources than on aiding toxic victims; and both emphasized the private nature of the problem. But in Michigan, the growing power of victims and their allies expanded the scope of the political issue and undermined the dissociation strategy; by making public institutions the focus of conflict, the victims increased the political costs of noninvolvement.

On the other hand, the regional administration in Lombardy did not begin by dissociating itself from the conflict. The incentives to become fully involved in managing the chemical disaster were perceived from the start as exceeding the potential costs. As a relatively new public institution, the region sought to use the disaster to demonstrate and build up its institutional capacity and to prove its superiority over the central authorities in Rome. The regional government quickly obtained emergency funds from the central government and became committed to managing all aspects of Seveso. The ability to obtain new funding, a significant difference from the other two cases, reflected a characteristic of the Christian Democratic party: its emphasis on the distribution of resources rather than the implementation of policy (Tarrow, 1980, p. 170; Pasquino, p. 90). In the Seveso case, good relations between Andreotti's new central government in Rome and Golfari's regional government in Milan, and among their respective factions within the Christian Democratic party, facilitated the shift of funds and responsibility to the regional administration. That transfer committed the Lombardy regional government to a policy of active intervention from the start and brought regional officials directly into managing social conflict around Seveso.

From the victims' perspective, a social institution that adopts a dissociation strategy becomes highly unresponsive. A pure dissociation strategy results in denying the resources of redress to victims: medical assistance, financial compensation, and other forms of welfare. The strategy represents an intentional form of secondary victimization. For victims and supporters to change that institution, they often need to use strategies of disruption, and increasingly higher levels of disruption.

Confrontation

Institutions pursue a strategy of confrontation when dissociation fails to shift blame and attention elsewhere, when the costs of dissociation are perceived to exceed the benefits. Confrontation can occur through coercion (the first dimension of power), the direct application of pressure on those perceived to be causing social conflict. But in our cases, confrontation occurred mostly in the symbolic realm (the third dimension of power). In this symbolic form, an institution seeks to present its side, often in official publications—media readily controllable by the institution—and through tactics of discrediting its opponents. Those tactics blame the victims, political groups, the public media, private companies, and public agencies not only for causing the conflict but also for creating the problem.

Both private and public institutions use discrediting tactics with remarkable similarities. Institutions of both types accuse victims of suffering from psychosomatic problems; they accuse victims' supporters and lawyers of political or economic profiteering or both; they accuse companies of forsaking their social responsibility; and they accuse public institutions of mishandling the problem. In sum, in a confrontation strategy, the institution seeks to blame other participants as the source of all problems, and it portrays itself as a victim of irresponsible social forces.

Some organizations engaged in a confrontation strategy with less passion than other organizations. Among private corporations, Givaudan and Michigan Chemical refrained from confrontation in public, as might be expected from their emphasis on dissociation. The other three companies actively engaged in confrontation, with varying degrees of discrediting tactics.

Givaudan and Michigan Chemical, for example, did not use official publications for confrontation. Givaudan's pamphlets instead followed its general strategy of presenting the company's positive accomplishments, of contesting only a few selected charges, of focusing on scientific rather than political problems (Givaudan). Michigan Chemical prepared no separate document that I could locate concerning its involvement in the PBB case, but it issued occasional press releases on the trial proceedings and included statements in annual reports on how the incident would not materially damage the company. The company's internal newsletter discussed aspects of the contamination but did not engage directly or publicly in discrediting tactics (*Crosscurrents*). In both companies, the strategy of dissociation dominated the strategy of confrontation. This de-emphasis on confrontation helped keep the companies out of social conflict.

Three companies used official corporate publications in a confrontation strategy to cope with expanding conflict. Farm Bureau Services used the monthly publication of the Michigan Farm Bureau to discredit other parties, especially in the controversy over lowering the PBB guidelines in 1977 (*Michigan Farm News*, 1977). The Kanemi Company used its internal newsletter in crude discrediting tactics to blame various participants for conflict associated with the Yusho case (Kanemi Sōko). Kanegafuchi Chemical, after losing its court cases in 1977 and 1978, prepared argumentative documents that complained in shrill tones about the unfairness of the decisions (Kanegafuchi Chemical, 1977, 1978). The company also persuaded its labor union to issue a pamphlet that repeated, in almost identical words, management's denial of responsibility and complaint of injustice (Kanegafuchi Kagaku Rōdō Kumiai). In all three instances, the documents portrayed the company as victim and other participants as victimizers.

An attorney for Farm Bureau Services neatly summed up the vision of corporation-as-victim: "The real message to others in our position is that Farm Bureau tried to be nice, and the federal people tried to indict them. The state people tried to turn it into a political issue. . . . This turning around was strictly due to [the] greed of four classes of animals: ambitious politicians, poor man's Woodward-and-Bernstein types, scientists on the make for a grant or luster to an unknown name, and overreaching claimants and counsels. My client was the target, the naive victim" (McIntyre interview). The effort to portray the corporation as victim entailed a symbolic inversion; it reconstructed an inverted reality to present the victimizer as victim. Such "dialectical reversals" generally play an important role in the social construction of ideologies (Jameson, pp. 309, 369–72).

Farm Bureau Services, indeed, was among the corporations more actively involved in confrontation. In March 1975, for example, when the independent trade journal *Michigan Farmer* devoted an entire issue to the PBB controversy, the top manager at Farm Bureau Services wrote a stinging letter of rebuttal (Armstrong, 1975). But joining in the public name-calling only dragged the company deeper into controversy and conflict. It did little to improve the company's public image and probably did more damage than good. As an attorney for Farm Bureau Services explained, the company's entrance into public debate and public controversy did not help his client (McIntyre interview), and it certainly did not help contain conflict.

Why did Farm Bureau Services cultivate so active an institutional presence in the conflict? A key reason was the central position of Farm Bureau Services within Michigan's political economy as the state's leading distributor of agricultural feed and as a subsidiary of the state's most powerful agricultural organization. That position gave the company a strong public presence and political role from the start. To many Michigan farmers, the company was not only a commercial feed distributor but also a subsidiary of a farmer cooperative, the Michigan Farm Bureau, which was supposed to serve its farmer-members, and therefore the company was a member of the Farm Bureau "family." The company's desire not to spoil its image as a provider of service to farmers—and its practical links to farmers—contributed to its more direct involvement in conflict and its strategy of confrontation. Unlike Michigan Chemical, Farm Bureau Services did not have the option of closing up shop and leaving the state.

Public administrations in Michigan and Lombardy used a strategy of confrontation to deal with conflict by producing official publications after the problems became political issues. The decisions to print and distribute official newsletters represented dissatisfaction with ordinary commercial mass media. Officials believed that greater control over the flow of information would reduce "distortion" and would tilt the conflict in favor of the public agency. These attempts to reach victims and others directly, to short-circuit the societal mediation of the mass media, provided the official story of the issue and emphasized what government was doing to provide redress.

In Japan, on the other hand, the government published as little information as possible. It issued no official public report on the Yusho incident, no public pronouncements of social policies adopted. Public agencies followed a basically conservative and closed approach to information. The government sought to reduce conflict not by providing information but by withholding it. That position reflected Japan's elitist bureaucracy, its stress on bureaucratic deliberation and decisions as closed and private affairs, and its almost compulsive resistance to directly releasing information. Instead of publishing its own information, the government used other institutions, especially the scientific Yusho Research Group, to release information in a strategy of diversion, as discussed below.

Public institutions also used discrediting tactics in a confrontation strategy to delegitimate their critics and their critics' claims. Like private institutions, public administrations caught in conflict sought to blame other participants and to portray their own organizations as victims of irresponsible social forces.

In all three cases, public officials sought to confront conflict by blaming the politicians, especially politicians of other parties. They accused other politicians of playing politics, of turning the issue into a political football, of *strumentalizzazione*. While these accusations were often in part true, the tactic is a prime example of how the authority of public agencies could be used to resymbolize problems and shift attention or blame elsewhere. This form of discrediting became part of interparty competition, with a party portraying its opponent as manipulating the issue for self-interest and political purposes, while proclaiming itself as dedicated solely to helping the victims, for public purposes. The portrayal of conflict in Manichean symbols masked, or attempted to mask, the political purposes of public policies and public administrations. In each case, the more the issue became incorporated into interparty competition, the more common became the charge of political manipulation as a discrediting tactic.

Blaming the victims for emotionalism or overreaction to contamination provided public administrations with another method of discrediting. In all three cases, public officials tended to blame difficult-to-diagnose problems on psychological difficulties of the victims themselves. The same form of blaming the victim occurs for occupational diseases and accidents, from public officials as well as private employers (Page and O'Brien, pp. 145–49). This blaming-the-victim syndrome occurred as a political response, an effort to deny sufferers an official or legitimate status as victims and to brand them instead as psychologically ill or as economic opportunists.

Public administrations also attempted to confront political conflict in contamination problems by blaming the media for sensationalism or distortion. The governments of Michigan and Lombardy, which followed relatively aggressive confrontation strategies, compiled and published collections of press reports, sometimes with official critiques, in efforts to demonstrate press distortions. As noted above, those two administrations also began publishing their own official newsletters.

Yet no public administration pursued with equal vigor a tactic of blaming the private companies, at least not in the public realm. Public administrations in all three cases filed court suits against the private companies involved, but those cases remained more in the courtroom than in the public media or political domain. While public administrations pushed to make the companies pay economic costs of the disaster, they did not push to make the companies pay their share of the political costs. As a result, the political costs remained more in the public administrative domain than in the private corporate domain. As the regional president of Lombardy pointed out, people forgot that the private company and not the public administration originally caused the disaster. But he did not connect ICMESA's public disappearing act with his administration's policy or with Givaudan's social and political strategy.

Electoral and political reasons worked to discourage public administrations from using tactics of blaming-the-private-company. In Michigan, the Republican governor sought to maintain credibility with the agriculture industry and tried not to alienate completely the Michigan Farm Bureau. In Japan, the Kanemi Company's president maintained close ties with important leaders of the Liberal Democratic party, which discouraged outright political criticism of the company. In Lombardy, the powerful economic interests around Milan influenced regional officials' tendency to downplay public castigation of private industry by politicians of the governing Christian Democrats. The moderate to conservative political ideology of these administrations contributed to a reluctance to put private industry on a political hot spot. The cases suggest that political leaders fear the direct and indirect consequences of blaming private companies. Put simply, administrations that do not blame the companies, for political and economic reasons, need to find others to blame—other politicians, the media, the victims.

In sum, both public and private institutions engage in a strategy of confrontation to cope with social conflict. For private companies, that strategy may serve to bolster the loyalty of employees to the organization, something quite needed in times of crisis. But confrontation can also decrease the company's public legitimacy, as discrediting tactics become seen as untrue, deceptive, and self-serving. As Sissela Bok has pointed out, lies to enemies not only have moral dimensions but also have practical consequences and can backfire (pp. 148–53). Similar problems exist for public institutions that adopt a strategy of confrontation. In addition, that strategy rarely results in a containment of conflict; indeed, confrontation usually produces an expansion of conflict. Public institutions, as part of the governmental domain, generally find it more difficult than private companies to resist confrontation. As the issue becomes part of existing political competition, public administrators find themselves pressed to defend policy as well as political positions. Discrediting tactics provide an easily accessible defense but do little for the victims' redress. What works for political or bureaucratic purposes of organizational survival may not be the best for social or moral purposes of providing redress.

Diversion

The strategy of diverting conflict away from one's own institution to other organizations, which then contain the conflict or become the focus of conflict, often occurs together with dissociation, as an institutional form of passing the buck. The use of other institutions for one's own defense is most effective if it occurs privately, behind the scenes. The strategy can involve deception as well as concessions, programs, or policies, which may or may not work, but which provide symbolic or material reassurances (Edelman, 1971).

Private corporations use various forms of diversion to reduce political attention focused on the company, although not all companies use this strategy to its fullest potential. In our cases, differences in degree of use among the companies can be explained by the company's position in the political economy.

ICMESA/Givaudan provides the clearest example of intentionally using a strategy of diversion. The company's outside consultant designed a strategy that stressed "the lessons of judo" and "the art of taking slaps," which transform attacks into strengths and contest only the most serious charges, with the least possible exertion. Saul D. Alinsky used strikingly similar words in recommending a protest strategy for relatively powerless groups: "The basic tactic of Have-Nots in warfare against the Haves is a mass political jujitsu: the Have-Nots do not rigidly oppose the Haves, but yield in such planned and skilled ways that the superior strength of the Haves becomes their own undoing" (p. 152). Givaudan's strategy sought to remove all private friction between the company and public authorities and at the same time to shift the management of contamination problems and the focus of social conflict from the private company to the public administration.

Givaudan's consultant admitted in an interview that an approach other than his diversion-and-dissociation strategy might also have been effective. But he thought the strategy he created "was the best for the interests of Givaudan, although not for the interests of truth" (Chiape interview). The consultant knew the processes of public administration and politics well, since he had earlier worked for the Lombardy region's first president. But he did not consider his advice to Givaudan, a plan to produce administrative and political chaos for the region, as "morally incorrect." He acted "as a professional" and would have counseled the region if asked. "In this case, the client was Givaudan. . . . The region had better weapons available than Givaudan. If they didn't use them, that's their fault" (Chiape interview).

Although it is not clear from the information available to me whether Northwest Industries and Michigan Chemical designed a strategy to shift the focus of public conflict to the public administration and to Farm Bureau Services, the companies' actions had that result. The companies may have been lucky, or they may have intentionally sought to use diversion and dissociation to insulate themselves from PBB social conflicts. Michigan Chemical, in this case, also had an advantage over Givaudan in being able to use another private company (Farm Bureau Services) as well as the state government for its strategy of diversion.

Farm Bureau Services, in contrast, adopted a position of public collaboration with state agencies instead of a strategy of institutional buck-passing. That public collaboration reflected the company's important position in the state's political economy—its close connections to Michigan Farm Bureau, the state Department of Agriculture, and Michigan

State University. But the active involvement of Farm Bureau Services with public authorities produced persistent embroilment in social conflict, a result similar to that predicted by the private consultant for Givaudan in the Seveso case. From the private corporation's perspective, that conflict complicated management, demoralized employees, gave the company a bad public image, and probably also reduced revenues. An alternative strategy for Farm Bureau Services could have diminished if not avoided the conflict and its consequences.

In Japan, the Kanemi Company did not attempt to shift responsibility to public institutions. The company was restricted in shifting more political attention to public authorities by the company's connections to the Liberal Democratic party. Kanegafuchi Chemical, on the other hand, began to promote government involvement in the conflict after the court decision in October 1977. Motivated by financial concerns, the company argued that central and local governments take responsibility for resolving Yusho problems, since they shared responsibility for causing the contamination, especially in the dark oil incident (Kanegafuchi Chemical, 1978). Both companies, however, sought to use public institutions in settlement-mediation efforts; the companies invoked the power and authority of central government officials in order to diffuse and divert some of the conflict. Each company also publicly blamed the other for causing the disaster, thereby adding to conflicts around each company.

Public administrations also rely on diversion strategies, but in contrast to private companies, they do not often combine diversion with dissociation. Public agencies often turn to organizations in the public sector, using standard administrative approaches such as expert commissions, public hearings, and investigatory committees; sometimes they create totally new organizational structures. Through these subordinate and specially mandated institutions, public agencies can divert social conflict but still maintain direct access to key dimensions of power: the ability to control the agenda and manipulate evocative symbols. In the cases, public institutions diverted the conflict, but not far enough. Still tied to the subordinate bodies, the institutions remained exposed to conflict and to pressure from victims and their allies for effective redress.

The use of a new structure to control a political issue is illustrated by the ICMESA case. As conflict in Seveso expanded, the region created a new institution, the special office. The regional president stated that this organization was intentionally designed to contain social conflict, reduce participation in the issue, and implement controversial plans (Golfari interview). The director of the special office, however, encountered enormous problems in making the office function and discovered to his dismay that the newly created institution attracted rather than quelled protest (Spallino interview). From the Lombardy regional government's perspective, the institution did contain conflict: most protest became focused on the special office in Seveso rather than on the regional offices in Milan. The example demonstrates that a structure designed to contain and reduce conflict for some in the public sector can create and increase conflict for others.

In the phase of a public issue, administrative agencies use scientific commissions to create legitimacy for government positions. As conflict heightens, they use these commissions as instruments in political struggle in a strategy of diversion. They increasingly seek experts not only with the appearance of objectivity (to ensure the public legitimacy of decisions) but also with the appropriate partisanship (to ensure the political acceptability of decisions). The demand for both public legitimacy and political acceptability typifies the processes of scientific advice and scientific disputes in general (Robinson; Nelkin, 1979b). In the phase of a political issue, public agencies increasingly emphasize the political role of expert commissions.

In Michigan, as political conflict around PBBs erupted in early 1976, the governor created an expert commission to review state policies and answer scientific questions. Two typical difficulties arose, one around the selection of the experts, the other around the final report of the commission. The governor first sought to select Michigan scientists of different political positions on PBBs. When conflicts arose among members of the panel, he then turned to out-of-state scientists and sought to isolate the process from politics. The final report unexpectedly criticized state policies on PBBs, concluding that existing measures might not protect the future health of Michigan residents, a conclusion that significantly expanded political conflict and debate. The governor's science advisor later admitted that "a mistake was made in paneling the commission without having someone from [the governor's] staff directly involved. Staff are aware of consequences, politically and organizationally" (Taylor interview). Greater political control of the commission could have produced a more politically acceptable report and thereby reduced the conflict that the governor subsequently confronted. But greater political control might also have reduced government attention to redress for toxic victims, which resulted from expanded conflict.

The Lombardy region set up expert commissions almost immediately after the Seveso disaster became a public issue. As in Michigan, the region at first used the commissions for legitimacy by selecting scientists to represent a broad spectrum of political viewpoints. But as conflict increased in society and in the commissions, the region reduced the political diversity among the commissions' scientists. The region pruned the commissions of more critical members, often on the left, to reduce dissension and the range of opinions. Later, when it established the special office, the region reorganized the commissions to make them even more closely aligned with regional politics. As controversy grew, the region imposed greater political control over scientific information. Indeed, in 1979, even obtaining an interview with a scientist heading an expert commission required approval by a politician, the director of the special office, who was closely aligned with the regional president.

Japan's public administration neither isolated its scientific commission from government agencies, as occurred in Michigan, nor incorporated its scientific commission into the administration, as occurred in Lombardy. Instead public officials created a separate expert commission, the Yusho Research Group, which maintained the appearance of objectivity through its basis in Kyūshū University but retained an approach to policy that supported government positions. That combination of scientific legitimacy and political cooperation provided the public administration with relatively effective control over the political issue.

Japan's public administration transformed university scientists into official providers of information, reducing conflict for public officials, but increasing conflict for Yusho Research Group doctors. For example, each year since 1969 the group has published an annual collection of scientific articles on Yusho in the *Fukuoka Acta Medica*, whereas no public administration has published an official review of policies on the poisoning. The strategy stressed the scientific and medical aspects and downplayed the social and policy aspects of Yusho. That emphasis reflected the government's inadequate redress policies, which left Yusho victims for a decade without any significant government assistance. The strategy also resulted in a public loss of legitimacy for the research group, especially in the minds of the victims. Eventually, because of the victims' antipathy toward the doctors, it became difficult or impossible for researchers to perform epidemiological studies.

Prefectural officials used the Yusho Research Group as a shield to protect government officials from protests. The experts not only decided the criteria for diagnosing Yusho as a disease but also examined patients to determine if they qualified. When patients complained to prefectural officials about not qualifying as a certified Yusho victim, public officials directed them back to the doctors. The ambiguous relationship between experts and bureaucrats created a situation of blurred responsibility, which allowed each to blame the other for problems and left the dissatisfied patients in the middle with no clear means of appeal. The government strategy transformed the physicians from givers of medical care into gatekeepers of official redress.

In a strategy of diversion, public hearings, in either an administrative or a legislative arena, are structured to present only one side of an issue and to provide participation with little influence on policy. Murray Edelman emphasized that the form of public hearing is often more important than the content. Such hearings provide a setting "heavily imbued with stylized and ritualistic components that justify policy to mass audiences" (1977, pp. 112–13). The appearance of an expert witness in such a setting is often more important than what the witness says; testimony can provide "cover" for already negotiated deals (p. 113). Public hearings are a prime stage for political theatrics, an institutional forum that permits ritualized conflict, generally with the public agency able to set the agenda and control key symbols. Hearings can divert conflict from other locations into environments that are less threatening to the public agency. Hearings may give the appearance of confrontation, but by restricting the boundaries of controversy, they usually end up containing the conflict.

In Michigan, the Department of Agriculture used a public administrative hearing in June 1976 in a political challenge to the governor's expert committee and its proposal to lower the PBB tolerance. The department translated its well-known bias on the standard into a hearing that heard mainly one side of the issue, its side. The public hearing remained firmly under the control of the Department of Agriculture and reflected the organizational strength of the agriculture establishment and the relatively weak position of the affected farmers.

Legislative hearings, however, can be more conflictual. Some legislative hearings in Michigan (such as those held by State Representative Albosta) helped expand conflict when they became the focus for complaints by the farmers. Other legislative hearings (such as those held by U.S. Senators Griffin and Riegle) served to dissipate and contain conflict. These senatorial hearings were designed to be bipartisan and "nonpolitical," to show attention to constituents by established politicians. The senators listened to all sides and created a "record"; they went through the motions of seeking to provide redress without actually doing so. By contrast, the oversight hearings organized by the U.S. House Subcommittee on Oversight and Investigation created conflict through a concerted attack on specific federal agencies and sometimes on individual bureaucrats. This conflict within the hearing arena was intended to promote redress for the victims.

In Italy and Japan, administrations did not hold public hearings as in Michigan, but they did argue their cases in regional and national legislatures. Those meetings took the form of staged encounters, which provided bureaucrats and politicians of the party in power with the opportunity to defend existing policies and gave victims and politicians of opposition parties the chance to attack. One ministry official in Japan reported that the hearings had little influence on policy content (Itō interview). But the hearings did provide symbolic support and assurances that the government and the opposition were both doing their best on the political issue, even when they were doing little.

In Italy, the national legislature used institutional diversion in its creation of a special investigatory commission on the ICMESA incident, which held private hearings and wrote an analytic report on the incident's causes and on institutional reforms. The commission published a final report but not a transcript of interviews or copies of internal documents. The high quality of the final report and the commission's unanimous approval of it eloquently demonstrated the reduced conflict between Christian Democratic and Communist parties in Italy's Seventh Legislature.

In Japan, the national legislature did not form a special investigatory commission on the Yusho poisoning and published only the transcripts of committee meetings for a limited circulation. No effort was made to collect or analyze or publish administrative documents on the Yusho case. The relative inactivity of Japanese legislators reflects the weak position of the nation's legislature with regard to policies proposed and formulated by the strong bureaucracy and the dominant party (Pempel). While the Japanese Diet is supposed to exercise final decision power, it has more often served as ratifier rather than decider because of the dominant position of the Liberal Democratic party (Richardson, p. 248). This role of the Diet may change, since the Liberal Democratic party's loss of its majority in the Upper House in the elections of 1989.

In conclusion, for private corporations, the strategy of diversion can be quite effective in insulating the organization from social conflict, especially when it is combined with a strategy of dissociation. For public institutions, however, the nature of public accountability, along with pressures of political organizations and social movements in alliances with victims' groups, make a diversion strategy less effective, since public sector institutions are often used as the vehicle for containment. In short, it is harder for public agencies to pass the buck effectively or completely. The public administration that involves its own institutions in diversion makes those institutions vulnerable to conflict. The institution designed to contain conflict can end up being used to expand conflict. From the victims' perspective, however, this is fortunate, because it keeps open the possibility of achieving redress through public institutions.

Resolution

Neither social conflict nor a political issue persists forever. With time, and with the responses of private companies and public agencies, the extent of institutional disruption declines and the number of groups involved in the issue decreases. As the last groups of victims receive some measure of redress, the conflict becomes reintegrated into society. The issue leaves the political agenda.

Victims rarely receive what they believe they deserve. Indeed, it is impossible to provide full redress to toxic victims. Monetary compensation cannot replace lost health or lost life or lost community, even though it is society's most common effort at redress. Most toxic victims receive only partial redress, and they often end up bearing the risks (however uncertain they may be) of long-term health hazards from toxic exposure. The disappearance of open social conflict represents a "resolution" of the issue but not necessarily a feeling of justice among the victims.

The failure of litigation can undermine the victims' movement with allies in the political realm, as occurred in the Yusho and Michigan cases. In the Yusho case, in 1987, nearly twenty years after the poisoning's outbreak, Japan's Supreme Court refused to rule on corporate and governmental responsibility and instead proposed final mediation on compensation for the victims. That action effectively ended the political issue. The prolonged conflict over Yusho for two decades reflected not only the victims' profound demand for redress but also the depth of organizational support provided by the victims' allies. The victims ended their battle without feeling fully vindicated, but they believed they had done all they could do.

In Michigan, the devastating defeat in the Tacoma trial in October 1978, only five years after the poisoning's start, led to small settlements in other outstanding trials for claims of both cattle damage and human illness. Following the statewide election in November, PBB action in the political realm largely disappeared. Some farmers affected by low levels of contamination, who applied for low-interest loans from the state, subsequently had their loans forgiven in the mid-1980s in a measure that amounted to additional compensation. That redress resulted from the continued activism of one farmer-victim and from bipartisan perception that forgiving the loans served the interests of each party. But many farmers similarly affected by PBBs felt abandoned by the Republican party, double-crossed by their farm organization, and rejected by the court system. The support they received from private lawyers and Democratic party politicians and staff had been no match for the superior resources—both economic and political—of the private companies.

In the ICMESA case, most victims did not become embroiled in complex litigation against Givaudan. Negotiated settlements provided most victims with monetary compensation for material losses. The company decided in its dissociation and diversion strategy to pay for damages in the private realm rather than contest the victims in litigation and the political domain. Protests by victims and supporters against the region eventually succeeded in changing policies so as to provide limited redress to some peripheral victims previously excluded. By the mid-1980s, disputes over ICMESA's dioxin no longer raised bitter social conflicts or appeared as a political issue. The physical landscape around Seveso still showed scars from the poisoning, and the inner feelings of victims still echoed with fears of future health problems, but the body politic was no longer torn apart regularly by ICMESA's dioxin.

Chapter 8

Politics, Policies, and Power

The cases in this book are only a sampling of chemical disasters. Each of the three countries has other chemical disasters in its recent history and many cases of toxic contamination. The United States has its Love Canals (Levine); Italy has its Manfredonias; Japan has its Minamatas (Huddle and Reich). Every industrializing nation confronts similar accidents and abuses of toxic chemicals. The chemical explosion at the Sandoz factory in Switzerland and its pollution of the Rhine River remind us that we have not seen the end of chemical disasters in advanced industrial societies. The world's worst industrial disaster at Bhopal, India, illustrates the potential for new and even more deadly catastrophes in industrializing nations (Morehouse and Subramanian).

Politics and Power in Advanced Industrial Societies

The political problems confronted by toxic victims in their struggle for redress belong to a broader category of obstacles faced by relatively powerless groups in society, illustrated especially by the difficulties in transforming nonissues into public and political issues. The cases confirm the importance of analyzing nonissues; they also show political strategies available to victims, public institutions, and private institutions that seek to use power to determine who gets what in advanced industrial societies.

Nonissues

Our cases provide ample evidence that nonissues exist in various societies and can produce devastating social and personal consequences.

254

This conclusion about nonissues is important for both academic and practical reasons. The conclusion first provides additional fuel to the academic debate about the limits of pluralism and the methods of analyzing power. It provides additional evidence that intentional obstacles can block the transformation of nonissues into public issues.

Political scientists in the United States have long debated the question of non-issues and the implications for pluralism. Schattschneider pointed out in 1960 how the "mobilization of bias" in organizations keeps some issues on and other issues off the agenda. Peter Bachrach and Morton Baratz then analyzed the "two faces of power" (1962), how power is used to affect society through decisions and through nondecisions. The other side of the debate has been argued often by Nelson Polsby, who in 1979 was still defending the conventional pluralist position, still calling for more empirical evidence that nonissues exist, and still doubting that a study of the mobilization of bias could provide meaningful analysis or useful conclusions (Polsby, 1979).

John Gaventa, in his study of power and powerlessness in Appalachia, argued strongly against conventional pluralism's resistance to nonissues. He approvingly cited Frederick W. Frey's two conditions under which nonissues might be expected to exist: "when: 1) glaring inequalities occur in the distribution of things avowedly valued by actors in the system, and 2) those inequalities do not seem to occasion ameliorative influence attempts by those getting less of those values" (Frey, p. 1097). Gaventa concluded that it is more important to understand why quiescence persists in a situation of potential conflict than to study why conflict occurs in a situation expected to be conflict-free (p. 26).

Gaventa proposed several approaches to the study of nonissues "to substantiate the expectation that B would have thought and acted differently, were it not for A's power" (p. 26). First, he recommended research in the community to discover how people perceive the use of power to maintain nonissues. Second, he called for analysis of patterns of communication and symbolization in order to assess connections between the ideologies or actions of the power holders and those of the powerless. He also proposed research of evolving social conditions, which allow for new actors or new circumstances to influence the emergence of issues from nonissues. Finally, he suggested comparative analysis of different groups faced with similar problems but varying power configurations. I have used such approaches throughout this book, placing it in line with other community-power studies critical of conventional pluralism.

This study uses the empirical method demanded by the defenders of conventional pluralism—the examination of "concrete decisions"—to criticize conventional views about the maintenance of nonissues and how issues get on the agenda. Nonissues are not a vaporous phenomenon. They can be demonstrated empirically by examining the processes by which they become public and political issues. To study that process, I looked at political behavior, but also at "the institutions, structures, and rules in which behavior is embedded," for those factors "constrain behavior, mediate between actors, set the boundaries for political arenas, and influence political agendas" (Newton, p. 545).

Nonissues result from both intentional and unintentional acts. The nonissues of this study did not result from a simple "conspiracy" of a few political actors, but rather from a complex confluence of organizational and human impulses, which I examined as "middle-level" hypotheses based in theories of human psychology and organizational behavior. Such complex origins do not make nonissues any less important for society. Social problems perceived and treated as private troubles raise critical questions about individual suffering and social justice. But the complexity demands attention to both intentional and unintentional factors, especially when actors seek to move a nonissue into the public or political domains.

Intentional acts to create and maintain nonissues occur for various social problems. Some individuals may suspect or recognize the social scope of a problem, yet intentionally act to keep the problem defined as private. In the three cases, officials in private companies responded in that way, as did some officials in public agencies, and even some victims. The general patterns transcend differences among the three societies. Some individuals perceive their personal and organizational interests to be served in keeping the problem confined to the private realm, and they act accordingly. When such an individual (or group of individuals) is able to influence or determine what an organization does or does not do, one can think about the organization itself as acting to keep a nonissue from becoming an issue.

Unintentional efforts to maintain nonissues generally occur more at the institutional level. In the three cases, public and private institutions lacked adequate systems to prevent, evaluate, or manage chemical disasters and toxic contamination. Loopholes existed in legal systems supposed to control hazardous substances and in administrative systems supposed to implement and enforce the laws. The cases support the general tendency of public organizations to be constrained by legal and institutional structures from effectively controlling the activities of private corporations (Stone, 1975).

Because of these intentional and unintentional acts, individual victims of social problems cannot easily transform nonissues into public issues. Victims confronted similar difficulties in all three societies, representing a general social phenomenon, what might be called the "enemy of the people" phenomenon. Those circumstances are dramatically illustrated in Henrik Ibsen's play about a doctor who discovered that the "curative" spring waters in the town's Health Institute were in fact contaminated and dangerous. The doctor tried to warn the townspeople about the public health hazard, but threats to health became perceived as threats to wealth, both personal and social. Instead of a protector of public health, the doctor became portrayed and perceived as *An Enemy of the People*.

Victims and Power

Toxic victims and potential victims confront the problems of a chemical disaster with less preparation, less resources, and less power than social institutions. Those toxic victims who begin as socially marginal, as well as politically powerless, confront even greater obstacles in their quest for redress. Toxic victims usually have no way of predicting they will suffer a chemical disaster. They learn about toxic contamination through suffering, trial and error, and their own struggle. They must learn about obtaining redress from the strategies of earlier toxic victims and from the struggles of other kinds of victims in society. As long as social institutions lack adequate policies to manage or prevent chemical disasters, and as long as power remains unequally distributed in society, potential victims need to understand the politics of toxic contamination, for it is through politics that the victims obtain redress.

Victims come to realize that they suffer not only from a social problem but also from a political problem. Their problems derive both from toxic contamination and from relative powerlessness. The victims' strategies must therefore take into account the political realities of the society, especially the roles of political parties and interest groups. The victims need to understand the structure of political competition and use it to design an effective strategy. In the United States, those strategies concentrate on mobilizing individual politicians and their staffs, on using constituency connections, and on activating independent environmental groups. In Italy, victims need strategies that stimulate political parties and their affiliated organizations. In Japan, victims' strategies also focus on parties and affiliated organizations, and require attention to conflict certain to occur among competing support groups. In all three societies, the strategies must develop allies in the mass media and must anticipate the responses of private and public organizations.

The individual can play a central role in transforming nonissues into public issues. The process, however, often requires a deep stubbornness and a willingness for self-sacrifice. But one individual can influence the responses of both private and public institutions, even from outside those organizations, as shown in our cases. The individual's capacity to exert influence, however, depends on the individual's material, organizational, and spiritual resources. This influence appears proportional to social class. In our three cases, for example, socially marginal victims suffered more and longer than victims with the connections and means to influence social institutions. One person can, however, overcome the disadvantages of social class to achieve significant influence on society.

As the three cases show, the sources of organizational assistance depend on the nature of the party system and the structure of political competition. In general, the existence of a party in full opposition facilitates the victims' search for assistance in raising an issue. But the existence of a high degree of competition among opposition parties and groups produces divisions and conflicts in the victims' search for political allies and organizational assistance. In Michigan, for example, victims relied on individual state legislators for assistance, utilizing the emphasis of U.S. politicians on serving their constituents more than their party. That pattern occurred even in a state noted for relatively strong party organizations and party discipline. In Italy, victims relied on party organizations or union organizations to serve as social arbitrators in obtaining compensation; this pattern reflected the dominant role of Italian parties in controlling social resources and the prominent social role of unions in the mid-1970s. In Japan, toxic victims relied on organizations linked to opposition parties or independent citizens' groups, encountering severe conflict among the various organizations.

Victims do not make their decisions on individual or group action solely on the basis of rational economic calculations. Mancur Olson argued that a dilemma of collective action occurs when an individual can gain the benefits of collective action without participating in that action—the problem of the free rider. That dilemma does not hold, however, when the acts of an additional individual do make a difference in total benefits obtained, as occurs for example with small groups or with specially active individuals. Moreover, not all individuals' decisions on whether to participate are based on estimates of what they expect to gain for the costs of participation. Other incentives prompt individuals to join in group action: anger against government and business, demands for moral retribution, and feelings of group solidarity. While these incentives may be nonrational from a strictly economic perspective, they are rational from a broader political perspective. The incentives create the group actions that compel social institutions to change policy and provide redress.

Social conflict resulting from political action can produce positive consequences for some groups. Indeed, the politics of redistributing resources in society commonly involves conflict among groups. Although groups in power seek to contain conflict perceived as threatening the status quo, groups out of power tend to favor conflict perceived as promoting a redistribution of resources. The pattern of social conflict depends not only on the nature of the social harm and on the structure of political competition but also on larger social values. The depth and persistence of social conflict in the Yusho case suggests that Japanese do not simply value consensus. They also value conflict as a means to make social decisions, as Frank Upham argued in an analysis of litigation in Japan (1987). The Yusho case illustrates the falsity of the conventional view of Japan as a society that has a "traditional preference for decision by consensus," in which "there is no open confrontation of groups or parties" (Ward, p. 49). The relatively slow buildup of conflict in Michigan indicates the existence of strong social controls over conflict, controls that only gradually broke down. The rise and fall of social conflict in Italy similarly does not fit a simple categorization of that society as dominated by conflict.

Public Institutions and Power

Public bureaucracies, when confronted with a chemical disaster, similarly need political strategies to cope with the consequences of toxic contamination. Public officials can use a strategy of passing the buck to private companies, aiming to make private corporations more responsible in the public sphere and to make companies share the political as well as the economic costs of disasters. Public officials can also follow a strategy of using, adapting, or creating public institutions, a strategy that tends to involve public agencies in social conflict but that also holds the potential of providing assistance to toxic victims. On the other hand, a strategy of dissociation tends to protect public officials from social conflict while effectively assisting private companies more than toxic victims. The choice of whose interests to protect—those of private companies, toxic victims, or public officials—represents a choice of political strategy by public institutions, but one that rarely is made explicitly or explained publicly.

Public officials can expect disasters to create social crises and crises to provide political opportunities, which lead to unplanned confrontation. For politicians in power, opportunities arise to solidify the grasp of power and improve the capabilities of public institutions. For politicians in opposition, opportunities arise to gain more power and to criticize the actions of public institutions and incumbent officials. In that way, the social conflicts associated with chemical disasters reflect the existing structure óf political competition in society. Crises provide the possibility for designing better policies to prevent and manage disasters and better methods to increase the public accountability of private organizations. But as Richard Sennett argued, the changes in social structure accompanying crises tend to be minimal and moreover tend to support the status quo more than to alter it (Sennett).

Public institutions often lack the organizational and the political capability to respond to the rapid and complex technological changes of the late twentieth century. These problems often remain as nonissues until a disaster occurs or until the problem becomes a political threat. The AIDS epidemic provides a tragic example of both the public health consequences of institutional lag in the face of a new infectious disease and the need for public and political conflict to get an issue on the agenda and produce appropriate changes in policy.

Randy Shilts documented these processes for AIDS in his book, *And the Band Played On.* "From 1980, when the first isolated gay men began falling ill from strange and exotic ailments, nearly five years passed before all these institutions—medicine, public health, the federal and private scientific research establishments, the mass media, and the gay community's leadership—mobilized the way they should have in a time of threat. . . . In those early years, the federal government viewed AIDS as a budget problem, local public health officials saw it as a political problem, gay leaders considered AIDS a public relations problem, and the news media regarded it as a homosexual problem that wouldn't interest anybody else" (Shilts, pp. xxii, xxiii). According to sociologists Charles Perrow and Mauro F. Guillén, these organizational failures arose only partly from normal bureaucratic problems or from economic and ideological factors; they attributed the failures to the "unique characteristics" of AIDS, especially the associated stigma, fear, and expense (p. 106). The chemical disasters of this book suggest that the case of AIDS may not be as unique as Perrow and Guillén believe.

Not just Italy, then, can be described as a *lentocrazia*, a stalled society. Michel Crozier said the same about France and the *société bloquée*: "French society is not the only stalled society. Blockages seem to be an essential characteristic of modern advanced societies. . . . Even in day-to-day transactions, people are trapped in vicious circles; whatever their intentions, the logic of the system distorts their activities and forces them to collaborate in preserving the model" (pp. v, vi).

For our three cases, the process of making social decisions lacked the speed necessary to meet the toxic problems of an industrial nation. One Italian critic, Giorgio Galli, called the ICMESA disaster a logical outcome of the *lentocrazia* and of its underlying assumption that in doing nothing, society would remain the same. That assumption was wrong, he stressed, since the situation did not remain static but changed for the worse. In that sense, not deciding to deal with chemicals in society represented a decision by omission instead of by action (Galli). Our case studies of Japan and the United States illustrated similar processes of the *lentocrazia* for the problems of toxic chemicals.

Once the chemical disasters became social crises, institutions did respond, often in misdirected, symbolic, or defensive ways, but sometimes with incremental changes that helped provide victims with redress (see below). While Japan's public administration is considered among the more efficient (as in the vision promoted by Ezra Vogel in his book *Japan as Number One*), the Yusho case demonstrates the failure of Japanese public administration in an outstanding way. Any assessment of Japanese government that does not take into account the *lentocrazia* and its detrimental social consequences presents a one-sided view. On the other hand, Lombardy is among the most advanced and efficient regional administrations in Italy, as demonstrated by an analysis of regional social policies (Putnam, Leonardi, and Nanetti). The inadequacies of Lombardy's public institutions in dealing with chemical problems indicate the even greater institutional lag that other regions in Italy must be experiencing.

The *lentocrazia* in all three societies involves confusion and conflict among central and peripheral public bodies. In all three cases, the disasters resulted from inadequate systems in the periphery to control chemicals and from confusion and division among narrowly focused public bureaucracies in both periphery and center. Local governments are well known for a lack of adequate resources to control complex technology (Tarrow, 1978, p. 7). In another area of environmental control, regional sewage planning, Bruce Ackerman and his coauthors found that "even moderately large local governments" lacked sufficient staff to carry out analysis or even to evaluate analysis performed by outside consultants (pp. 309–10). Local administration suffers from a central paradox, pointed out by Douglas Yates: dependence on higher-level governments restricts the ability to take independent policy initiatives; and independence of higher-level governments weakens the ability to ensure competence, control, and fairness in the implementation of national programs (1977, p. 6). Local governments are not only weak technically but also vulnerable politically. Our cases illustrated the limitations of municipalities in controlling complex industrial operations and in recognizing and responding to chemical disasters. Once the disasters became public issues, however, differences appeared in the roles of center and periphery, reflecting divergent structural and historical backgrounds as well as the characters of the particular peripheral regions.

Michigan's state government maintained the most control over the contamination problems but used the federal government as a source of technical assistance and of public legitimation. That approach reflects an evolution of the American federal system over the past two centuries, involving a marked transformation in the relationship between center and periphery. The layer-cake model has been increasingly replaced by a marble-cake model. The functional specialists in state and federal governments, especially in the twentieth century and particularly since the New Deal, have devised various systems of communication and collaboration. Federal programs have been designed to encourage the pattern of state control and federal assistance and legitimation.

The Lombardy regional executive, on the other hand, used the central government as a source of emergency funds, as a source for some technical assistance, but rarely for public legitimation, seeking to build up Lombardy's own technical services and legitimacy and to avoid the generally discredited central authorities. That approach reflects Italy's groping toward a federal structure since the 1970s (Gourevitch). But the transformation is a slow process. The center is jealous of its old prerogatives and power. The periphery is determined to make real progress in the development of regional and local autonomy. Those at the periphery, however, go to great extremes in not asking the center for assistance. They also avoid using the center as a source of public legitimation, because the center is commonly viewed as inefficient, ineffective, and corrupt.

The Fukuoka and Nagasaki authorities followed yet another approach, using the central government as a major source of legitimacy, technical assistance, and general policy decisions. Their actions reflect the continuing dominance of Japan's center, despite postwar reforms that have given prefectures increased autonomy. As reflected in the Yusho case, the central government in Japan retains an image of legitimacy and a grasp on power that do not exist in either the United States or Italy. On the other hand, local governments in Japan have provided an important source of legislative innovation in such areas as environmental ordinances, freedom of information, and affirmative action (Reed; Reich, 1983), and legislative reform at the local level has sometimes compelled similar advances at the central level.

The chemical disasters, in exposing nonissues and the *lentocrazia*, raise basic questions of social structure and social justice. As Hugh Stretton wrote about environmental issues, "They are conflicts about what to use, what to produce and how to pay for it—conflicts between people competing in familiar ways for rival values or for shares of scarce goods" (p. 4). His solution, a form of social democracy, called for "the development of new styles of public and private management. . . . To get an efficient flow of business, officials will have to be trusted again—but with some new arrangements to encourage them to behave in trustworthy ways" (pp. 292–93).

The creation and persistence of nonissues in these cases reflected structural gaps in the distribution of power and information in society between workers and managers, between consumers and producers, and between community and factory. For chemical disasters, a concentration of information and power in some groups (managers, producers, factory) increased the damage to other groups (workers, consumers, community). The case studies demonstrated that the public institutions responsible for mediating those gaps in social structure lacked the political capability to respond to the rapid, extraordinarily complex social and technological changes of the late twentieth century.

But public institutions are not static organizations. They adapt to social and technological changes, as documented in the case studies. Those adaptations often improve the political capability of institutions, so that problems that previously required politicization can be handled within institutions. Institutions follow a learning curve, which allows them over time to resolve problems with less social conflict and less political controversy. There seems to be a trade-off between the political capability of public institutions and the political activation of social victims (Reich, 1984b). As mentioned in chapter 1, the three cases of this study were chosen because they moved from private to public to political. In short, they all involved public institutions with low political capability for the problems of chemical disasters, which therefore required a high level of political activation for the victims to achieve redress.

Private Corporations and Power

Private corporations confronted by toxic contamination can limit their liabilities by following a strategy of diversion combined with a strategy of dissociation. That combination can increase the company's ability to control the scope of the issue and to control the publication and meaning of information. While seeming counterintuitive to most corporate officials, the strategy of institutional buck passing includes the passing of information to public officials about suspected contamination problems. Making public officials disclose the information can help reduce the effects of contamination as well as limit public criticism of the private company. Moreover, in most cases, the data become public anyway, regardless of efforts to keep company records private.

Officials in private corporations can expect increased public regulatory control over corporate activities in chemicals through both legal and political means. One corporate manager, observing this general trend, recommended that company officials adopt what he called the "TV test." He urged, "Don't do anything you wouldn't be willing to explain on TV. Most of the actions business is being criticized for would not have occurred if those responsible thought that their actions would one day become public." In his view, the TV test is better than simply conforming with current legal decisions, since it anticipates the direction in which the law is moving (Miller). In other words, corporate actions and information once defined as solely private matters are becoming increasingly redefined as public concerns and as subject to public scrutiny. Legal criteria alone cannot provide corporate executives with appropriate strategies to cope with social and political problems.

Private corporations can be expected to design strategies to protect their own interests, which may mean strategies to evade more effectively their social responsibilities. We can expect private companies to attempt to pass along the economic and political costs of chemical disasters to public organizations and to private individuals—a pattern illustrated in the three cases. Moreover, large corporations not only recognize the need for a social and political strategy but also have sufficient resources to design and implement an effective one.

These considerations raise broader questions of private corporations and public responsibility. In areas beyond toxic contamination, private corporations pose major obstacles to victims of social problems. Private corporations, according to the law, are "fictitious persons." But, as one analyst noted, this concept ignores "the privileged position of business," for corporations "are taller and richer than the rest of us and have rights that we do not have. Their political impact differs from and dwarfs that of the ordinary citizen" (Lindblom, p. 5). As we observed in all three cases, the special influence of private corporations on public affairs contributes to the difficulties in transforming a social problem from nonissue to public or political issue.

Corporations differ in their use of power, as shown by the responses of the firms in the three cases. The Kanemi Company showed the greatest social irresponsibility in its initial denial and cover-up of the problem and in its later resistance to compensating victims. A private corporation can maintain a social problem as a nonissue by combining a rigid policy to protect corporate profits and growth with inadequately developed corporate strategies. The corporate ability to understand and respond to complex social phenomena depends on its structure of interests and its structure of cognitions.

The Yusho case also demonstrates the social tragedies and dysfunctions resulting from Japanese organizational mechanisms that enhance loyalty to the company and that promote cooperation between public and private organizations. This negative example, and others, raise serious questions about the recommendations of those American authors who urge U.S. firms to follow Japanese corporate practices intended to improve management and productivity but who have not considered the potential negative social consequences of those same practices (E. Vogel; Pascale and Athos).

Corporate responses were critically influenced by cultural variables. The Kanemi Company's use of token monetary compensation (the solatium) and the company president's repeated public apologies represent specific Japanese forms that are socially expected and demanded in situations of corporate wrongdoing. One cannot easily imagine either the formal solatium or the public apology in an American or an Italian context. At a more general level of analysis, however, companies in the three countries did seek private out-of-court settlements with victims, especially once the problems became public issues. But the means they used necessarily depended on the particular cultural and social context.

Might the responses of other corporations in these three societies be more forthcoming, more attentive to the problems of redress? Although this question is difficult to answer in a definite way, several points are worth noting. For Japan, the history of pollution has been marked by corporate intransigence and callousness toward victims, until a court rules against the offending company, precisely as occurred in the Yusho case. For Italy, the branch of the major multinational Givaudan probably responded in a more sophisticated manner than a typical Italian firm might have, certainly with greater attention than ICMESA did before 1969, when the company ignored its pollution problems almost completely. For the United States, the reliance of the companies on their legal staffs and on insurance policies represents a fairly typical response.

This discussion points to the difficulties that private corporations have in confronting the concept of social responsibility. The cases suggest that corporate actors behave first with a narrow sense of corporate responsibility and give only secondary attention to a broad social responsibility. But as noted below, the incentives for corporate social responsibility can be changed through public policy. If victims gained easier access to corporate resources for redress, then the actions of corporations could change as well. Increasing the economic and political costs of socially irresponsible acts would encourage a recalculation of corporate responsibility, making private organizations redefine their private interests to include larger areas of the public welfare. Changing the social structure of incentives for corporate behavior, however, runs directly into the problems of redistributing power in society and the dominant tilt of influence toward the private sector.

Disasters provide an opportunity, and sadly may even be necessary, for realigning the mobilization of bias and for redefining the dominant symbols in ways to tip the balance of power, at least momentarily, in the direction of public interests and in favor of the relatively powerless. Crises not only provide opportunities for viewing the ordinary processes of society, they also provide the powerless with brief chances for reforming those processes, for redirecting social resources in more equitable ways.

Policies for Chemical Disasters

Policy for chemical disasters can focus on managing the consequences of disasters after they occur or on preventing the release of toxic chemicals. One broad issue for public and private institutions is which line of policy receives priority. Two distinct types of argument exist on this question.

The first argument is economic and uses cost-benefit analysis. This line of reasoning suggests that prevention might not always be the best policy. Some chemical leaks may not be worth trying to prevent if the costs of prevention are great and the benefits of prevention are small. Making that judgment about costs and benefits involves a host of problems, including those inherent in cost-benefit analysis (Fischhoff) as well as the uncertainties of toxic contamination. The costs of prevention (for example, in equipment purchases) are usually more tangible and easier to estimate than the benefits of prevention (for example, in avoiding cleanup costs). Economic rationality also entails assumptions about the equivalence of providing redress and avoiding damage and about the monetization of injury as an acceptable form of making one whole again. The consequences of nonprevention are often difficult to predict, since the effects of a chemical on the human body, natural environment, and social harmony may be unknown, uncertain, or intangible. For the three cases in this book, however, the costs of nonprevention, that is, the costs of the chemical disasters, certainly exceeded the costs of those actions that could have prevented the disasters.

The second type of argument assesses the ability of modern society to prevent catastrophes. If it is impossible to prevent catastrophes, then policies should focus on ways to manage their effects, though prevention should not be ignored. There is considerable debate on the ability to prevent catastrophes. Some (such as Joseph G. Morone and Edward J. Woodhouse) argue that the United States has done quite well at preventing catastrophes and advocate a policy focusing on prevention. Charles Perrow, on the other hand, in his book *Normal Accidents*, concluded that failures are inevitable in the components of high-risk industrial systems and that catastrophic consequences occur from the unanticipated interaction of multiple failures. He characterized his review of high-risk industries as a "dismal and dismaying travelogue," with a discouraging record in each case (p. 351). "At each turn, even in the best of industries, we found rampant attribution of operator error to the neglect of errors by the Great Designers and the Centralized Managers. We found organizations that could not carry the burden of error-free operation, and sometimes seemed insensitive to the damage they did or could do. We may be thankful for the regulatory agencies, but too often they were shown to be ineffective, sometimes natteringly so, sometimes even criminally indifferent or co-conspirators" (p. 351).

Perrow did not accept the existing structure of industrial systems, nor did he offer a simple prescription for methods to improve risk management in society. He concluded that "much could be done in all these systems to increase safety" (p. 350) and suggested ways of thinking to reduce the likelihood of failures in components and in systems. He stressed that different types of industrial systems pose different risks of catastrophe, leading one to broader judgments and specific strategies about high-risk systems: "abandon this, it is beyond your capabilities; redesign this, regardless of short-run costs; regulate this, regardless of the imperfections of regulation" (p. 351).

Perrow's concern with the systemic obstacles to prevention shapes my comments on both types of policies, those to manage and those to prevent chemical disasters. I advocate a strategy that recognizes that *some* chemical disasters are inevitable. While not ignoring measures to prevent disaster, public policy must seek to provide effective care, compensation, and cleanup. Central to such a strategy is the likelihood that an effective public policy of disaster management will encourage better private efforts at prevention. The higher the cost to corporations of providing redress, and the more likely that they will be required to pay full redress, the more likely that private firms will adopt more effective prevention measures. Assuring adequate redress for the victims serves not only the goals of equity but also the objectives of prevention.

Policies to Manage Chemical Disasters

Most social institutions and most toxic victims are not prepared to cope with a chemical disaster (chapters 6 and 7). Social institutions are not usually structured to deal adequately with the normal control of toxic chemicals; few are prepared to manage a crisis situation. A disaster highlights the weaknesses in dealing with both ordinary and extraordinary circumstances. A disaster exposes the institution's lack of preparation and planning for dealing with low-probability high-cost events.

Each of the three areas of redress—care, compensation, and cleanup—can be used as an indicator to evaluate disaster-management policies. How effectively do public policies ensure each type of redress? Questions concerning the three areas rarely develop as limited technical decisions, and often become social controversies, because of problems of available information, model choice, data collection, data analysis, and interpretation. These problems can easily combine with threats to important public and private interests, increasing the potential to become major political issues. This mixture of scientific and political problems occurs commonly in environmental controversies (Socolow) and in other technical conflicts (Nelkin, 1979a), as evidenced, for example, in the debates around DDT (Dunlap). The uncertainties of scientific questions promote the intervention of social values, especially when proposed answers threaten the economic or political interests of powerful social institutions. Our cases showed how various social groups interacted to produce political conflict around the uncertainties and controversies of toxic contamination.

Chemical disasters create strong organizational incentives for both public agencies and private firms to design plans for the management of potential chemical disasters, similar to the contingency plans for natural disasters (Haas, Kates, and Bowden; Kates). Those plans may include strategies for addressing the substantive policy questions of care, compensation, and cleanup, as discussed below. But such efforts, to be complete, also need to provide strategies for managing the political consequences of chemical disasters.

Various efforts have been made to improve chemical disaster planning in the United States. Industry groups, for example, have established a computerized chemical emergency information system, called CHEMTREC, to provide data on toxic chemicals in transportation accidents (Rawls). Title III of the Superfund Act Reauthorization Amendments in 1986 required planning by companies, communities, and states for the consequences of toxic substance releases. Whether these requirements will be effectively implemented remains to be seen, but they do represent a trend toward more systematic efforts at chemical disaster planning. It is also uncertain whether these efforts will address the problems of slow or invisible toxic contamination, such as the PBB poisoning of Michigan, or will encompass the full range of policies for care, compensation, and cleanup.

Care. Problems associated with receiving care are often immediate and troubling for victims, in part because exposure to toxic chemicals carries important psychological consequences as well as potential physical impairments. On this point, Henry M. Vyner concluded: "A new, but steadily growing, body of research is beginning to indicate that exposure to the invisible environmental contaminants can also cause the development of psychological trauma amongst exposed individuals" (1988b, p. 1102). The psychological dimensions of toxic contamination complicate enormously the problems of giving care. Exposed individuals typically confront multiple uncertainties about exposure, evacuation, dose, latency, etiology, diagnosis, prognosis, treatment, and finances (p. 1098). These uncertainties provide a foundation for other adverse psychological responses. Vyner recommended that public policy address the psychological trauma associated with toxic exposure. Health-care systems need to promote the patient's "vigilance needs," which require providing patients "with a sense of control over their health and with the knowledge that they have a clearly defined path of action whenever a new health problem arises" (p. 1102).

Policies for the care of toxic victims often encounter conflicts between the public and private medical sectors. The ability of public authorities to provide adequate care for the victims of toxic contamination depends to a great degree on the broader health system. In the United States, public health departments confront a long-standing tension in the structural dichotomy between the public and private sectors. Private medicine has traditionally opposed public provision of medical services, except under extraordinary circumstances. While health departments are responsible for the epidemiologic investigation of environmental problems (about which private medicine has little interest), they lack routine mechanisms to provide victims of chemical contamination with health services (which would compete with the interests of private medicine). Deciding what role public health departments take toward victims of toxic contamination remains a source of continuing tension.

In Japan and Italy, on the other hand, toxic victims can receive care through the national health system, with most costs paid by insurance or the government. These safety nets can reduce the pressure to create public policies for the care of toxic victims. But as we saw in the cases, the services normally provided are not adequate or appropriate for chemical disasters. In such situations, public officials must address the need for special clinics for toxic victims. In both the Yusho and ICMESA cases, public authorities did create special services, after increasing degrees of political pressure, in efforts to provide what the victims considered adequate care.

The provision of public services, either in normal or special clinics, raises important economic as well as political questions. To the extent that such services are provided solely with funds from public coffers or victims' pockets, the companies responsible are relieved of the costs of care. In addition, other firms' incentives to avoid creating similar public harms are reduced. In our three cases, public authorities made efforts through legal action and direct negotiation to recover the costs of public services provided. Whether the funds collected actually covered the full outlay of public services remains uncertain, although it seems unlikely.

Conflicts around personal choice arose in all three cases. In Japan, many victims rejected the doctors in the special Yusho clinic and preferred their own physicians. Some victims developed an alternative Yusho clinic. In Italy, the question of abortion and personal choice propelled the dioxin issues into the national political arena. In Michigan, the discovery of high levels of PBBs in breast milk of mothers raised questions about the safety of breast feeding. Policymakers in all three cases had to consider what degree of public intervention was appropriate in basic decisions about private lives. The Michigan case showed the great reluctance of public authorities in the United States to intervene in what are perceived as private individual decisions; officials opted to provide information to mothers through tests of PBB levels in breast milk but to leave decisions about whether to breast feed to individuals. The Italian case demonstrated a willingness for the state to intervene—even on the extremely controversial subject of abortion—in providing information, counseling, and services.

So that public authorities can identify who deserves public care, they must develop criteria for certifying official patients. The three cases represent a range of possibilities in drawing boundaries. In Michigan, the authorities stated in effect that no one deserved care, a clear and simple policy. Although the state eventually provided some medical assistance for PBB farmers, problems of access to health services remained for people without health insurance (MDPH, 1986). In Japan, public officials used a restrictive definition of official Yusho patients based on specific symptoms, producing continual conflict over efforts to enlarge those boundaries. And in Seveso, the officials used an inclusive definition of official patients based on geographic criteria; their policy produced less conflict on this point than in the Yusho case, although controversy did occur over where to draw the geographic lines. In the Yusho and Seveso cases, victims who met the official criteria gained access to health services without the normal copayment obligations.

Our cases suggest that any policy that draws lines around official victims (no lines, symptom-based lines, or geography-based lines) is likely to produce conflict in a case of toxic contamination. A strategic choice of public policy would include consideration of the kind of conflict likely to occur and its implications. A policy inevitably addresses the uncertainty of who deserves care by applying social values related to state intervention: no state provision of public care, limited provision, or inclusive provision. In our cases, these policies and their values both structure the forms of social conflict and are affected by social conflict.

Policies for care of toxic victims in the United States have been promoted through the 1986 amendments to the Superfund Act. In particular, the amendments expanded the role of the Agency for Toxic Substances and Disease Registry in ways that could assist persons who are seeking care. "Among the most significant additions are the agency's responsibilities for officially listing which chemicals are 'hazardous substances,' gathering data on the health effects of exposure to the substances, preparing toxicological profiles, responding to requests from individuals and from licensed physicians for health assessments, conducting required health assessments in the areas surrounding cleanup sites, disseminating information to state governments, the medical community and affected communities on the dangers of exposure, conducting periodic medical testing for populations at risk and providing treatment to injured individuals, reporting to Congress on the results of health assessments, and running health surveillance programs in areas where there exists a significantly increased risk of adverse health effects in humans" (Soble and Brennan, p. 1065). If fully implemented, this policy could go a long way toward meeting the needs for care, through both information and services, for toxic victims in the United States.

Compensation. Who has the authority to decide how victims are compensated? The conventional response is to leave the problem to the courts, private damage suits, and traditional tort law. The difficulties with that approach are enormous: the reluctance of courts to enter complex scientific disputes, differences between legal and scientific theories of causation, and the burden of proof that must be shouldered by the victims (Environmental Law Institute). Lawyer Stephen M. Soble summed up the problems: "Too often victims of toxic substance poisoning are also victims of the legal process. Victims seeking compensation through general tort remedies frequently face long and costly litigation, uncertain of the outcome. Alternatively, victims who are entitled to recovery under existing forms of non-judicial compensation often find their actual recoveries to be pitifully small" (Soble, p. 703).

Japan's administrative compensation system for pollution-related diseases provided an alternative to litigation (first as a limited system in 1969 and then in expanded form in 1973). The Japanese system emerged from political conflict over pollution in the 1960s and a series of successful lawsuits for toxic victims in the early 1970s (Aronson). Industry in Japan supported the legislation for administrative compensation as a means to protect companies against disruptions from social conflict and the liabilities of tort litigation. The enactment of Japan's compensation law thus supports social theory that views policy as institutional responses to contain protest and conflict.

The first fundamental problem of design encountered by the Japanese system was deciding who should be compensated. This problem for policymakers resembled the uncertainties about care discussed above. Japan's system combined two types of criteria: geographic boundaries and individual symptoms. For health damage related to general air pollution (class I areas), the system used strict geographic boundaries and loose individual symptoms, while for health damage related to toxic substances (class II areas), the system used loose geographic boundaries and strict individual symptoms.

As in our three cases and other environmental conflicts, the way of drawing the lines determined the form of controversy. For Japan's compensation system in the 1970s, criticism of air-pollution criteria focused mainly on the drawing of geographic boundaries for designated pollution areas, while criticism of the toxic pollution criteria focused on the diagnosis of individual symptoms for certified pollution victims (Aronson). Controversy in a compensation system is probably inevitable; but the form of controversy can be chosen or shaped by the design of the system, as Peter S. Barth showed in his study of workers' compensation.

Design decisions about which polluting sources pay and about the criteria used for assessing payment levels contributed to the Japanese system's demise. In the 1980s, the political winds shifted in Japan and the compensation system was reformed in ways that effectively eliminated relief for new cases of health damage related to air pollution. Japan's top industrial lobbying group, Keidanren (the Federation of Economic Organizations in Japan), became the lead group lobbying to revise the compensation system and reduce the economic burden on industry, arguing that sulfur dioxide levels (used to assess industry payments to the compensation fund) had substantially declined and could no longer be considered the primary cause of nonspecific respiratory diseases. Various citizen groups and victims' groups opposed the revisions (Awaji and Tsukatani). But, despite scientific evidence suggesting continued health problems associated with air pollution, Japan's Environment Agency followed the recommendations of an expert committee to take steps that effectively abolished the compensation system for new cases of respiratory diseases.

The demise of Japan's compensation system illustrates that a system built through politics can also be demolished through politics. The legislative changes to dismantle the compensation system provided for continued compensation payments to existing official patients, thereby defusing potential opposition from this group of beneficiaries. The transformation reflected the declining power of Japan's antipollution movement, the slowdown in overall Japanese economic growth, and the changing economic structure. Japan's progressive system to assist persons with nonspecific respiratory diseases associated with air pollution was scuttled out of fear that the compensation would drag the heavy industries into even slower growth and a worse competitive disadvantage. In the 1970s, political pressure compelled big business in Japan to shoulder social welfare costs associated with economic growth; in the 1980s, industry pressure compelled the government to reverse this redistributive effort and return the costs of illness associated with air pollution to individuals.

Over the past decade, the U.S. Congress has repeatedly confronted the question of compensating toxic victims, considering various bills but refusing to pass a comprehensive compensation law. In 1977 and 1978, Congress rejected proposals for a comprehensive administrative system as amendments to the Toxic Substances Control Act. Congress rejected similar proposals in 1980 when it passed the Superfund Act (Comprehensive Environmental Response, Compensation, and Liability Act) and again on several occasions in its reauthorization in the mid-1980s. Measures for compensation were removed from the original Superfund Act in order to assure passage and were replaced with the request for a report on existing remedies. The report of a study group was submitted to Congress in 1983, with two policy measures to facilitate compensation: creation of a national administrative fund and expansion of state laws to give broader causes of action for toxic victims (Grad). Congress has yet to act on these recommendations.

Soble and Brennan reviewed the failures of Congress to construct a national policy on compensating toxic victims and concluded: "Congress continues to rely only upon controlling industrial chemicals, leaving victims to fend for themselves in state courts with nonuniform and often unfair and unfavorable standards for adjudication and recovery for toxic torts" (p. 1061). "The basic policy choice for Congress is whether it will react sometime in the future to an escalating battle between victims and the chemical industry as litigants, or whether it will choose to govern by developing a positive legal regime which is comprehensive, rational and fair and which brings the best scientific evidence to bear on decisions regarding safety, prevention and treatment" (p. 1062). Although some states have moved in the direction of administrative systems for toxic victims (*Columbia Journal of Environmental Law*), policymakers at the national level have refused to do more than hold hearings and request studies. Another major chemical disaster in the United States, or a series of successful and expensive toxic lawsuits, may be required to move national policymakers to take action on an administrative system for compensation.

Some authors have argued, on the other hand, that an administrative compensation system is not necessary and that the existing tort system in the United States is providing toxic victims adequate redress (Anderson). Leslie Boden and his coauthors pointed out that decisions about causation in court are made by lay judges and juries, with considerable flexibility to rely on commonsense experience. They criticized proposed reforms of the tort system, warning that "any reforms should be approached with caution, since our ability to predict the effects of such reforms is limited" (p. 1027). They concluded, "While [these reforms] may reduce the evidentiary burden on plaintiffs, they do not appear to eliminate the problem of demonstrating causation" (p. 1028).

How effectively did public policies compensate the victims of our three cases? The Italian administration did best, as Lombardy effectively used central government funds to begin compensating victims and to provide alternative housing and other needs. This solution resulted from a particular political moment in which the new national government was supported by the Italian Communist party and ties between the Christian Democratic factions in power in Rome and in Milan were relatively close. Subsequently, the Lombardy government initiated efforts to collect those funds from Givaudan. The Japanese administration, on the other hand, did worst, insisting on private company action and a court decision, which was not reached until 1976, eight years after the poisoning, and which left most victims with no significant compensation until then. The Michigan administration followed a policy between the extremes adopted in Italy and in Japan, first encouraging the parties to settle out of court and later pushing the companies to put together additional funds for settlement.

Cleanup. The conflicts over cleanup in all three cases involved typical siting controversies for hazardous activities and hazardous wastes: the "not in my backyard" phenomenon and the "if it's so safe, put it in the city" response. The controversies raised uncertainties about the safety of underground water (endangered by burial) and the possibility of air pollution (from incineration). Protests in court and in direct action compelled politicians and bureaucrats to explain, justify, and improve the technical decisions. Our three cases also reflect broader problems with policies for toxic cleanup, especially the ways in which such controversies illustrate how "the perceived interests of the local community clash with the broader public interests of the state" (Bowman, p. 240).

In the Michigan case, state officials defined burial as the correct scientific solution for cleanup, and they decided on a site without consulting local politicians or local residents. Officials sought to portray burial as a narrow technical solution in order to give legitimacy to their choice of policy. But their emphasis on rapid burial had an important political dimension: to remove contaminated cattle from the public sphere as fast as possible. As Dorothy Nelkin noted about technical decisions in general, the definition of inherently political problems as technical problems reflects a greater value placed on efficiency than on democracy. "Technical planning limits public choice and threatens the widely held assumption that people should be able to influence decisions that affect their lives" (Nelkin, 1979b, p. 20). The efforts to bypass the social and political problems of cleanup delayed disposal and increased the burden on farmers and, ironically, exacerbated the social and political problems of cleanup. People refused to consider burial as simply a technical solution to cleanup and refused to let public officials ignore the social and political dimensions.

One political solution for the problem of toxic cleanup is a policy of exporting the problem. In the Michigan case, officials decided to export remaining carcasses, which could not be buried, out of the state, thereby exporting the problem to someone else's (already contaminated) backyard in Nevada. In Seveso, barrels of toxic waste traveled around Europe in search of a final resting place. These examples illustrate a broader policy of efforts to export toxic waste across international boundaries. These efforts, however, have run into problems as increasing numbers of potential recipients refuse to accept the wastes.

In general, the export of toxic wastes represents a "one-directional hazard," where the risk creators are in one country and the pollution victims are in one or more other countries (Majone, p. 43). While international regulatory policy is trying to come to terms with hazardous exports (including wastes and other substances), basic political and equity problems remain (Majone, pp. 42–44). Governments are reluctant to relinquish areas of national sovereignty to international agencies or agreements. Moreover, when the maldistribution of economic benefits and environmental damages between risk creators and pollution victims corresponds to a maldistribution of political power across nations, international regulatory policy becomes even more difficult to design or implement effectively. These political and equity problems across national boundaries parallel the problems between subnational governments, as in the Michigan case of PBBs. No one at the receiving end likes one-directional hazards.

Yet the United States has made important advances in the past decade in its efforts to construct a policy of cleanup for toxic contamination. The Superfund Act of 1980 marked a key legislative attack on identifying toxic waste sites and promoting their detoxification. But the problems confronting this task are enormous. The U.S. Office of Technology Assessment, in a study of Superfund in 1985, concluded that the Environmental Protection Agency had seriously underestimated the number of hazardous waste sites requiring cleanup and the costs of cleanup. The Office of Technology Assessment predicted that ten thousand sites or more could require Superfund cleanup attention, in contrast to the Environmental Protection Agency's estimate of two thousand, and that the costs could rise to $100 billion, far beyond the initial authorization of $1.6 billion for five years. The report also criticized the Environmental Protection Agency's policy of transferring wastes from the hazardous site to a disposal site, which tended to increase both risks and costs (U.S. Office of Technology Assessment, 1985).

Economic and technical constraints, along with political conflicts, have continued to confront Superfund. The reauthorization of Superfund in 1986 provided $9 billion for the next five years to deal with the most dangerous sites. Although new technologies are being developed for destroying or detoxifying hazardous wastes—including high-temperature incineration, solidification, and biodegradation with microbes—the innovations tend to be high-cost as well (*New York Times*, 23 December 1987). But beyond the economic and technical problems, "responsible hazardous waste management is primarily a function of political activity" (Bowman, p. 246). As might be expected, Superfund cleanup priorities have been set according to "squeaky wheel" principles, according to one analysis, as communities have competed for their piece of the limited Superfund pie (Landy, p. 68).

Policies to Prevent Chemical Disasters

There are two general categories of actions to prevent the consequences of chemical disasters. Actions can be directed at the early discovery of chemical spills in order to prevent them from expanding into more complex disasters. Other actions can be directed at eliminating the causes of spills in processing and production. Both levels of action are needed. The rare accident provides a window to observe normal processes, and preventing the rare accident requires changing the normal processes.

In chapter 5, I reviewed four types of policy actions that could hasten the discovery of toxic contamination: improved detection systems, improved understanding of the problem, improved internal communication, and corrective policies for external coordination. What policies might help prevent the initial release of toxic chemicals? One must stress that our three cases resulted from processing errors, from faulty management and operations—as occurred also at Three Mile Island and Chernobyl. How can such internal problems be prevented? In this section, I review preventive strategies based on government regulation, industrial self-regulation, and changes to the structure of incentives.

All three countries in this book have laws related to toxic chemicals, but with various weaknesses. The United States and Japan have pursued policies on new toxic chemicals more actively than has Italy. Japan passed its Chemical Substances Control Law in 1973, and the United States passed the Toxic Substances Control Act in 1976. Italy's National Health Reform Law, which passed in 1979, contains a clause on chemical controls, but only in the 1980s did Italy enact separate statutes for this purpose, including one based on a European Commission directive. Italy does not yet have a general environmental control law, whereas Japan's Basic Law for Environmental Pollution was passed in 1967 and the U.S. National Environmental Policy Act was passed in 1969. Moreover, Italy established a central government agency for the environment only in the mid-1980s—still considered ineffective at best (Amendola, p. 18)—long after the United States had established its Environmental Protection Agency (1970) and Japan its Environment Agency (1971).

These legislative acts on toxic chemicals reflect broad differences in governmental structures. Italy's past approach, as well as its current approach under the National Health Reform Act, stresses decentralized regulation, with control at the level of the region or the local health unit. The decentralized approach can lead to wide disparities among regions. The United States, under the Reagan administration, attempted a similar program of decentralization with its "new" federalism that emphasized the role of states. Shifting the focus of regulation from Washington to the states could result in a situation similar to that of Italy, with great regional disparities emerging not only in the quality of services but also in the level of standards. President Reagan's new federalism typically opposed efforts by the states to create more stringent rules, for example, on the control of pesticides or on the right of workers to know about hazards. This effort reflects a concern of industry that deregulation at the federal level would result in confusing, overlapping, and duplicate regulations at the local and state levels, making it more difficult for industry to influence the regulatory process and more difficult to comply with the final laws and rules.

More fundamentally, these regulatory approaches are not directed at preventing the kinds of processing errors that caused the spills of our three cases. In the United States, the Toxic Substances Control Act (like Japan's Chemical Substances Control Law and Europe's Sixth Amendment) is a birth-control statute, in that the law is designed to prevent new substances from being manufactured and marketed unless they meet certain criteria. Another major U.S. statute in this area, the Resource Conservation and Recovery Act (RCRA), is a waste-disposal law that sets standards for the handling, transportation, and disposal of certain classes of wastes. Neither statute regulates how companies conduct their chemical processes. Conceivably, all the companies in our cases could have been in compliance with TSCA and RCRA and still have caused the same chemical disasters.

In recent years, public institutions have given greater attention to the control of chemicals by passing new laws, changing public agencies, and shifting organizational goals. These government activities include efforts to reduce the manufacture and use of existing toxic substances, to review new chemicals prior to manufacture, and to regulate the use of allowed chemicals at both central and local government levels. Whether these policies will prevent new disasters from occurring, however, is far from certain. At the least, the measures could decrease the severity of chemical contamination that does occur, reducing damage to human health and the environment. A comparison of efforts in Europe and the United States shows remarkable similarities in the objectives and outcomes of chemical regulation but striking differences in the national processes of regulation (Brickman, Jasanoff, and Ilgen, p. 301).

In addition to TSCA and RCRA, the United States has other statutes that provide opportunities for government to regulate the chemical operations of private firms: the Clean Air Act, the Clean Water Act, the Superfund Act, the Occupational Safety and Health Act, the Federal Insecticide, Fungicide, and Rodenticide Act, as well as state laws. I cannot review the requirements and problems with these statutes here. Suffice it to say that legislation allows agencies to regulate some processes, but agencies cannot possibly regulate all processes. Laws can require private firms to install the best available technology for chemical controls, but agencies cannot effectively inspect the operations to ensure that the equipment is used properly and consistently. The limits of direct government intervention in altering private actions are well recognized.

One alternative to government regulation of industry is to depend on industry's self-regulation. That approach, as illustrated by the three cases, has not worked well in the past, owing to industry's self-interest as well as to problems in detection, understanding, and internal communication. If one were to depend on industry's self-regulation, an ethical code for chemical controls might be necessary, with explicit means to ensure accountability, to determine violations, and to provide punishments. So far no such code exists. And one cannot be optimistic about the effectiveness of an international code on corporate behavior, judging from experiences with existing codes for multinational enterprises engaged in pharmaceutical marketing and infant-formula sales.

Another approach to reducing errors in processing would be to change the structure of incentives for corporations and their managers. Economists and lawyers tell us that corporations will engage in appropriate preventive measures if it is cheaper to prevent an accident than it is to pay the costs of redress when an accident occurs (Calabresi). Various policies might be adopted to increase the probability that firms would pay the costs of redress. For example, the use of strict liability by which a corporation would be financially responsible for damages without proof of negligence could help internalize the costs of harm for the firms involved. Another policy to change the incentives for corporate behavior would be to enforce criminal penalties for managers in firms that cause toxic contamination.

Such proposals, however, are not likely to be well received in corporate circles. When plaintiffs frequently win in the tort system, for instance, corporations can respond by seeking to change the rules so that the system spreads the costs of risk away from the individual firms or the industry. Such pressures were behind the administrative compensation system for pollution-related damages in Japan (Aronson) and have given rise to industry requests for an asbestos-compensation system in the United States. An administrative compensation system or an insurance system that is broadly financed could serve this goal of corporate protection.

Insurance policies in this situation can have a pernicious effect, in what economists refer to as "moral hazard." As Kenneth Arrow explained, "The insurance policy might itself change incentives and therefore the probabilities upon which the insurance company has relied. Thus, a fire insurance policy for more than the value of the premises might be an inducement to arson or at least to carelessness" (p. 142). Insurance for chemical spills could have the counterproductive effect of reducing corporate efforts at prevention by reducing the anticipated costs of toxic contamination.

Arrow's solution to this hazardous moral hazard is "coinsurance" to make the company partly responsible for the damages. "If a complete absence of risk-shifting is bad because it inhibits the undertaking of risky enterprises and if total risk-shifting is bad because it reduces the incentives for their success, then it is reasonable to suggest that partial risk-shifting might be best" (p. 143). Indeed, a recent proposal for using insurance in compensation of victims of major industrial risks provided for a five-tranche system, including some company responsibility for damages (Smets). The more likely that companies will be held liable for environmental damages, the more likely that this insurance market will develop (Katzman). Structuring this system to work effectively requires careful consideration of who bears the burden of proof, how information is distributed, and what level of care is expected.

Reducing the victims' barriers to redress could create incentives for private companies to take more care in production. This approach has particular appeal to economists who believe that organizations respond reasonably well to changes in the market environment. Public policy could be used to change the signals to companies about the probability of bearing costs for certain actions, and thereby encourage more prevention and caution. Experience with occupational illnesses, however, indicates that a compensation system alone does not create adequate incentives to achieve a healthy and safe workplace, especially for diseases with long latency periods (Barth, p. 260). One major reason is that profit-maximizing companies compare the current costs of investing in preventive measures against the future marginal benefits of lower compensation costs. These calculations and the discounting process yield a present value of future claims that is quite small, making the prospect of compensation itself relatively ineffective as an incentive.

The empowerment of the victims vis-à-vis the victimizers, however, could create significant incentives to change corporate behavior. In short, if corporations are reasonably certain that they will pay not only economic but also political costs in providing redress to victims of chemical disasters, managers might be persuaded to take extra measures in production and processing to prevent the occurrence of toxic contamination. For this reason, as well as for reasons of equity, better mechanisms are needed to empower the victims of toxic contamination, including peripheral victims, and to assure they receive full redress. Broad-based political pressure for payment of full redress is likely to induce private firms to invest more in prevention.

In two of the three countries, social change supported the empowerment of toxic victims in the 1980s. The United States experienced the rise of a grass-roots movement on toxic chemicals, exemplified by the Citizen's Clearinghouse for Hazardous Wastes and the National Toxics Campaign (Russell). In Italy, local environmental groups spread through the country, and the political spectrum became decidedly "greener," with the Greens winning thirteen seats in Parliament in 1987 (Flavin). Japan, on the other hand, in the 1980s underwent a reversal of earlier trends, perhaps best reflected in the demise of the compensation law.

Changing the system of incentives for corporations will not be a simple task. Charles Perrow, in his study of normal accidents, stressed that the systems that produce industrial catastrophes are human constructions; as such, they are theoretically amenable to change. But he also issued stern cautions: "These systems are human constructions, whether designed by engineers and corporate presidents, or the result of unplanned, unwitting, crescive, slowly evolving human attempts to cope. Either way they are very resistant to change. Private privileges and profits make the planned constructions resistant to change; layers upon layers of accommodations and bargains that go by the name of tradition make the unplanned ones unyielding. But they are human constructions, and humans can destruct them or reconstruct" (p. 351).

Once again, we are confronted by a dilemma. Incremental tinkering may be achievable but may have little impact, while significant change confronts deep political obstacles. As Perrow concluded, "Ultimately, the issue is not risk, but power; the power to impose risks on the many for the benefit of the few" (p. 306). Disaster may provide the opportunity and the impetus for institutional change, allowing for a redistribution of power and a transformation of policy. But the existing distribution of power creates formidable blockages to social change that might benefit the relatively powerless. To obtain even incremental changes in laws, institutions, and goals, more social crisis may be required. To prevent more chemical disasters, paradoxically and tragically, we may need more chemical disasters.

References

Ackerman, Bruce, Susan Rose-Ackerman, James W. Sawyer, Jr., and Dale W. Henderson. 1974. *The Uncertain Search for Environmental Quality*. New York: Free Press.

Aftosmis, J. G., R. Culik, K. P. Lee, H. Sherman, and R. S. Waritz. 1972. "Toxicology of brominated biphenyls: I. Oral toxicity and embryotoxicity." *Toxicology and Applied Pharmacology* 22:316.

Akagi Katsuo. 1973. "Kanemi raisu oiru chūdoku jiken (Part 1)." *Shokuhin Eisei Kenkyū* 23:405–18.

Alberoni, Francesco. 1977. *Movimento e Istituzione*. Bologna: Il Mulino.

Albosta, Donald J. N.d. (ca. September 1975). Press release on irresponsible acts by Governor Milliken and Departments of Public Health and Agriculture on PBB problems. Office of State Representative Albosta. Lansing, Mich.

———. 3 September 1975. Press release on Governor Milliken's veto of Bill 5033. Lansing, Mich.

———. 16 October 1975. Press release on organizational meeting of PBB Committee. Lansing, Mich.

Alinsky, Saul D. 1969 (1946). *Reveille for Radicals*. New York: Vintage Books.

Allport, Gordon W. 1958. "Traits due to victimization." Chap. 9 in *The Nature of Prejudice*, 138–58. Garden City, N.Y.: Doubleday/Anchor Books.

Allum, P. A. 1973. *Italy—Republic without Government?*. New York: W. W. Norton.

Amendola, Gianfranco. 1985. *In Nome del Popolo Inquinato*. Milan: Franco Angeli Libri.

Anderson, H. A., E. C. Holstein, S. M. Daum, L. Sarkozi, and I. J. Selikoff. 1978. "Liver function tests among Michigan and Wisconsin farmers." *Environmental Health Perspectives* 23:333–39.

Anderson, H. A., R. Lilis, I. J. Selikoff, K. D. Rosenman, J. A. Valciukas, and S. Freedman. 1978. "Unanticipated prevalence of symptoms among dairy farmers in Michigan and Wisconsin." *Environmental Health Perspectives* 23:217–26.

Anderson, Robert C. 1985. "Toxic compensation bills." *Environmental Health Perspectives* 62:365–71.

Anderson, Robert M., Robert Perrucci, Dan E. Schendel, and Leon E. Trachtman. 1980. *Divided Loyalties: Whistle-Blowing at BART.* West Lafayette, Ind.: Purdue University Press.

Andō Masahiro. 1978. "Kanemi Yushō jiken Kokura hanketsu ni tsuite." *Shokuhin Eisei Kenkyū* 28:581–90.

Ansoff, H. Igor. 1969. "The process of strategic change." In H. Igor Ansoff, ed., *Business Strategy: Selected Readings.* New York: Penguin.

Armstrong, Donald R. 1975. "PBB: Farm Bureau responds." Letter to the editor. *Michigan Farmer,* 5 April.

———. 1977. Prepared statement in *Hearings before the U.S. Senate Subcommittee on Science, Technology, and Space, Committee on Commerce, Science, and Transportation,* 28, 29, and 30 March 1977, Part 1, 1627–34. Washington, D.C.: U.S. Government Printing Office.

Aronson, Bruce. 1988. "Compensation of pollution-related health damage in Japan." In Michael R. Reich, ed., *Social Policy for Pollution-related Diseases,* special issue of *Social Science & Medicine* 27:1043–52.

Arrow, Kenneth J. 1974. "Insurance, risk, and resource allocation." Chap. 5 in *Essays in the Theory of Risk-Bearing,* 134–43. New York: Elsevier.

Assennato, Giorgio. "Health Surveillance in TCDD Clean-up Workers." Ph.D. diss. Baltimore: Johns Hopkins School of Public Health.

Awaji Takehisa and Tsukatani Tsuneo. 1988. "Current problems and prospects with the Japanese compensation system for pollution-related health damage." In Michael R. Reich, ed., *Social Policy for Pollution-related Diseases,* special issue of *Social Science & Medicine* 27:1053–60.

Bachrach, Peter, and Morton Baratz. 1962. "The two faces of power." *American Political Science Review* 56:947–52.

———. 1975. "Power and its two faces revisited: A reply to Geoffrey Debnam." *American Political Science Review* 69:900–904.

Badaracco, Joseph L., Jr. 1985. *Loading the Dice: A Five-Country Study of Vinyl Chloride Regulation.* Boston: Harvard Business School Press.

Barkun, Michael. 1977. "Disaster in history." *Mass Emergencies* 2:219–31.

Barone, Michael, Grant Ujifusa, and Douglas Matthews, comps. 1979. "Michigan." In *The Almanac of American Politics, 1980,* 415–54. New York: E. P. Dutton.

Barth, Peter S., with H. Allan Hunt. 1980. *Workers' Compensation and Work-related Illnesses and Diseases.* Cambridge: MIT Press.

Bauer, Raymond A., with Richard S. Rosenbloom and Laure Sharp, and the assistance of others. 1969. *Second-Order Consequences: A Methodological Essay on the Impact of Technology.* Cambridge: MIT Press.

Bekesi, J. G., J. F. Holland, H. A. Anderson, A. S. Fishbein, W. Rom, M. S. Wolff, and I. J. Selikoff. 1978. "Lymphocyte function of Michigan dairy farmers exposed to polybrominated biphenyls." *Science* 199:1207–9.

Berger, Peter L., and Thomas Luckmann. 1966. *The Social Construction of Reality.* New York: Doubleday.

Berger, Samuel R. 1971. *Dollar Harvest: The Story of the Farm Bureau.* Lexington, Mass.: Heath Lexington Books.

Berlin, A., A. Buratta, and M.–Th. Van der Venne, eds. 1976. *Proceedings of the Expert Meeting on the Problems Raised by TCDD Pollution.* 30 September and 1 October 1976. Commission of the European Communities, Ministero della Sanità, Regione Lombardia, and Istituto Superiore di Sanità. Milan.

Bernstein, Marver. 1955. *Regulating Business by Independent Commission.* Princeton: Princeton University Press.

Bertazzi, Pier Alberto, Carlo Zocchetti, Angela C. Pesatori, Stefano Cuercilena, Maurizio Sanarico, and Laura Radice. 1989. "Ten-year mortality study of the population involved in the Seveso incident of 1976." *American Journal of Epidemiology* 129:1187–1200.

Bisanti, L., F. Bonetti, F. Caramaschi, G. Del Corno, C. Favaretti, S. Giambelluca, E. Marni, E. Montescarchio, V. Puccinelli, G. Remotti, C. Volpato, and E. Zambrelli. 1978. "Experiences of the accident of Seveso." Regione Lombardia, Ufficio Speciale per i programmi della L.R. 17.1.1977, no. 2. In *Proceedings of the 6th European Teratology Society,* Budapest, 4–7 September 1978. Budapest: Kiado Publications.

Boden, Leslie I., J. Raymond Miyares, and David Ozonoff. 1988. "Science and persuasion: Environmental disease in U.S. courts." In Michael R. Reich, ed., *Social Policy for Pollution-related Diseases,* special issue of *Social Science & Medicine* 27:1019–29.

Boeri, R., B. Bordo, P. Crenna, G. Filippini, M. Massetto, and A. Zecchini. 1978. "Preliminary results of a neurological investigation of the population exposed to TCDD in the Seveso region." *Rivista Patologia Nervosa e Mentale* 99:110–27.

Bok, Sissela. 1978. *Lying: Moral Choice in Public and Private Life.* New York: Random House.

Bowman, Ann O'M. 1983. "The politics of hazardous waste regulation: Theoretical and practical implications." In James P. Lester and Ann O'M. Bowman, eds., *The Politics of Hazardous Waste Management,* 234–51. Durham: Duke University Press.

Brandt, Allan M. 1986. "AIDS: From social history to social policy." *Law, Medicine & Health Care* 14:231–42.

Brickman, Ronald, Sheila Jasanoff, and Thomas Ilgen. 1985. *Controlling Chemicals: The Politics of Regulation in Europe and the United States.* Ithaca: Cornell University Press.

Brunn, Stanley D., and Dennis Koons. 1977. "Selecting a burial site for contaminated cattle: Kalkaska County vs. The State of Michigan." *East Lakes Geographer* 12:58–70.

Budd, M. L., N. S. Hayner, H. E. B. Humphrey, J. R. Isbister, H. Price, M. S. Reizen, G. van Amburg, and K. R. Wilcox, Jr. 1978. "Polybrominated biphenyl exposure—Michigan." *Morbidity and Mortality Weekly Report* 27:115–16, 121.

Burnham, Walter Dean. 1967. "Party systems and the political process." In William Nesbit Chambers and Walter Dean Burnham, eds., *The American Party Systems: Stages of Political Development,* 238–76. New York: Oxford University Press.

Calabresi, Guido. 1970. *The Cost of Accidents.* New Haven: Yale University Press.

Cantrell, J. S., N. C. Webb, and A. J. Mabis. 1967. "Search for the chick edema factor." *Chemical & Engineering News* 30 January, 10.

Caramaschi, F., G. Del Corno, C. Favaretti, S. E. Giambelluca, E. Montesarchi, and G. M. Fara. 1981. "Chloracne following environmental contamination by TCDD in Seveso, Italy." *International Journal of Epidemiology* 10:135–43.

Cardin, Michael, and Lawrence B. Brilliant. 1979. "The search for effective state decision-making about toxic substances: Michigan's Toxic Substance Control Commission Act." *Wayne Law Review* 25:1217–49.

Casarett, Louis J. 1975. "Origin and scope of toxicology." In Louis J. Casarett and John Doull, eds., *Toxicology: The Basic Science of Poisons*, 3–10. New York: Macmillan.

Cavallaro, Aldo. 1979. "Contributo del Laboratorio Provinciale di Igiene e Profilassi al riscontro del TCDD nell'ambiente." In Regione Lombardia—Ufficio Speciale de Seveso, ed., *Disastro ICMESA*, 223–41. Milan: Franco Angeli.

Cerruti, Giovanni. 1976. "Cento giorni alla diossina." In Marco Martorelli, ed., *ICMESA: Una Rapina di Salute, di Lavoro e di Territorio*, 7–16. Milan: Mazzotta.

Chemical & Engineering News. 1985. "Court drops, reduces Seveso dioxin charges." 27 May, 7–8.

Chemical Industries Association. *Basic International Chemical Industry Statistics, 1963–1986.* London: Chemical Industries Association.

Chemical Week. 1982. "After a big overhaul, Velsicol is stepping out." 24 March, 26–27.

Chen, Edwin. 1979. *PBB: An American Tragedy.* Englewood Cliffs, N.J.: Prentice-Hall.

Clausen, A. W. 1981. "Voluntary disclosure, an idea whose time has come." In Thornton Bradshaw and David Vogel, eds., *Corporations and Their Critics: Issues and Answers to the Problems of Corporate Social Responsibility*, 61–70. New York: McGraw-Hill.

Climo, Martha. 1980. Office of Communication Services, Michigan Department of Public Health. Letter to author, 20 August.

Cloward, Richard A., and Frances Fox Piven. 1975. *The Politics of Turmoil: Essays on Poverty, Race, and the Urban Crisis.* New York: Random House, Vintage Books.

Cobb, Roger W., and Charles D. Elder. 1972. *Participation in American Politics: The Dynamics of Agenda-Building.* Baltimore: Johns Hopkins University Press.

Cobb, Roger, Jennie-Keith Ross, and Marc Howard Ross. 1976. "Agenda building as a comparative political process." *American Political Science Review* 70:126–38.

Cohn, Murray Steven. 1985. "Description of a carcinogenic risk assessment used in a regulatory proceeding: Formaldehyde." In David G. Hoel, Richard A. Merrill, and Frederica P. Perera, eds., *Risk Quantitation and Regulatory Policy*, 269–77. Cold Spring Harbor, N.Y.: Cold Spring Harbor Laboratory.

Columbia Journal of Environmental Law. 1985. "Developments in victim compensation legislation: A look beyond the Superfund Act of 1980." 10:271–94.

Commissione Parlamentare di Inchiesta sulla Fuga di Sostanze Tossiche Avvenuta il 10 Luglio 1976 nello Stabilimento ICMESA e sui Rischi Potenziali per la Salute e per l'Ambiente Derivanti da Attivita Industriali (Legge 16 Guigno 1977, n. 357). 25 July 1978. Acts of Parliament, Seventh Legislature, Rome.

Commoner, Barry. 1973. "Workplace burden." *Environment*, July/August, 15–20.

Conti, Laura. 1977. *Visto da Seveso, l'Evento Straordinario e l'Amministrazione Ordinaria.* Milan: Feltrinelli.

Corbett, Thomas H. 1977. *Cancer and Chemicals.* Chicago: Nelson-Hall.

Corbett, Thomas H., A. R. Beaudoin, R. G. Cornell, M. R. Anver, R. Schumacher, J. Endres, and M. Szwabowska. 1975. "Toxicity of polybrominated biphenyls (Firemaster BP-6) in rodents." *Environmental Research* 10:390–96.

Cordle, F. 1977. Testimony in *Hearings before the U.S. House of Representatives Subcommittee on Oversight and Investigation, Committee on Interstate and Foreign Commerce*, 2 and 3 August 1977, 109–394. Washington, D.C.: U.S. Government Printing Office.

Cott, Arthur. 1986. "The disease-illness distinction: A model for effective and practical integration of behavioural and medical sciences." In Sean McHugh and T. Michael Vallis, eds., *Illness Behavior: A Multidisciplinary Model*, 71–99. New York: Plenum Press.

Courter, Paul, and Dick Lehnert. 1975. "PBB: Answers taking shape." *Michigan Farmer*, 1 November, 10, 12, 14–15.

Coye, Molly Joel. 1979. "Crisis: Control in the workplace, a review of three major works in occupational health." *International Journal of Health Services* 9:169–83.

Coyer, Brian W., and Don S. Schwerin. 1978. "PBB and the election: Milliken vulnerable on PBB." *Michigan Farmer*, 21 October, 4–5.

——. 1981. "Bureaucratic regulation and farmer protest in the Michigan PBB contamination case." *Rural Sociology* 46:703–23.

Creech, J. L., Jr., and M. N. Johnson. 1974. "Angiosarcoma of liver in the manufacture of polyvinyl chloride." *Journal of Occupational Medicine* 16:150–51.

Crenson, Matthew A. 1971. *The Un-Politics of Air Pollution: A Study of Non-Decision Making in the Cities*. Baltimore: Johns Hopkins University Press.

Crim, Bobby. 8 August 1977. Press conference transcript. Michigan House of Representatives. Lansing, Mich.

Crosscurrents. 1979. 1 (January). Published by Public Affairs Department of Velsicol Chemical Corporation.

Crozier, Michel. 1973 (1970 in French). *The Stalled Society*. New York: Viking Press.

Cyert, Richard M., and James G. March. 1963. *A Behavioral Theory of the Firm*. Englewood Cliffs, N.J.: Prentice-Hall.

Dahl, Robert A. 1961. *Who Governs?* New Haven: Yale University Press.

——. 1982. *Dilemmas of Pluralist Democracy: Autonomy vs. Control*. New Haven: Yale University Press.

Dahrendorf, Ralf. 1959. *Class and Class Conflict in Industrial Society*. London: Routledge and Kegan Paul.

Dando Shigemitsu. 1965. *Japanese Criminal Procedure*. Trans. by B. J. George, Jr. South Hackensack, N.J.: Fred B. Rothman.

dell'Anno, P. 1976. *The Law and Practice Relating to Pollution Control in Italy*. London: Graham and Trotman.

Deutsch, Morton. 1973. *The Resolution of Conflict: Constructive and Destructive Processes*. New Haven: Yale University Press.

De Vos, George, and Wagatsuma Hiroshi. 1966. *Japan's Invisible Race: Caste in Culture and Personality*. Berkeley: University of California Press.

DeVries, Walter, and V. Lance Farrance. 1972. *The Ticket Splitter: A New Force in American Politics*. Grand Rapids, Mich.: Eerdmans.

Disaster Research Center. 1979. "Selected second phase observations and findings." *Unscheduled Events* 13(3):4–7.

Doniger, David D. 1978. *The Law and Policy of Toxic Substances Control: A Case Study of Vinyl Chloride.* Baltimore: Johns Hopkins University Press.

Douglas, Mary. 1973. *Natural Symbols: Explorations in Cosmology.* New York: Vintage Books.

Douglas, Mary, and Aaron Wildavsky. 1982. *Risk and Culture.* Berkeley: University of California Press.

Downs, Anthony. 1967. *Inside Bureaucracy.* Boston: Little, Brown.

Drayton, William. 1980. "Economic law enforcement." *Harvard Environmental Law Review* 4:1–40.

Dunbar, Willis Frederick. 1970. *Michigan: A History of the Wolverine State.* 2d ed. Grand Rapids, Mich.: Eerdmans.

Dunlap, Thomas R. 1981. *DDT: Scientists, Citizens, and Public Policy.* Princeton: Princeton University Press.

Easton, David. 1965. *A Framework for Political Analysis.* Englewood Cliffs, N.J.: Prentice-Hall.

Eckstein, Harry. 1988. "A culturalist theory of political change." *American Political Science Review* 82:789–804.

Edelman, Murray. 1971. *Politics as Symbolic Action: Mass Arousal and Quiescence.* New York: Academic Press.

———. 1977. *Political Language, Words That Succeed and Politics That Fail.* New York: Academic Press.

———. 1988. "Skeptical studies of language, the media, and mass culture." *American Political Science Review* 82:1333–39.

Egginton, Joyce. 1980. *The Poisoning of Michigan.* New York: W. W. Norton.

Eisenberg, Leon. 1986. "The genesis of fear: AIDS and the public's response to science." *Law, Medicine & Health Care* 14:243–49.

Ellis, John. 1975. *The Social History of the Machine Gun.* New York: Pantheon.

Enloe, Cynthia H. 1975. *The Politics of Pollution in a Comparative Perspective: Ecology and Power in Four Nations.* New York: David McKay.

Environmental Law Institute. 1980. *Six Case Studies of Compensation for Toxic Substances Pollution: Alabama, California, Michigan, Missouri, New Jersey, and Texas.* Prepared under the supervision of the Congressional Research Service, Library of Congress. Washington, D.C.: U.S. Government Printing Office.

Epstein, Edward Jay. 1975 (1967). *Between Fact and Fiction: The Problem of Journalism.* New York: Vintage.

Erikson, Kai T. 1976. *Everything in Its Path: Destruction of Community in the Buffalo Creek Flood.* New York: Simon and Schuster.

Fagen, Richard R. 1966. "Mass mobilization in Cuba: The symbolization of struggle." *Journal of International Affairs* 20:254–71.

Fara, G. 1977. "Rapporto Preliminare sullo State di Salute nella Zona Inquinata da TCDD." Presented at Convegno di Seveso, 28 May 1977, Comitato di Coordinamento tra i CSZ di Brianza e di Seveso 1–2–3.

Farber, T. M., and A. Baker. 1974. "Microsomal enzyme induction by hexabromobiphenyl." *Toxicology and Applied Pharmacology* 29:102.

Ferrara, Marcella. 1977. *Le Donne di Seveso.* Rome: Editori Riuniti.

Fischhoff, Baruch. 1977. "Cost-benefit analysis and the art of motorcycle maintenance." *Policy Sciences* 8:177–202.

Flavin, Christopher. 1989. "Italia verde." *World Watch*, September/October, 5–6.

Food and Drug Administration. 4 June 1974. "Initial notification of class II voluntary recall." Teletype. Detroit, Mich.

Food and Drug Administration. 1977. "FDA enforcement actions against Michigan Chemical Company." In *Hearings before the U.S. House of Representatives Subcommittee on Oversight and Investigation, Committee on Interstate and Foreign Commerce*, 2 and 3 August 1977, 386. Washington, D.C.: U.S. Government Printing Office.

Franck, S. A. 1985. "Economic trends and the chemical industry." In European Chemical Marketing Research Association, ed., *The Changing Face of Chemicals: Proceedings of the International Conference of the European Chemical Marketing Research Association, 14–16 October 1985*, 45–85. Leverkusen, Germany: Bayer.

Frankl, Victor E. 1963 (1959). *Man's Search for Meaning: An Introduction to Logotherapy*. New York: Pocket Books.

Frey, Frederick W. 1971. "Comment: On issues and non-issues in the study of power." *American Political Science Review* 65:1081–1101.

Fujii Mitsuru and Michael R. Reich. 1988. "Rising medical costs and emerging efforts to reform Japan's health insurance system." *Health Policy* 9:9–24.

Fujiwara Kunisato. 1977. *PCB Osen no Kiseki*. Tokyo: Ishiyaku Shuppan.

Fukada Shinsuke. 1970. *Ningen Fushoku*. Tokyo: Shakai Shinpō Sha.

Fukuhara Susumu. 1968. "Toki no ugoki: 'Haigō shiryō no hinshitsu kanri' ni kansuru tsūtastu." *Shiryō Kensa*, July, 1–3.

Fukuoka Kanemi Yushō Genkoku Junbi Shomen. 14 April 1976. Isharyō Seikyū (Kanemi Yushō) Jiken Hanketsu, Besshi Junbi Shomen, Fukuoka Chihō Saibansho Daini Minjibu.

Galimberti, Mario. 1977. "I giorni della tragedia." In Mario Galimberti, Giacomo Citterio, and Luigi Losa, comps., *Seveso: La Tragedia della Diossina*, 13–78. Besana Brianza: Edizioni GR.

Galli, Giorgio. 1976. "La lezione della ICMESA." *Giornale della Lombardia*, September.

Gamson, William A. 1968. *Power and Discontent*. Homewood, Ill.: Dorsey Press.

Gans, Herbert J. 1979. *Deciding What's News*. New York: Pantheon.

Gaventa, John. 1980. *Power and Powerlessness: Quiescence and Rebellion in an Appalachian Valley*. Urbana: University of Illinois Press.

Geertz, Clifford. 1973. "Thick description: Toward an interpretive theory of culture." Chap. 1 in *The Interpretation of Cultures*, 3–30. New York: Basic Books.

Gente. 1976. "Documenti inediti: In due lettere al Sindaco l'ICMESA continua a mentire." 11 August.

Gibney, Frank. 1975. *Japan: The Fragile Superpower*. Tokyo: Charles E. Tuttle.

Gilfillan, S. C. 1965. "Lead poisoning and the fall of Rome." *Journal of Occupational Medicine* 7:53–60.

Gitlin, Todd. 1980. *The Whole World is Watching: Mass Media in the Making and Unmaking of the New Left*. Berkeley: University of California Press.

Giunta Municipale di Seveso. 1977. "Posizione sulla situazione della bonifica, 21 April 1977." In Mario Galimberti, Giacomo Citterio, and Luigi Losa, comps., *Seveso: La Tragedia della Diossina*, 210. Besana Brianza: Edizione GR.

Givaudan Corporation. 1980. *Seveso Backgrounder*. Geneva: Givaudan Corporation.

Gleisen, Victor. 5 March 1976. Testimony at Michigan House of Representatives Special Committee to Investigate PBB, Cadillac hearing. Transcript, pp. 97–98.

Goffman, Erving. 1963. *Stigma: Notes on the Management of Spoiled Identity.* Englewood Cliffs, N.J.: Prentice-Hall.

Golfari, Cesare. 9 December 1976. Intervento alla seduta del consiglio regionale del 9 dicembre 1976 sull' "Attuazione del D.L. 10 agosto 1976, n. 542— intervento per l'inquinamento da sostanze tossiche veraficatosi in provincia di Milano il 10 luglio 1976."

Gotō Masayasu and Higuchi Kentaro. 1969. "Yushō (enka bifenīru chūdoku shō) no hifu kagakuteki shōkoron." *Fukuoka Igaku Zasshi* 60:409–31.

Gourevitch, Peter. 1978. "Reforming the Napoleonic state: The creation of regional governments in France and Italy." In Sidney Tarrow, Peter J. Katzenstein, and Luigi Graziano, eds., *Territorial Politics in Industrial Nations,* 28–63. New York: Praeger.

Grad, Frank P. 1982. *Injuries and Damages from Hazardous Wastes: Analysis and Improvement of Legal Remedies.* Washington, D.C.: U.S. Government Printing Office.

Graziano, Luigi. 1980. "On political compromise: Italy after the 1979 elections." *Government and Opposition* 15:190–207.

Grigliè, Remo, and Gianni Brera. 1978. *La Grande Brianza.* Milan: Istituto Editoriale Regioni Italiane.

Gross, Bertram. 1964. *The Managing of Organizations.* New York: Free Press.

——. 1980. *Friendly Fascism: The New Face of Power in America.* New York: M. Evans.

Gruppo P. I. A., B. Mazza, and V. Scatturin. 1976. "Icmesa: Come e perchè." *Sapere,* November/December, 10–36.

Haas, J. Eugene, Robert W. Kates, and Martyn J. Bowden, eds. 1977. *Reconstruction Following Disaster.* Cambridge: MIT Press.

Hah, Chong-do, and Christopher C. Lapp. 1978. "Japanese politics of equality in transition: The case of the Burakumin." *Asian Survey* 18:487–504.

Halbert, Frederic. 1977. Testimony in *Hearings before the U.S. Senate Subcommittee on Science, Technology, and Space, Committee on Commerce, Science, and Transportation,* 28, 29, and 30 March 1977, Part 1, 36–142. Washington, D.C.: U.S. Government Printing Office.

Halbert, Frederic, and Sandra Halbert. 1978. *Bitter Harvest.* Grand Rapids, Mich.: Eerdmans.

Haley, John O. 1978. "The myth of the reluctant litigant." *Journal of Japanese Studies* 4:359–90.

——. 1982. "Sheathing the sword of justice in Japan: An essay on law without sanctions." *Journal of Japanese Studies* 8:265–81.

Hall, Peter M. 1973. "A symbolic interactionist analysis of policies." In Andrew Effrat, ed., *Perspectives in Political Sociology,* 35–75. New York: Bobbs-Merril.

Hamilton, Alice, and Harriet L. Hardy. 1974. *Industrial Toxicology.* 3d ed. Acton, Mass.: Publishing Sciences Group.

Hancock, M. Donald. 1983. "Comparative public policy: An assessment." In Ada W. Finifter, ed., *Political Science: The State of the Discipline,* 283–308. Washington, D.C.: American Political Science Association.

Hara Ichirō. 1969. "Aru kondensa seizō kōjō ni okeru enka bifenīru sagyō no kenkō kanri." *Osakafu Kōeiken Kenkyū Hōkoku (Rōdō Eisei Hen)* 7:26–31.

Hardin, Russell. 1982. *Collective Action*. Baltimore: Johns Hopkins University Press, for Resources for the Future.

Hardy, Harriet L. 1965. "Beryllium poisoning—lessons in control of man-made disease." *New England Journal of Medicine* 273:1188–99.

Haynes, William. 1954. "Michigan Chemical Corporation." In *American Chemical History*. Vol. 4, *Company Histories*, 275–76. New York: Van Nostrand.

Hazama Yoshiyuki, ed. 1969. *Kore ga Yushō Da*. Kitakyūshū: Kanemi Raisu Oiru Higaisha o Mamoru Kai.

Heath, Robert L., and Richard Alan Nelson. 1986. *Issues Management: Corporate Public Policymaking in an Information Society*. Beverly Hills: Sage.

Heberle, Rudolf. 1968. "Types and functions of social movements." In David L. Sills, ed., *International Encyclopedia of the Social Sciences*. Vol. 14, 438–44. New York: Macmillan.

Heller, Joseph. 1979. *Good as Gold*. New York: Simon and Schuster.

Herxheimer, K. 1899. "Uber chlorakne." *Munchener Medizinische Wochenschrift* 46:278–82.

Hikōsojinra dairijin bengoshi. 15 March 1980. *Ikensho*. Isharyō seikyū kōso jiken. Fukuoka Kōtō Saibansho Dai Yon Minjibu.

Hill and Knowlton International. 1983. *Seveso Information Backgrounder*. Prepared for Givaudan. April.

Hilltop Research. 1970. *Acute Toxicity and Irritation Studies on Firemaster BP-6 Hexabromobiphenyl for Michigan Chemical Corporation*. Miamiville, Ohio: Hilltop Research.

Hoel, David G., Richard A. Merrill, and Frederica P. Perera, eds. 1985. *Risk Quantitation and Regulatory Policy*. Cold Spring Harbor, N.Y.: Cold Spring Harbor Laboratory.

Hoeting, Alan L. 20 February 1975. "Briefing Memorandum—Michigan PBB problem." Detroit Office, Food and Drug Administration.

Hoffman, Paul. 1977. Testimony in *Hearings before the U.S. Senate Subcommittee on Science, Technology, and Space, Committee on Commerce, Science, and Transportation*, 31 March 1977, Part 2, 1373–93. Washington, D.C.: U.S. Government Printing Office.

Holden, Constance. 1980. "Love Canal residents under stress." *Science* 208:1242–44.

Homberger, E., G. Reggiani, J. Sambeth, and H. K. Wipf. 1979. "The Seveso accident: Its nature, extent, and consequences." *Annals of Occupational Hygiene* 22:327–71.

Huddle, Norie, and Michael Reich. 1987 (1975). *Island of Dreams: Environmental Crisis in Japan*. Rochester, Vt.: Schenkman Books.

Ibsen, Henrik. 1923. *An Enemy of the People*. Trans. by Eleanor Marx-Aveling. New York: Brentanos.

ICMESA. 28 March 1975. "Relazione tecnica." Meda. Submitted to Regione Lombardia Comitato Contro L'Inquinamento Atmosferico. In "Dossier Icmesa," *Acqua Aria*, October 1976, 599–602.

Inglehart, Ronald. 1988. "The renaissance of political culture." *American Political Science Review* 82:1203–30.

Insight Team of *The Sunday Times* of London. 1979. *Suffer the Children: The Story of Thalidomide*. New York: Viking Press.

Isbister, Jack. 29 April 1975. "FDA PBB investigations." Memorandum. Michigan Department of Public Health, Lansing, Mich.

Isono Naohide. 1975. *Kagaku Busshitsu to Ningen: PCB no Kako, Genzai, Mirai.* Tokyo: Chūkō Shinsho.

Jackson, Robert J. 1976. "Crisis management and policy-making: An exploration of theory and research." In Richard Rose, ed., *The Dynamics of Public Policy: A Comparative Analysis,* 209–35. Beverly Hills: Sage.

Jackson, T. F., and F. L. Halbert. 1974. "A toxic syndrome associated with the feeding of polybrominated biphenyl-contaminated protein concentrate to dairy cattle." *Journal of the American Veterinary Medicine Association* 165:437–39.

James, Judson L. 1969. *American Political Parties: Potential and Performance.* New York: Pegasus.

Jameson, Frederic. 1971. *Marxism and Form.* Princeton: Princeton University Press.

Japan v. Katō. Hanrei Jihō. No. 885, 17, Fukuoka District Court, Kokura Branch. 24 March 1978.

Jensen, Michael C. 1981. "Business and the press." In Thornton Bradshaw and David Vogel, eds., *Corporations and their Critics: Issues and Answers to the Problems of Corporate Social Responsibility,* 49–60. New York: McGraw-Hill.

Jensen, Soren. 1972. "The PCB story." *Ambio* 1:123–31.

Kaga Takashi. 1972. *PCB Osen no Kyōfu.* Tokyo: Karin Kikaku.

Kamino Ryūzō. 1972. "Ningen bubetsu to no tatakai." In Ishi Hiroyuki, ed., *PCB: Jinrui o Kuu Bunmei no Senpei,* 13–39. Tokyo: Asahi Shinbunsha.

———. 1973. *Ondo no Tami: Kanemi Yushō Kanja no Kiroku.* Tokyo: Kyōbun Kan.

Kamps, L. R., W. J. Trotter, S. J. Young, L. J. Carson, J. A. G. Roach, J. A. Sphone, J. T. Tanner, and B. McMahon. 1978. "Polychlorinated quaterphenyls identified in rice oil associated with Japanese 'Yusho' poisoning." *Bulletin of Environmental Contamination and Toxicology* 20:589–91.

Kanegafuchi Chemical Corporation. 1977. *Kanemi Yushō Fukuoka Hanketsu ni Tsuite.* October.

———. 1978. *Kanemi Yushō Jiken ni Tsuite.* November.

———. 1979. *Junbi Shomen (Dai San Kai).* Isharyō seikyū kōso jiken. Fukuoka Kōtō Saibansho Dai Yon Minjibu. 23 October.

Kanegafuchi Kagaku Rōdō Kumiai. 1978. *Kanemi Yushō Jiken ni Tsuite: Watakushi Tachi no Kihonteki Kangaekata.* 20 January.

Kanemi Sōko Company. 1974. "Yushō jiken no genjō." *Himugashi (Shanaihō),* June.

"Kanemi Yushō misoshō higaisha kyūsai ni kan suru kakunin sho." 6 July 1978. Mimeo of agreement. Osaka.

Kashimoto T., Miyata H., and Kunita N. 1981. "The presence of polychlorinated quaterphenyls in the tissues of Yusho victims." *Food and Cosmetics Toxicology* 19:335–40.

Kashimoto T., Miyata H., Kunita N., Tung T. C., Hsu S. T., Chang K. J., Tang S. Y., Ohi G., Nagakawa J., and Yamamoto S. 1981. "Role of polychlorinated dibenzofuran in Yusho (PCB poisoning)." *Archives of Environmental Health* 36:321–26.

Kates, Robert W. 1977. "Major insights: A summary and recommendations." In J. Eugene Haas, Robert W. Kates, and Martyn J. Bowden, eds., *Reconstruction Following Disaster,* 261–93. Cambridge: MIT Press.

Katō I., Toyota M., Haraguchi T., Tōda T., Morishima A. 1978. "Shokuhin jiko, yakugai to higaisha no kyūsai." Roundtable discussion. *Jiyurisuto*, 15 January, 17–38.

Katō Kunioki. 1977. "Yushō gen-in jiko toshite no furanji roshutsu setsu ni tsuite." *Hōritsu Jihō*, April, 47–55.

Katō Yachiyo. 1979. "Watakushi ga daita kazukazu no gimon: Damasu beki toki ga ari kataru beki toki ga aru." *Yushi*, October, 38–43.

———. 1989. *Kanemi Yushō Saiban no Ketchaku*. Tokyo: Saiwai Shobō.

Katsuki Shibanosuke. 1969. "Jogen." *Fukuoka Igaku Zasshi* 60:403–7.

Katzman, Martin T. 1985. *Chemical Catastrophes: Regulating Environmental Risk through Pollution Liability Insurance*. Homewood, Ill.: Richard D. Irwin.

Katznelson, Ira, and Mark Kesselman. 1975. *The Politics of Power: A Critical Introduction to American Government*. New York: Harcourt Brace Javonovich.

Kawashima Takeyoshi. 1963. "Dispute resolution in contemporary Japan." In Arthur Taylor von Mehren, ed., *Law in Japan: The Legal Order in a Changing Society*, 41–72. Cambridge: Harvard University Press.

Kelman, Steven. 1981. *Regulating America, Regulating Sweden: A Comparative Study of Occupational Safety and Health Policy*. Cambridge: MIT Press.

Kiefer, David M. 1987. "United States, profits up sharply as producers reap benefits of restructuring." *Chemical & Engineering News*, 14 December, 26–28.

Kikuchi Masahiro, Mikagi Yoshiharu, Hashimoto Michio, and Kojima Tōru. 1971. "Iwayuru Yushō kanja no 2 hōken rei." *Fukuoka Igaku Zasshi* 62:89–103.

Kim, Hong N. 1966. "Deradicalization of the Japanese Communist party under Kenji Miyamoto." *World Politics* 28:273–99.

Kimbrough, Renate. 1972. "Toxicity of chlorinated hydrocarbons and related compounds." *Archives of Environmental Health* 25:125–31.

———. 1979. "The carcinogenic and other chronic effects of persistent halogenated organic compounds." In William J. Nicholson and John A. Moore, eds., *Health Effects of Halogenated Aromatic Hydrocarbons*. Vol. 320, 415–18. New York: New York Academy of Sciences.

Kimbrough, Renate D., D. F. Groce, M. P. Korver, and V. W. Burse. 1981. "Induction of liver tumors in female Sherman strain rats by polybrominated biphenyls." *Journal of the National Cancer Institute* 66:535–42.

Kingdon, John W. 1984. *Agendas, Alternatives, and Public Policies*. Boston: Little, Brown.

Kleinman, Arthur. 1986. "Illness meanings and illness behaviour." In Sean McHugh and T. Michael Vallis, eds., *Illness Behavior: A Multidisciplinary Model*, 149–60. New York: Plenum Press.

Kōga K., Watanabe H., Mochida Y., and Hiratsuka S. 1970. "Aru kome nuka abura fukusanbutsuchū no dokusei busshitsu ni kan suru kenkyū." *Nihon Chikusan Gakkaihō* 41:336–42 and 439–44.

———. 1971. "Aru kome nuka abura fukusanbutsuchū no dokusei busshitsu ni kan suru kenkyū." *Nihon Chikusan Gakkaihō* 42:16–24.

Kohanawa Mokoto. 1974. "Niwatori (PCB konnyū) daku oiru chūdoku jiken o kaerimite." *Kagaku* 44:117–19.

Kolbye, Albert C., Jr. 1972. "Food exposures to polychlorinated biphenyls." *Environmental Health Perspectives* 1:85–88.

——. 25 June 1975. "Low level polybrominated biphenyl contamination in dairy herds." Memorandum. Food and Drug Administration, Washington, D.C.

——. 7 March 1977a. "Statement on polybrominated biphenyls (PBBs)." Presented before the Michigan House of Representatives Committee on Public Health, Lansing, Mich.

——. 1977b. Testimony in *Hearings before the U.S. Senate Subcommittee on Science, Technology, and Space, Committee on Commerce, Science, and Transportation*, 31 March 1977, Part 2, 953–1367. Washington, D.C.: U.S. Government Printing Office.

Krier, James E., and Edmund Ursin. 1977. *Pollution and Policy: A Case Essay on California and Federal Experience with Motor Vehicle Pollution 1940–1975.* Berkeley: University of California Press.

Krimsky, Sheldon, and David Ozonoff. 1982. *Genetic Alchemy: The Social History of the Recombinant DNA Controversy.* Cambridge: MIT Press.

Kubota v. Kanemi Sōko K.K.. Hanrei Jihō. No. 866, 21, Fukuoka District Court, Kokura Branch. 5 October 1977.

Kuratsune Masanori. 1972. "Yushō jiken no keika to kyōjun—ekigakusha no tachiba kara." In Ishi Hiroyuki, ed., *PCB: Jinrui o Kuu Bunmei no Senpei*, 40–65. Tokyo: Asahi Shinbunsha.

——. 1989. "Jogen (1)." *Fukuoka Igaku Zasshi* 80:179–80.

Kuratsune Masanori and Raymond E. Shapiro, eds. 1984. *PCB Poisoning in Japan and Taiwan: Progress in Clinical and Biological Research.* Vol. 137. New York: Alan R. Liss.

Kuratsune Masanori, Yoshimura T., and Matsuzaka J. 1972. "Epidemiologic study on Yusho, a poisoning caused by ingestion of rice oil contaminated with a commercial brand of polychlorinated biphenyls." *Environmental Health Perspectives* 1:119–28.

Laitin, David D., and Aaron Wildavsky. 1988. "Political culture and political preferences." *American Political Science Review* 82:589–96.

Lancet. 1977. "Polybrominated biphenyls, polychlorinated biphenyls, pentachlorophenyl—and all that." 2 July, 19–21.

Landy, Marc. 1986. "Cleaning up Superfund." *Public Interest* 85:58–71.

Lange, Peter. 1980a. "Crisis and consent, change and compromise: Dilemmas of Italian communism in the 1970s." In Peter Lange and Sidney Tarrow, eds., *Italy in Transition: Conflict and Consensus*, 110–32. London: Frank Cass.

——. 1980b. "Preface." In Peter Lange and Sidney Tarrow, eds., *Italy in Transition: Conflict and Consensus*, 1–5. London: Frank Cass.

LaPalombara, Joseph. 1960. *Guide to Michigan Politics.* 2d ed. East Lansing, Mich.: Bureau of Social and Political Research, Michigan State University.

——. 1964. *Interest Groups in Italian Politics.* Princeton: Princeton University Press.

Laseter, John L. 1979. "Approaches to monitoring organic environmental contaminants in food." In U.S. Office of Technology Assessment, U.S. Congress, *Environmental Contaminants in Food*, 187–94. Washington, D.C.: U.S. Government Printing Office.

Lawrence, Paul R., and Jay W. Lorsch. 1969. *Organization and Environment.* Homewood, Ill.: Richard D. Irwin.

Leeblow, Jim. 6 March 1976. Testimony at Michigan House of Representatives Special Committee to Investigate PBB, Cadillac hearing. Transcript, 30.

Leiserson, Michael. 1970. "Coalition government in Japan." In Sven Groennings,

E. W. Kelley, and Michael Leiserson, eds., *The Study of Coalition Behavior: Theoretical Perspectives and Cases from Four Continents*, 80–102. New York: Holt, Rinehart and Winston.

——. 1973. "Political opposition and political development in Japan." In Robert Dahl, ed., *Regimes and Opposition*, 341–98. New Haven: Yale University Press.

Lerner, Melvin J. 1970. "The desire for justice and reactions to victims." In J. Macaulay and L. Berkowitz, eds., *Altruism and Helping Behavior: Social Psychological Studies of Some Antecedents and Consequences*, 205–29. New York: Academic Press.

Lester, James P., and Ann O'M. Bowman, eds. 1983. *The Politics of Hazardous Waste Management*. Durham, N.C.: Duke University Press.

Levine, Adeline Gordon. 1982. *Love Canal: Science, Politics, and People*. Lexington, Mass.: Heath.

Lifton, Robert Jay. 1967. *Death in Life: Survivors of Hiroshima*. New York: Random House.

——. 1976. "Nuclear energy and the wisdom of the body." *Bulletin of the Atomic Scientists*, September, 16–20.

——. 1979. *The Broken Connection: On Death and the Continuity of Life*. New York: Simon and Schuster.

Lifton, Robert Jay, Katō Shūichi, and Michael R. Reich. 1979. *Six Lives/Six Deaths: Portraits from Modern Japan*. New Haven: Yale University Press.

Lindblom, Charles E. 1977. *Politics and Markets: The World's Political-Economic Systems*. New York: Basic Books.

Lipsky, Michael. 1968. "Protest as a political resource." *American Political Science Review* 62:1144–58.

Lockard, Duane. 1969. *The Politics of State and Local Government*. 2d ed. New York: Macmillan.

Losa, Luigi. 1977. "Le considerazioni della tragedia." In Mario Galimberti, Giacomo Citterio, and Luigi Losa, comps., *Seveso: La Tragedia della Diossina*, 157–201. Besana Brianza: Edizioni GR.

Lowi, Theodore J. 1967. "Party, policy, and constitution in America." In William Nesbit Chambers and Walter Dean Burnham, eds., *The American Party Systems: Stages of Political Development*, 238–76. New York: Oxford University Press.

Luken, Thomas A. 1977. Opening Statement in *Hearings before the U.S. House of Representatives Subcommittee on Oversight and Investigation, Committee on Interstate and Foreign Commerce*, 2 and 3 August 1977, 49–50. Washington, D.C.: U.S. Government Printing Office.

Lukes, Steven. 1974. *Power: A Radical View*. London: Macmillan.

Lundquist, Lennart. 1980. *The Hare and the Tortoise: Clean Air Policies in the United States and Sweden*. Ann Arbor: University of Michigan Press.

McCune, E. L., J. E. Savage, and B. L. O'Dell. 1962. "Hydropericardium and ascites in chicks fed a chlorinated hydrocarbon." *Poultry Science* 4:295–99.

McHugh, Sean, and T. Michael Vallis, eds. 1986. *Illness Behavior: A Multidisciplinary Model*. New York: Plenum Press.

MacMahon, Brian. 1979. "Strengths and limitations of epidemiology." In National Research Council, *The National Research Council in 1979: Current Issues and Studies*, 91–104. Washington, D.C.: National Academy of Sciences.

McNeill, William H. 1976. *Plagues and Peoples*. Garden City, N.Y.: Anchor Press/Doubleday.

McNelly, Theodore. 1960. *Contemporary Government of Japan*. Boston: Houghton Mifflin.

MAF. *See* Ministry of Agriculture and Forestry.

Maffii, Neva Agazzi. 1983. "Un giorno alle case IACP." *Scienza Esperienza*, June, 31.

Majone, Giandomenico. 1985. "The international dimension." In Harry Otway and Malcolm Peltu, eds., *Regulating Industrial Risks: Science, Hazards, and Public Protection*, 40–56. London: Butterworths.

Marconi, Pio. 1981. "I partiti nella crisi delle istituzioni." *Mondoperaio*, February, 109–15.

Martorelli, Marco, ed. 1976. *ICMESA: Una Rapina di Salute, di Lavoro e di Territorio*. Milan: Mazzotta.

Maugh, Thomas H., II. 1978. "Chemicals: How many are there?" *Science* 199:162.

May, George. 1973. "Chloracne from the accidental production of tetrachlorodibenzodioxin." *British Journal of Industrial Medicine* 30:276–83.

MDA. *See* Michigan Department of Agriculture.

MDPH. *See* Michigan Department of Public Health.

Mechanic, David. 1974. "Discussion of research programs on relations between stressful life events and episodes of physical illness." In Barbara Snell Dohrenwend and Bruce P. Dohrenwend, eds., *Stressful Life Events: Their Nature and Effects*, 87–97. New York: Wiley.

Meester, Walter D. 6 June 1975. "Critique on the Michigan Department of Public Health study on the short-term effects of PBB on health." Mimeo. Blodgett Memorial Medical Center, Grand Rapids, Mich. In *Hearings before the U.S. Senate Subcommittee on Science, Technology, and Space, Committee on Commerce, Science, and Transportation*, 28, 29, and 30 March 1977, Part 1, 527–29. Washington, D.C.: U.S. Government Printing Office.

Melucci, Alberto. 1977. *Sistema Politico: Partiti e Movimenti Sociali*. Milan: Feltrinelli.

MEO. *See* Michigan Executive Office.

Mercer, H. Dwight. 1975. "Kolbye memo, dated June 25, 1975, re: Michigan PBB problem." Memorandum. Food and Drug Administration, Washington, D.C.

Mercer, H. Dwight, Richard M. Teske, Robert J. Condon, Allen Furr, Gavin Meerdink, William Buck, and George Fries. 1976. "Herd health status of animals exposed to polybrominated biphenyl." *Journal of Toxicology and Environmental Health* 2:235–49.

MHW. *See* Ministry of Health and Welfare.

Michigan Chemical Corporation. 1971. *Firemaster BP-6: Health and Safety*. Chicago: Michigan Chemical Corporation.

——. 1973. *BP-6 Plant Housekeeping Inspection, Inter-Office Memorandum*. Saint Louis, Mich.: Michigan Chemical Corporation.

Michigan Department of Agriculture (MDA). 13 May 1974. Press release, Bureau of Consumer Protection, on exclusion from the market of milk from fifteen Michigan dairy farms. Lansing, Mich.

——. 17 May 1974. Bureau of Consumer Protection, *PBB Feed Contamination Conference*, W. C. Geagley Laboratory. Transcript of tapes 1–4. Lansing, Mich.

———. 7 November 1974. Press release, Bureau of Consumer Protection, on lowered federal guidelines for PBB residues. Lansing, Mich.

———. 3 December 1974. Press release, B. Dale Ball, director, on PBB contamination guidelines and quarantine regulations. Lansing, Mich.

———. 10 July 1975. *Summary of Testimony Presented at the Public Hearing on Proposed Regulation No. 554, Polybrominated Biphenyls, Held 29 May 1975.* Lansing, Mich.

Michigan Department of Public Health (MDPH). 25 July 1974. Press release, Office of Information and Education, on first phase of investigations into possible harmful effects of PBB in humans. Lansing, Mich.

———. 14 February 1975. Press release, Office of Information and Education, on no evidence of any human ill effects from exposure to milk or beef contaminated by PBB. Lansing, Mich.

———. 19 August 1976. Press release, Office of Information and Education, on PBB and PCB contamination of breast milk samples from nursing mothers in Michigan. Lansing, Mich.

———. 15 October 1976. Press release, Office of Information and Education, on follow-up study on exposure of general population to PBB and on contamination of breast milk of nursing mothers. Lansing, Mich.

———. 4 November 1976. Press release, Office of Information and Education, on continuing to recommend that mothers not living on contaminated farms who choose to breast feed should continue to do so, despite trace amounts of PBB in their breast milk. Lansing, Mich.

———. 1986. *PCB/PBB Health Study News* 5 (Spring). Center for Environmental Health Sciences, Division of Epidemiological Studies.

Michigan Executive Office (MEO). 27 March 1975. Press release and report on the press and PBB. Michigan State Government Task Force. Lansing, Mich.

———. 2 September 1975. Press release on governor's veto of Bill 5033 to provide low-interest loans to PBB farmers. Lansing, Mich.

———. 5 March 1976. Press release on governor's five-point "action plan" for PBB. Lansing, Mich.

———. 22 June 1976. Press release on decision of Michigan Agriculture Commission to maintain current PBB levels. Lansing, Mich.

———. 4 January 1977. "Transcript of Governor Milliken's news conference: Joint news conference held with House Speaker Bobby Crim, Dr. Maurice Reizen, and Dr. Irving Selikoff, on PBB." Lansing, Mich.

———. 8 August 1977. Press release on governor's response to "selected chronology" issued by House Speaker Bobby Crim. Lansing, Mich.

Michigan Farm News. 1974. "The contamination thing: What happened?" Interview with Donald Armstrong, Farm Bureau Services executive vice-president. 1 July, 1.

———. 1977. "PBB politics threaten all of agriculture." April, 1–2.

Miller, Arjay. 1980. "A director's questions." *Wall Street Journal*, 18 August.

Miller, F. DeWolfe, Lawrence B. Brilliant, and Richard Copeland. 1984. "Polybrominated biphenyls in lactating Michigan women: Persistence in the population." *Bulletin of Environmental Contamination and Toxicology* 32:125–33.

Mills, C. Wright. 1959. *The Sociological Imagination.* New York: Oxford University Press.

Milnes, M. H. 1971. "Formation of 2,3,7,8-tetrachlorodibenzodioxin by thermal decomposition of 2,3,5-trichlorophenate." *Nature* 232:395–96.

Ministero della Sanità. 14 September 1977. Minutes of meeting to review Seveso situation. Rome.

Ministry of Agriculture and Forestry (MAF). 29 October 1968. "Daku abura ni yoru niwatori no chūdoku jiko ni tsuite." Mimeo. Tokyo.

Ministry of Health and Welfare (MHW). 23 December 1968. "Komenuka abura chūdoku jiken kyōgi kaigi shiryō." Mimeo. Kankyō Eisei Kyoku, Shokuhin Eisei Ka, Tokyo.

———. 20 March 1969. "Dai 4 kai yushō taisaku honbu kaigi, kaigi shiryō." Mimeo. Kankyō Eisei Kyoku, Shokuhin Eisei Ka, Tokyo.

Mishan, E. J. 1971. "Evaluation of life and limb: A theoretical approach." Journal of Political Economy 79:687–705.

Miyata H., Murakami Y., and Kashimoto T. 1978. "Investigation of polychlorinated quaterphenyl in Kanemi rice oils caused the 'Yusho.'" Shokuhin Eiseigaku Zasshi 19:233–35.

Mocarelli, P., A. Marocchi, P. Brambilla, P. Gerthoux, D. S. Young, and N. Mantel. 1986. "Clinical laboratory manifestations of exposure to dioxin in children: A six-year study of the effects of an environmental disaster near Seveso, Italy." Journal of the American Medical Association 256:2687–95.

Mocarelli, P., F. Pocchiari, N. Nelson, and L. Needham. 1988. "Livelli di 2,3,7,8-tetraclorodibenzo-p-diossina (TCDD) nel siero di soggetti esposti a Seveso: Rapporto preliminare." Bollettino Epidemiologico Nazionale, 31 October, 1–3.

Molotch, Harvey. 1970. "Oil in Santa Barbara and power in America." Sociological Inquiry 40:131–44.

Morehouse, Ward, and M. A. Subramanian. 1986. The Bhopal Tragedy. New York: Council on International and Public Affairs.

Morone, Joseph G., and Edward J. Woodhouse. 1986. Averting Catastrophe: Strategies for Regulating Risky Technologies. Berkeley: University of California Press.

Motz, Mrs. Jerry. 29 May 1975. Testimony at Public Hearing on Proposed Regulation No. 554, Polybrominated Biphenyls, State of Michigan, Department of Agriculture. Lansing, Mich. Transcript, 113.

———. 1 March 1976. Testimony at Michigan House of Representatives Special Committee to Investigate PBB, Grand Rapids hearing. Transcript, 16–24.

Mueller, Claus. 1973. The Politics of Communication: A Study in the Political Sociology of Language, Socialization, and Legitimation. New York: Oxford University Press.

Murray, Michael A. 1975. "Comparing public and private management: An exploratory essay." Public Administration Review 35:364–71.

Nagashima Atsushi, assisted by B. J. George, Jr. 1963. "The accused and society: The administration of criminal justice in Japan." In Arthur Taylor von Mehren, ed., Law in Japan: The Legal Order in a Changing Society, 297–323. Cambridge: Harvard University Press.

Nagayama J., Masuda Y., and Kuratsune M. 1975. "Chlorinated dibenzofurans in Kanechlors and rice oils used by patients with 'Yusho.'" Fukuoka Igaku Zasshi 66:593–99.

National Academy of Sciences, Steering Committee on Identification of Toxic and Potentially Toxic Chemicals for Consideration by the National Toxicology Program. 1984. Toxicity Testing: Strategies to Determine Needs and Priorities. Washington, D.C.: National Academy Press.

Nature. 1980. "Seveso director killed by political group." 283:614.

Nelkin, Dorothy, ed. 1979a. *Controversy: The Politics of Technical Decisions.* Beverly Hills: Sage.

——. 1979b. "Science, technology, and political conflict: Analyzing the issues." In Dorothy Nelkin, ed., *Controversy: The Politics of Technical Decisions,* 9–22. Beverly Hills: Sage.

Nelkin, Dorothy, and Michael S. Brown. 1984. *Workers at Risk: Voices from the Workplace.* Chicago: University of Chicago Press.

Nelson, Barbara J. 1984. *Making an Issue of Child Abuse: Political Agenda Setting for Social Problems.* Chicago: University of Chicago Press.

Neustadt, Richard. 1970. *Alliance Politics.* New York: Columbia University Press.

Newton, K. 1979. "The language and the grammar of political power: A comment on Polsby." *Political Studies* 27:542–47.

Nishimura Mikio. 1972. "Hachi kagetsu no kūhaku o tsuikyū suru." In Ishi Hiroyuki, ed., *PCB: Jinrui o Kuu Bunmei no Senpei,* 65–94. Tokyo: Asahi Shinbunsha.

Noguchi v. Kanemi Sōko K.K.. Hanrei Jihō. No. 881, 17, Fukuoka District Court, Kokura Branch. 3 March 1978.

Nomura Shigeru. 1953. "Kurorufenoru chūdoku ni kan suru kenkyū." *Rōdō Kagaku* 29:474–83.

Nomura Shigeru and Arimatsu Tokuhide. 1969. "Hōkōzoku enka tanka suiso to hifu shōgai." *Seikatsu Eisei* 13:11–19.

Norris, J. M., J. W. Ehrmantraut, C. L. Gibbons, R. J. Kociba, B. A. Schwetz, J. Q. Rose, C. G. Humiston, G. L. Jewett, W. B. Crummett, P J. Gehring, J. B. Tirsell, and J. S. Brosier. 1973. "Toxicological and environmental factors involved in the selection of a decabromodiphenyl oxide as a fire retardant chemical." *Applied Polymers Symposium* 22:195–219. Presented at the Conference on Polymeric Materials for Unusual Service Condition, Ames Research Center, NASA, Moffett Field, Calif., 29 November 1972.

Northwest Industries, Inc. 1976. *Annual Report 1975.* Chicago: Northwest Industries, Inc.

Olson, Mancur, Jr. 1971 (1965). *The Logic of Collective Action: Public Goods and the Theory of Groups.* Cambridge: Harvard University Press.

Organization for Economic Cooperation and Development (OECD). 1979. *Technology on Trial: Public Participation in Decision-Making Related to Science and Technology.* Paris: OECD.

Page, Joseph A., and Mary-Win O'Brien. 1973. *Bitter Wages: Ralph Nader's Study Group Report on Disease and Injury on the Job.* New York: Grossman Publishers.

Parks, Marion, ed. 1976. *The Chemical Cloud that Fell on Seveso: Review and Selected Translations from the Italian Press, July 10, 1976 through September 10, 1976.* Trans. by Elizabeth Sullan. Washington, D.C.: Rachel Carson Trust for the Living Environment.

Parsons, Talcott. 1979. "Definitions of health and illness in the light of American values and social structure." In E. Gartly Jaco, ed., *Patients, Physicians, and Illness,* 120–44. 3d ed. New York: Free Press.

Pascale, Richard, and Anthony Athos. 1981. *The Art of Japanese Management.* New York: Simon and Schuster.

Pasquino, Gianfranco. 1980. "Italian Christian Democracy: A party for all seasons?" In Peter Lange and Sidney Tarrow, eds., *Italy in Transition: Conflict and Consensus,* 88–109. London: Frank Cass.

PBB News. 1980. 4 (March). Michigan Department of Public Health, Division of Environmental Epidemiology.

PBB/PCB News. 1988. 7 (Fall). Michigan Department of Public Health, Center for Environmental Health Sciences.

PBB Scientific Advisory Panel. *Report to William G. Milliken, Governor, State of Michigan, on PBB.* 24 May 1976. Lansing, Mich.

Pearson, Jessica. 1980. "Hazard visibility and occupational health problem solving: The case of the uranium industry." *Journal of Community Health* 6:136–47.

Pempel, T. J. 1974. "The bureaucratization of policymaking in postwar Japan." *American Journal of Political Science* 18:647–64.

Penning, C. H. 1930. "Physical characteristics and commercial possibilities of chlorinated diphenyl." *Industrial Engineering and Chemistry* 22:1180–82.

People of the State of Michigan v. Northwest Industries, Inc., Northwest Chemco, Inc., Velsicol Chemical Corporation, Michigan Chemical Corporation, Michigan Salt Company, Michigan Farm Bureau, and Farm Bureau Services, Inc. Demand for Trial by Jury. Circuit Court, Ingham County, Mich. 22 February 1978.

Perrow, Charles. 1984. *Normal Accidents: Living with High-Risk Technologies.* New York: Basic Books.

Perrow, Charles, and Mauro F. Guillén. 1990. *The AIDS Disaster: The Failure of Organizations in New York and the Nation.* New Haven: Yale University Press.

Peters, John M. 1978. "The Kepone episode: Another warning." *New England Journal of Medicine* 298:277–78.

Petrella, Enrico. 1976. "Fatti, misfatti e cronaca." *Medicina al Servizio delle Masse Popolari,* September, 2–4.

Piven, Frances Fox, and Richard A. Cloward. 1977. *Poor People's Movements: Why They Succeed, How They Fail.* New York: Pantheon.

Plumlee, John P., and Kenneth J. Meier. 1978. "Capture and rigidity in regulatory administration." In Judith V. May and Aaron Wildavsky, eds., *The Policy Cycle,* 215–34. Beverly Hills: Sage.

Pocchiari, F., V. Silano, and G. Zapponi. 1986. "The chemical risk process in Italy: A case study, the Seveso accident." *Science of the Total Environment* 51:227–35.

Polsby, Nelson W. 1963. *Community Power and Political Theory.* New Haven: Yale University Press.

——. 1979. "Empirical investigation of the mobilization of bias in community power research." *Political Studies* 27:527–41.

Pototschnig, Umberto. 1978. "Limiti della legislazione vigente e nuove strategie di intervento." In *Dalla Lotta all'Inquinamento alla Tutela Pubblica dell'Ambiente,* Proceedings of a conference sponsored by the Lombardy Regional Assembly, 4–5 March 1977, 27–44. Milan: Giuffre.

Primack, Joel, and Frank von Hippel. 1974. *Advice and Dissent: Scientists in the Political Arena.* New York: Basic Books.

Provincia di Milano. 1976. *Documentazione Riguardante l'Inquinamento Causato dallo Stabilimento ICMESA di Meda.* Rip III, Ecologia, Igiene e Sanità. Milan. September.

Przeworski, Adam, and Henry Teune. 1970. *The Logic of Comparative Inquiry.* New York: Wiley.

Il Punto a Seveso: Bolletino Ufficiale per le Popolazioni Colpite. No. 1, 4 March 1977, to no. 12, 10 June 1977. Reprinted in "A un anno di Seveso." *Ecologia, Acqua, Aria,* July/August 1977, 419–41.

Putnam, R., R. Leonardi, and R. Y. Nanetti. 1983. "Explaining institutional success: The case of Italian regional government." *American Political Science Review* 77:55–74.

Quartiere Case IACP—Baruccana, via Cavalla. 15 February 1978. "In defesa della salute." Mimeo. Seveso.

Ramazzini, Bernardino. 1914 (1700, 1713). *De Morbis Artificum Diatriba* (Diseases of workers). New York: New York Academy of Medicine and Harper Publishing Co.

Rawls, Rebecca L. 1980. "Chemical transport—coping with disasters." *Chemical & Engineering News,* 24 November, 2–30.

Reed, Steven R. 1982. "Is Japanese government really centralized?" *Journal of Japanese Studies* 8:133–64.

Reggiani, G. 1978. "Medical problems raised by the TCDD contamination in Seveso, Italy." *Archives of Toxicology* 40:161–88.

Regione Lombardia (RL). N.d. Assessorato alla Sanità, "Evento ICMESA, Cronistoria 1976." Mimeo. Milan.

———. 21 July 1976. Comunicato ufficiale della regione su Seveso, n. o, "Un piano di intervento nella zona contaminata da sostanze tossiche." Milan.

———. 27 July 1976. Comunicato ufficiale della regione su Seveso, n. 2. Milan.

———. 29 July 1976. Comunicato ufficiale della regione su Seveso, n. 3. Milan.

———. 23 August 1976. Assessorato alla Sanità, "Inquinamento provocato dalla nube tossica fuoriuscita dallo stabilimento dell'ICMESA di Meda il 10 luglio 1976: Relazione sulla situazione, sui provvedimenti assunti e sulle proposte di intervento." Report. Milan.

———. 20 September 1976. Comunicato ufficiale della regione su Seveso, n. 67, "Riuniti in regione i rappresentanti degli evacuati di Seveso." Milan.

———. 11 October 1976. Comunicato ufficiale della regione su Seveso, n. 79, "Il punto sugli interventi nella zone inquinata di Seveso." Milan.

———. 20 October 1976. Comunicato ufficiale della regione su Seveso, n. 83, "Commissione Medico-Epidemiologica." Milan.

———. 6 December 1976. Comunicato ufficiale della regione su Seveso, n. 96, "Interventi prioritari approvati dai rappresentanti della regione, della provincia e del comune di Seveso." Milan.

———. 1 February 1977. Comunicato ufficiale della regione su Seveso, n. 1, "Esame dermatologico per gli scolari di Seveso." Milan.

———. 11 February 1977. Comunicato ufficiale della regione su Seveso, n. 6, "Richiesta dal presidente della regione l'intervento dell'esercito per Seveso." Milan.

———. 2 June 1977. Giunta Regionale, excerpts in *Il Punto a Seveso,* no. 12, 10 June 1977. Milan.

———. 1 February 1979. Il Presidente della giunta, "Risposta del presidente della giunta regionale Dr. Cesare Golfari alla interperllanza dei consiglieri regionali Laura Conti e Natale Contini del 9/1/1979 e Franco Petenzi del 19/1/1979." Milan.

———. 3 February 1979. Ufficio Speciale, comunicato stampa, on congenital malformations. Milan.

Reich, Michael R. 1981. "Japan: A review of the Chemical Substances Control Law." *East Asian Executive Reports* 3(5):3, 18–23.

———. 1983. "Environmental policy and Japanese society." Parts 1 and 2. *International Journal of Environmental Studies* 20:191–98 and 199–207.

———. 1984a. "Crisis and routine: Pollution reporting by the Japanese press." In George De Vos, ed., *Institutions for Change in Japanese Society*, 148–65. Berkeley: Institute for East Asian Studies.

———. 1984b. "Mobilizing for environmental policy in Italy and Japan." *Comparative Politics* 16:379–402.

Reich, Michael R., and Jaquelin K. Spong. 1983. "Kepone: A chemical disaster in Hopewell, Virginia." *International Journal of Health Services* 13:227–46.

Reisch, Marc S. 1987. "Top 50 chemicals production steadied in 1986." *Chemical and Engineering News*, 13 April, 20–23.

Reizen, Maurice S. 26 August 1976. "Advisory re: PBB in human milk." Michigan Department of Public Health. Memorandum to Michigan physicians. Lansing, Mich.

Richardson, Bradley M. 1977. "Policymaking in Japan: An organizing perspective." In T. J. Pempel, ed., *Policymaking in Contemporary Japan*, 239–68. Ithaca: Cornell University Press.

Riegle, Donald W. 1977. Opening Statement in *Hearings before the U.S. Senate Subcommittee on Science, Technology, and Space, Committee on Commerce, Science, and Transportation*, 28, 29, and 30 March 1977, Part 1, 1–6. Washington, D.C.: U.S. Government Printing Office.

Risebrough, R. W., P. Reiche, S. G. Herman, D. B. Peakall, and M. N. Kirven. 1968. "Polychlorinated biphenyls in the global ecosystem." *Nature* 220:1098–102.

RL. *See* Regione Lombardia.

Roach, J. A. G., and I. H. Pomerantz. 1974. "The finding of chlorinated dibenzofurans in a Japanese PCB sample." *Bulletin of Environmental Contamination and Toxicology* 12:338–42.

Roberts, Marc J., and Jeremy S. Bluhm. 1981. *Choices of Power: Utilities Face the Environmental Challenge*. Cambridge: Harvard University Press.

Robinson, David Z. 1980. "Politics in the science advising process." *Technology in Science* 2:153–63.

Rohlen, Thomas P. 1976. "Violence at Yoka High School: The implications for Japanese coalition politics of the confrontation between the Communist party and the Burakumin Liberation League." *Asian Survey* 16:682–99.

Russell, Dick. 1990. "The rise of the grass-roots toxics movement." *Amicus Journal* (Winter), 18–21.

Ryan, William. 1976. *Blaming the Victim*. Rev. ed. New York: Vintage Books.

Sapere. 1976. "Un crimine di pace." November/December.

Schacht, Henry B., and Charles W. Powers. 1981. "Business responsibility and the public policy process." In Thornton Bradshaw and David Vogel, eds., *Corporations and their Critics: Issues and Answers to the Problems of Corporate Social Responsibility*, 23–32. New York: McGraw-Hill.

Schattschneider, E. E. 1960. *The Semisovereign People: A Realist's View of Democracy in America*. New York: Holt, Rinehart and Winston.

Scheingold, Stuart A. 1974. *The Politics of Rights: Lawyers, Public Policy, and Political Change.* New Haven: Yale University Press.

Schmidt, Alexander M. 26 March 1975. "Current status of PBB contamination of animal feed in the state of Michigan." Memorandum. Food and Drug Administration, Washington, D.C.

Schmittle, S. C., H. M. Edwards, and D. Morris. 1958. "A disorder of chickens probably due to a toxic feed—preliminary report." *Journal of American Veterinary Medical Association* 132:216–19.

Schur, Edwin M. 1980. *The Politics of Deviance: Stigma Contests and the Uses of Power.* Englewood Cliffs, N.J.: Prentice-Hall.

Schwartz, L. 1936. "Dermatitis from synthetic resins and waxes." *American Journal of Public Health* 26:586–92.

Seidman, Harold. 1970. *Politics, Position, and Power: The Dynamics of Federal Organization.* New York: Oxford University Press.

Selikoff, Irving J. 4 January 1977. *PBB Health Survey of Michigan Residents: November 4–10, 1976, Initial Report of Findings.* New York: Environmental Sciences Laboratory, Mt. Sinai School of Medicine.

———. 1980. "Asbestos-associated disease." In John M. Last, ed., *Public Health and Preventive Medicine,* 568–98. 11th ed. New York: Appleton-Century-Crofts.

Selikoff, Irving J., and Henry A. Anderson. 30 September 1979. *A Survey of the General Population of Michigan for Health Effects of Polybrominated Biphenyl Exposure.* Report to the Michigan Department of Public Health. New York: Environmental Sciences Laboratory, Mt. Sinai School of Medicine.

Sennett, Richard. 1980. *Authority.* New York: Alfred A. Knopf.

Settimana 3: Notizario dell'Ufficio Speciale della Regione Lombardia. No. 0, January 1978; no. 12, June 1979; no. 17, November 1979; no. 58, April 1983.

Shakai Rōdō Iin Kaigiroku. 30 October 1968. 59th Diet, 1–26.

———. 12 November 1968. 59th Diet, 1–8.

Shaver, Kelly G. 1975. *An Introduction to Attribution Processes.* Cambridge: Winthrop.

Shilts, Randy. 1988. *And the Band Played On: Politics, People, and the AIDS Epidemic.* New York: Penguin.

Shrivastava, Paul. 1987. *Bhopal: Anatomy of a Crisis.* Cambridge: Ballinger.

Sigal, Leon V. 1973. *Reporters and Officials.* Lexington, Mass.: Heath.

Silano, V. 1981. "Case study: Accidental release of 2,3,7,8-tetrachlorodibenzo-*p*-dioxin (TCDD) at Seveso, Italy." In *Planning Emergency Response Systems for Chemical Accidents,* 167–203. Copenhagen: World Health Organization Regional Office for Europe.

Simon, Herbert A. 1976 (1945). *Administrative Behavior.* 3d ed. New York: Free Press.

Smets, Henri. 1988. "Major industrial risks and compensation of victims: The role for insurance." In Michael R. Reich, ed., *Social Policy for Pollution-related Diseases,* special issue of *Social Science & Medicine* 27:1085–95.

Smith, Barbara Ellen. 1981. "Black lung: The social production of disease." *International Journal of Health Services* 11:343–59.

Smith, Elton R. 1977. "We must help tell PBB's other side." *Michigan Farm News,* April, 2.

Soble, Stephen M. 1977. "A Proposal for the administrative compensation of victims of toxic substance pollution: A model act." *Harvard Journal on Legislation* 14:683–824.

Soble, Stephen M., and Janis H. Brennan. 1988. "A review of legal and policy issues in legislating compensation for victims of toxic substance pollution." In Michael R. Reich, ed., *Social Policy for Pollution-related Diseases*, special issue of *Social Science & Medicine* 27:1061–70.

Socolow, Robert H. 1976. "Failures of discourse." In Harold A. Feiveson, Frank W. Sinden, and Robert H. Socolow, eds., *Boundaries of Analysis: An Inquiry into the Tocks Island Dam Controversy*, 9–40. Cambridge: Ballinger.

Sontag, Susan. 1978. *Illness as Metaphor*. New York: Farrar, Straus and Giroux.

Spallino, Antonio. 1 March 1979. Press conference. Ufficio Speciale di Seveso.

Stieber, Carolyn. 1970. *The Politics of Change in Michigan*. East Lansing: Michigan State University Press.

Stockwin, J. A. A. 1975. *Japan: Divided Politics in a Growth Economy*. New York: W. W. Norton.

Stone, Christopher D. 1975. *Where the Law Ends: The Social Control of Corporate Behavior*. New York: Harper and Row.

———. 1980. "The place of enterprise liability in the control of corporate conduct." *Yale Law Journal* 90:1–77.

Stone, Deborah A. 1984. *The Disabled State*. Philadelphia: Temple University Press.

Stretton, Hugh. 1976. *Capitalism, Socialism, and the Environment*. New York: Cambridge University Press.

Szasz, Thomas. 1970. *The Manufacture of Madness*. New York: Harper and Row.

Roy M. Tacoma and Marilyn K. Tacoma v. Michigan Chemical Corporation, The Michigan Salt Company, Velsicol Chemical Corporation, Farm Bureau Services, Inc., and Falmouth Cooperative Company, No. 2933. Circuit Court, Wexford, Mich. Unreported opinion. 26 October 1978.

Takabatake Michitoshi. 1975. "Citizens' movements: Organizing the spontaneous." *Japan Interpreter* 9:313–23.

Takamatsu Makoto. 1974. "Kanemi Yushō no tsuiseki: Gotō Tama no Ura cho no baai." *Kagaku* 44:755–62.

Takamatsu Makoto, Mizoguchi Keiko, Hirayama Hachirō, and Inoue Sumiko. 1979. "Gyomin Yushō kanja no ketsueki chū PCB nōdo to gasu kuromatoguramu patān no tokuchōteki shoken." *Fukuoka Igaku Zasshi* 70:223–28.

Takeshita Yasuhiko. 1969. "Kome to kome nuka abura kōgyō." *Seikatsu Eisei* 13:20–27.

Tannock, Ian F. 1987. "Treating the patient, not just the cancer." *New England Journal of Medicine* 317:1534–35.

Tarrow, Sidney. 1978. "Introduction." In Sidney Tarrow, Peter J. Katzenstein, and Luigi Graziano, eds., *Territorial Politics in Industrial Nations*, 1–27. New York: Praeger.

———. 1980. "Italy: Crisis, crises, or transition?" In Peter Lange and Sidney Tarrow, eds., *Italy in Transition: Conflict and Consensus*, 166–86. London: Frank Cass.

Terracini, Benedetto. 1977. "Le risposte non date alle popolazioni colpite sulle conseguenze del TCDD." *Sapere*, October/November, 36–43.

Teske, Richard H., and D. J. Wagstaff. 25 October 1974. "Review of the current status of polybrominated biphenyls (PBB) toxicosis in dairy cattle in Michigan." Memorandum. Food and Drug Administration, Washington, D.C.

Thomas, Mrs. Wayne. 29 May 1975. Testimony at Public Hearing on Proposed Regulation No. 554, Polybrominated Biphenyls, State of Michigan, Department of Agriculture. Lansing, Mich. Transcript, 73.

Thurston, Donald R. 1973. *Teachers and Politics in Japan*. Princeton: Princeton University Press.

Tocqueville, Alexis de. 1969 (1850). *Democracy in America*. Trans. by George Lawrence. Ed. by J. P. Mayer. Garden City, N.Y.: Anchor Books.

Tognoni, Gianni, and Bianca Torri. 1978. "L'aborto a Seveso." *Inchiesta*, January/February, 61–72.

Totokawa H., Nakatani K., and Kitamura K. 1969. "Keratosis follicularis?" *Seinichi Hifu* 11:341. Presented at Dai 26 Kai Nihon Hifu Kagaku Kai, Oita Chihō Kai, 7 September 1968.

Totten, George O. 1973. "The people's parliamentary path of the Japanese Communist party." Parts 1 and 2. *Pacific Affairs* 46:193–217 and 384–406.

Totten, George O., and Kawakami Tamio. 1964. "Gensuikyo and the peace movement in Japan." *Asian Survey* 4:833–41.

Toxic Substances Strategy Committee. 1980. *Toxic Chemicals and Public Protection: A Report to the President*. Washington, D.C.: U.S. Government Printing Office.

Trombley, Louis. 5 March 1976. Testimony at Michigan House of Representatives Special Committee to Investigate PBB, Cadillac hearing. Transcript, 9–12.

Upham, Frank K. 1976. "Litigation and moral consciousness in Japan: An interpretative analysis of four Japanese pollution suits." *Law & Society Review* 10:579–619.

———. 1987. *Law and Social Change in Postwar Japan*. Cambridge: Harvard University Press.

Urabe H. 1974. "Yushō to PCB ni kan suru kenkyū hōkoku: Jogen." *Fukuoka Igaku Zasshi* 65:1–4.

Urabe H., Koda H., and Asahi M. 1979. "Present state of Yusho patients." *Annals of New York Academy of Science* 320:273–76.

U.S. General Accounting Office. 8 June 1977. *Federal Efforts to Protect Consumers from Polybrominated Biphenyl Contaminated Food Products*. Report of the Comptroller General of the United States. Washington, D.C.: U.S. Government Printing Office.

U.S. Occupational Safety and Health Administration. 21 July 1978. "Access to employee exposure and medical records." Proposed rule. *Federal Register* 43:31371.

U.S. Office of Technology Assessment, U.S. Congress. 1979. *Environmental Contaminants in Food*. Washington, D.C.: U.S. Government Printing Office.

———. 1985. *Superfund Strategy*. Washington, D.C.: U.S. Government Printing Office.

Vogel, David. 1978. *Lobbying the Corporation*. New York: Basic Books.

———. 1986. *National Styles of Regulation: Environmental Policy in Great Britain and the United States*. Ithaca: Cornell University Press.

Vogel, Ezra. 1979. *Japan as Number One: Lessons for America*. Cambridge: Harvard University Press.

Vyner, Henry M. 1988a. *Invisible Trauma: The Psychosocial Effects of the Invisible Environmental Contaminants*. Lexington, Mass.: Lexington Books.

——. 1988b. "The psychological dimensions of health care for patients exposed to radiation and the other invisible environmental contaminants." In Michael R. Reich, ed., *Social Policy for Pollution-related Diseases*, special issue of *Social Science & Medicine* 27:1097–103.

Wade, Nicholas. 1979. "Viets and vets fear herbicide health effects; Vietnamese official brings concerns to Washington." *Science* 204:817.

Wagatsuma Hiroshi. 1972 (1966). "Postwar political militance." In George De Vos and Wagatsuma Hiroshi, eds., *Japan's Invisible Race: Caste in Culture and Personality*, 68–87. Rev. ed. Berkeley: University of California Press.

Waldvogel, Guy. 23 July 1976. President du Conseil d'Administration, ICMESA, letter to Monsieur Ufficiale Sanitario di Seveso e di Meda. Meda. English translation in E. Homberger, G. Reggiani, J. Sambeth, and H. K. Wipf. 1979. "The Seveso accident: Its nature, extent, and consequences." *Annals of Occupational Hygiene* 22:368.

Wallen, I. Eugene. 1977. Testimony in *Hearings before the U.S. House of Representatives Subcommittee on Oversight and Investigation, Committee on Interstate and Foreign Commerce*, 2 and 3 August 1977, 53–109. Washington, D.C.: U.S. Government Printing Office.

Ward, Robert E. 1967. *Japan's Political System*. Englewood Cliffs, N.J.: Prentice-Hall.

Welborn, John A., Richard Allen, Gary Byker, Alvin DeGrow, John Hertel, Robert Noordhoek, and Dennis Koons. July 1975. *The Contamination Crisis in Michigan: A Report from the Senate Special Investigating Committee on Polybrominated Biphenyls*. Lansing, Mich.: Michigan State Senate.

Whiteside, Thomas. 1979. *The Pendulum and the Toxic Cloud: The Course of Dioxin Contamination*. New Haven: Yale University Press.

Wilson, Dana. December 1978. *The Workers: Effects of PBB on Michigan Chemical Corporation Workers*. Lansing, Mich.: House Special Committee Examining the Effects of PBB on Michigan Chemical Corporation Workers.

Wilson, James Q. 1961. "The strategy of protest: Problems of Negro civic action." *Journal of Conflict Resolution* 3:291–303.

Wolf, Charles, Jr. 1988. *Markets or Governments: Choosing between Imperfect Alternatives*. Cambridge: MIT Press.

Wolff, M. S., H. A. Anderson, and I. J. Selikoff. 1982. "Human tissue burdens of halogenated aromatic chemicals in Michigan." *Journal of the American Medical Association* 247:2112–16.

Yamaryō Tomoko, Miyazaki T., Masuda Y., and Nagayama J. 1979. "Pori enka bifenīru no kanetsu ni yoru pori enka kuōtāfenīru no seisei." *Fukuoka Igaku Zasshi* 70:88–92.

Yates, Douglas. 1977. *The Ungovernable City: The Politics of Urban Problems and Policy Making*. Cambridge: MIT Press.

——. 1982. *Bureaucratic Democracy: The Search for Democracy and Efficiency in American Government*. Cambridge: Harvard University Press.

Young, A. L., J. A. Calcagni, C. E. Thalken, and J. W. Tremblay. 1978. *The Toxicology, Environmental Fate, and Human Risk of Herbicide Orange and Its Associated Dioxin*. Springfield, Va.: National Technical Information Service.

Yushōhan. 20 November 1968. *Yushōhan Kenkyū Chūkan Hōkokusho*. Mimeo. Ministry of Health and Welfare. Tokyo.

Zola, Irving. 1986. "Illness behavior: A political analysis." In Sean McHugh and

T. Michael Vallis, eds., *Illness Behavior: A Multidisciplinary Model*, 213–18. New York: Plenum Press.
Zuiderveen, Garry. 29 May 1975. Testimony at Michigan Department of Agriculture hearing. Lansing, Mich.

Interviews

Interviews were conducted by the author in the language of the interviewee. Excerpts from interviews in Italian and Japanese were translated by the author. Below are listed those interviews cited in the text.

Argiuolo, Amedeo, former worker at ICMESA, member of Consiglio di Fabbrica. 2 April 1979.
Armstrong, Donald R., executive vice-president, Farm Bureau Services. 30 August 1978.
Blake, Maddelena, resident of Baruccana section of Seveso. 18 February 1979.
Brambilla, Mrs. S., resident of Seveso in Case Fanfani. 1 March 1979.
Chiape, Aldo, management consultant. 26 March 1979.
Chiappini, Antonio, former worker at ICMESA, secretary for People's Scientific and Technical Committee. 12 February 1979.
Cordle, Frank, epidemiologist, U.S. Food and Drug Administration. 3 November 1978.
Corna, Giacomo, resident of zone A, member of town council in Seveso. 22 March 1979.
Fujiwara Benshi, activist in Tama no Ura. 22 September 1979.
Galli, Aldo, Italian Communist party activist in Seveso. 24 March 1979.
Gatzmeyer, Norman, veterinarian, Laboratory Division of Michigan Department of Agriculture. 22 August 1978.
Golfari, Cesare, president of the Lombardy Region. 29 March 1979.
Gotō Masayasu, dermatologist, formerly at Kyūshū University. 19 September 1979.
Green, Al, dairy farmer. 29 July 1978.
Halbert, Frederic, dairy farmer. 9 August 1978.
Haraguchi Torio, lawyer for Yusho victims in Fukuoka trial. 10 September 1979.
Higuchi Kentarō, professor of dermatology, Kyūshū University. 19 September 1979.
Humphrey, Harold, environmental epidemiologist, Michigan Department of Public Health. 16 August 1978.
Itō Rentarō, assistant director, Food Sanitation Division, Japanese Ministry of Health and Welfare. 11 October 1979.
Jūta Sueno, Yusho victim, Kitakyūshū. 6 September 1979.
Kamino Ryūzō, Yusho victim, Tagawa. 16 September 1979.
Kaneda Hiroshi, activist in Fukuoka. 13 September 1979.
Kashimoto Takashi, chief of Laboratory of Food Chemistry, Osaka Prefecture Institute of Public Health. 4 September 1975.
Kōga Kiyomi, researcher at feed company that used Kanemi dark oil. 28 August 1979.
Kohanawa Makoto, scientist at National Animal Health Research Laboratory. 22 August 1979.

Kōno Hiroaki, photojournalist who covered Kanemi Yusho. 22 August 1979.
Koons, Dennis, former staff for Michigan State Senator John Welborn. 15 August 1978.
Kuratsune Masanori, epidemiologist, Kyūshū University. 8 September 1979.
Kuroiwa Tomoyuki, Yusho victim, Naru. 21 September 1979.
Liberti, Irma, resident of Seveso in zone R. 11 February 1979.
McIntyre, Kenneth, attorney for Farm Bureau Services. 3 August 1978.
Menaspà, Mrs. B., former resident of zone A. 24 February 1979.
Menaspà, Sergio, former resident of zone A. 24 February 1979.
Miller, Pat, dairy farmer. 23 July 1978 and October 1989.
Motta, Mrs. Carlo, former resident of zone A. 25 February 1979.
Nebbia, Giorgio, environmentalist, Rome. 28 December 1978.
Nishimura Mikio, *Asahi Shinbun* reporter. 17 August 1979.
Poland, Alan, professor, University of Wisconsin School of Medicine. 14 September 1978.
Rivolta, Vittorio, health commissioner for Lombardy Region. 21 March 1979.
Sagady, Alex, director, Environmental Action of Michigan. 24 August 1978.
Spallino, Antonio, commissioner, Special Office of Seveso. 8 March 1979.
Taylor, William, science advisor to Governor Milliken. 4 August 1978.
Tsukauchi Katsuyuki, member of Kitakyūshū City Council, Japanese Communist party. 6 September 1979.
Uejima Kazuyoshi, member of Fukuoka Prefecture Assembly, Japanese Socialist party. 6 September 1979.
Ujino Kazuyuki, Yusho victim, Kitakyūshū. 15 September 1979.
Watanuki Reiko, science writer. August 1979.
Wilcox, Kenneth, chief, Bureau of Disease Control and Laboratory Services, Michigan Department of Public Health. 22 August 1978.

Newspapers

Asahi Shinbun
Avanti
Avvenire
Boston Globe
Cadillac Evening News (Cadillac, Mich.)
Il Cittadino della Domenica
Corriere della Sera
Corriere d'Informazione
Detroit Free Press
Detroit News
Il Giornale
Il Giorno
Grand Rapids Press
Japan Times
Mainichi Shinbun
Il Manifesto
Le Monde
New York Times
Nihon Keizai Shinbun

Paese Sera
Pioneer
La Repubblica
La Stampa
Stampa Sera
State Journal (Lansing, Mich.)
Sunday Times (London)
L'Unità
Washington Post

Index

Library of Congress Cataloging-in-Publication Data

Reich, Michael R., 1950–
  Toxic politics : responding to chemical disasters / Michael R. Reich.
    p. cm.
  Includes bibiliographical references and index.
  ISBN 0-8014-2434-8 (alk. paper)
    1. Crimes against the environment—Political aspects. 2. Hazardous substances—
Accidents—Political aspects. I. Title.
HV6401.R45 1991
363.73'84—dc20   91-55054